Natural Language Processing

Semantic Aspects

Haldane: *Where is the* **bedeutung** *of a proposition in your system, Turing? It is worth talking in terms of a universal grammar in mind, but it is also possible to construct meaningless propositions.*
Turing: *For example?*
Haldane: *"Red thoughts walk peacefully".*

. . . . . . *This is an example of a proposition which is formulated correctly, according to the rules of the English grammar. If your theory is correct, then I was able to construct such a proposition, since I was able to activate the English version of the universal grammar in my mind, the semantic content of which equals to zero. Where can I find in your theory that this proposition makes no sense?*
Turing: *It is very simple, I do not know.*

—from the book *The Cambridge Quintett,* John Casti, 1998

...however, the "system" is (as regards logic) a free play with symbols according to (logically) arbitrarily given rules of the game. All this applies as much (and in the same manner) to the thinking in daily life as to a more consciously and systematically constructed thinking in the sciences.

—Albert Einstein, *On Remarks of B. Russell's Theory of Knowledge,* Ideas and Opinions, New York, 1954

Natural Language Processing
Semantic Aspects

Epaminondas Kapetanios
University of Westminster
Faculty of Science and Technology
London, UK

Doina Tatar
Babes-Bolyai University
Faculty of Mathematics and Computer Science
Cluj-Napoca, Romania

Christian Sacarea
Babes-Bolyai University
Faculty of Mathematics and Computer Science
Cluj-Napoca, Romania

CRC Press
Taylor & Francis Group
Boca Raton   London   New York

CRC Press is an imprint of the
Taylor & Francis Group, an **informa** business

A SCIENCE PUBLISHERS BOOK

CRC Press
Taylor & Francis Group
6000 Broken Sound Parkway NW, Suite 300
Boca Raton, FL 33487-2742

© 2014 Copyright reserved
CRC Press is an imprint of Taylor & Francis Group, an Informa business

Library of Congress Cataloging-in-Publication Data

Kapetanios, Epaminondas.
  Natural language processing : semantic aspects / Epaminondas
Kapetanios, Doina Tatar, Christian Sacarea.
       pages cm
  Includes bibliographical references and index.
  ISBN 978-1-4665-8496-9 (hardback)
 1. Natural language processing (Computer science) 2. Semantic
computing. I. Tatar, Doina. II. Sacarea, Christian. III. Title.

  QA76.9.N38K36 2013
  006--dc23

                                                          2013035172

Visit the Taylor & Francis Web site at
http://www.taylorandfrancis.com

CRC Press Web site at
http://www.crcpress.com

Science Publishers Web site at
http://www.scipub.net

Preface

Communication has always played a pivotal role in the evolution of human culture, societies and civilisation. From symbols, cave paintings, petroglyphs, pictograms, ideograms and alphabet based forms of writing, to computerised forms of communication, such as the Web, search engines, email, mobile phones, VoIP Internet telephony, television and digital media, communication technologies have been evolving in tandem with shifts in political and economic systems.

Despite the emergence of a variety of communication means and technologies, natural language, or ordinary language, signed, written or spoken, remained the main means of communication among humans with the processing of natural language being an innate facility adhered to human intellect. With the rise of computers, however, natural language has been contrasted with artificial or constructed languages, such as Python, Java, C++, a computer programming language, or controlled languages for querying and search, in that natural languages contribute to the understanding of human intelligence. Nonetheless, the rise of social networks, e.g., Facebook, Twitter, did not replace natural language as one of the main means of information and communication in today's human civilisation.

What proved, however, to be an innate ability of humans from an early age to engage in speech repetition, language understanding and, therefore, so quickly acquire a spoken vocabulary from the pronunciation of words spoken around them, turned out to be a challenge for computing devices as symbol manipulators following a predefined set of instructions. Given the overwhelming amount of text based data and information generated and consumed daily, which is also boosted by the speedy pace of technological advances, effective solutions have been sought after, in accordance with many academic, industrial and scholar activities, in order to computationally improve natural language processing and understanding with the purpose to make meaningful information and communication stand out in an amalgamation of humans and machines.

Hence, it is not surprising that many books and research publications, around this topic, saw the light of this world amid all technological advances, which have been primarily geared towards faster communication rather than a qualitative one. The main tenor, what so ever, has been given through the lenses of text analytics, text mining, natural languages and information systems, information retrieval, as well as knowledge representation, artificial intelligence and machine learning. In all these contributions, a rather mathematical and statistical approach to natural language processing and understanding, than a pure linguistics based one, prevails, particularly when it comes to semantics and pragmatics based processing of text based data and information.

As having a long standing contribution to computational natural language processing and understanding, as well as to the underpinning mathematics, the authors epitomise, in this book, their experience and knowledge in a series of classic research areas at the cross-roads of natural language processing with information retrieval and search, text mining, knowledge representation, formal concept analysis and further mathematical aspects, with a particular emphasis on semantic aspects. To this extent, the book is, by no means, an exhaustive reference list of related work, or a handbook for natural language processing, however, it does provide a roadmap for those aspiring to contribute to world knowledge in the area through the lenses of semantic computing. Besides, the book aspires to guide all those academics, scholars and researchers, who wish to engage in this world knowledge contribution, through the major challenges and bumpy road ahead, as well as through a methodological baseline based on algorithmic and mathematical thinking, which underpins any serious attempt in computational approaches.

In this context, the first part of the book (Part I) introduces the reader into the main key challenges when it comes to representing and extracting meaning with such a symbol based system called natural language. The second part (Part II) discusses those mathematical aspects, which are considered fundamental for semantics based natural language processing. From a didactical point of view, some traditional mathematical concepts such as Lattice Theory, upon which Formal Concept Analysis is based, are being addressed in order to provide a second thought about the recurrent problems mainly caused by the shortly lived memory of classical studies in the field. Part III embarks on the knowledge representation aspects related with natural language processing in the flavour of measuring similarity among words, the pivotal role of semantics in query languages, as well as attempts to specify universal grammar for natural languages in the context of multi-lingual querying. Finally, part IV discusses knowledge extraction aspects related with natural language processing such as word sense disambiguation, text segmentation and summarisation, text entailment, named entity recognition.

Epaminondas Kapetanios
Doina Tatar
Christian Săcarea

Contents

PART IV: Knowledge Extraction and Engineering for NLP

PART I
Introduction

CHAPTER 1

The Nature of Language

1.1 Syntax versus Semantics

It has been claimed many times in the past that humans are, somehow, born for grammar and speech as an innate ability to see the structure underlying a string of symbols. A classic example is the ease with which children pick up languages, which, in turn, has undergone evolutionary pressure. Without language, knowledge cannot be passed on, but only demonstrated. For instance, chimps can show the offsprings processes but cannot tell their about them, since a demonstration is required. Languages are, therefore, brought into connection with information, sometimes quite crucial. Language can help you to make plans. Many of the Spanish conquistadores who conquered Mesoamericans could not read, but their priests could. Moreover, being able to record language provides access to thousands of years of knowledge.

Generation and recognition of sentences pose two main problems for the concept of language as an assembly of a set of valid sentences. Though most textbooks deal with the understanding of the recognition of languages, one cannot ignore understanding the generation of language, if we aspire to understand recognition seriously.

A language can be described as a series of simple syntactic rules. For instance, English is a language defined with some simple rules, which are more loose than strict. This fact, however, may also highlight that a language can be hard to define with a only series of simple syntactic rules. Let us assume the following sentences:

- John gesticulates
- John gesticulates vigorously
- The dog ate steak
- The dog ate ravenously

There are semantic rules (rules related to the meanings of sentences) in addition to the syntactic rules (rules regarding grammar). The rule usually specified, are strictly syntactic and, at least for computer languages, the easiest to formulate. The semantic rules, however, are notoriously difficult to formulate and are anchored in one's brain subconsciously, associating concepts with words and structuring the words into phrases and groups of phrases, which convey the meanings intended.

At a syntactic level and working towards some grammatical patterns or rules in English, one might be doing this consciously. There will always be a person or thing (a *subject*) and a verb describing an action (a *verb phrase*) in almost every language. In addition, there will sometimes be an *object* that the subject acts upon. In order to reflect on these abstract structures, one might find oneself using some other symbols acting as containers or patterns for sentences with a similar structure. For instance, one may end up with something like:

- *Subject* gesticulates
- *Subject* gesticulates vigorously
- *Subject* ate steak
- *Subject* ate ravenously

Next, abstract the verb phrases:

- *Subject VerbPhrase*
- *Subject VerbPhrase*
- *Subject VerbPhrase* steak
- *Subject VerbPhrase*

Finally, abstracting away the objects, we may end up with something like:

- *Subject VerbPhrase*
- *Subject VerbPhrase*
- *Subject VerbPhrase Object*
- *Subject VerbPhrase*

It is now easy to spot two main types of sentences that underpin the lexical-syntactic meanings of these four sentences:

- *Subject VerbPhrase*
- *Subject VerbPhrase Object*

You may also break down subject or verb phrases by having emerging sub-structures such as *noun (e.g., John)* and *determiner-noun (e.g., The dog)* for subject phrases, or *verb* (e.g., ate) and *verb-adverb* (e.g., ate ravenously) for verb phrases. Subsequently, you may end up with a finite language defined by the following rules of grammar:

1. A **sentence** is a subject followed by a verb phrase, optionally followed by an object.
2. A **subject** is a noun optionally preceded by a determiner.
3. A **verb** phrase is a verb optionally followed by an adverb.
4. A **noun** is John or dog.
5. A **verb** is gesticulates or ate.
6. An **adverb** is vigorously or ravenously.
7. An **object** is steak.
8. A **determiner** is The.

Despite the fact that the structure of these rules might seem to be right, here is exactly where the problem lies with the meaning of the sentences and the semantic rules associated with it. For example, the rules may allow you to say, "dog gesticulates ravenously," which is perfectly meaningless and a situation, which is frequently encountered as specifying grammars.

Having taken a look at how easily things might become quite complex when we need to define semantic rules on top of the syntactic ones, even with such a finite language, one can imagine that defining a strict grammar, i.e., including semantic rules, is almost impossible. For instance, a book on English grammar can easily become four inches thick. Besides, a natural language such as English is a moving target. For example, consider the difference between Elizabethan English and Modern English.

Again, as one discovers meta-languages in the next section one can bear in mind, that there is sometimes a gap between the language one means and the language one can easily specify. Another lesson learned is that using a language like English, which is neither precise nor terse, to describe other languages and, therefore, use it as a meta-language, one will end up with a meta-language with the same drawbacks.

The manner in which computer scientists have specified languages has been quite similar and is continuously evolving. Regardless of the variety and diversity of computer languages, semantic rules have rarely been an integral part of the language specification, if at all. They are mostly syntactic rules, which dominate the language specification. Take, for instance, context-free grammar (CFG), the first meta-language used extensively and preferred by most computer scientists. CFG specifications provide a list of rules with left and right hand sides separated by a right-arrow symbol. One of the rules is identified as the *start rule* or *start symbol*, implying that the overall structure of any sentence in the language is described by that rule. The left-hand side specifies the name of the substructure one is defining and the right hand side specifies the actual structure (sometimes called a *production*): a sequence of references to other rules and/or words in the language vocabulary.

Despite the fact that language theorists love CFG notation, most language reference guides use BNF (Backus- Naur Form) notation, which is really just a more readable version of CFG notation. In BNF, all rule names are surrounded by <...> and Æ is replaced with "::=". Also, alternative productions are separated by '|' rather than repeating the name of the rule on the left-hand side. BNF is more verbose, but has the advantage that one can write meaningful names of rules and is not constrained vis- à-vis capitalization. Rules in BNF take the form:

> *<rulename> ::= production 1*
> *| production 2*
> ...
> *| production n*

Using BNF, one can write the eight rules used previously in this chapter as follows:

> *<Sentence> ::= <Subject> <VerbPhrase> <Object>*
> *<Subject> ::= <Determiner> <Noun>*
> *<VerbPhrase> ::= <Verb> <Adverb>*
> *<Noun> ::=* John | dog
> *<Verb> ::=* gesticulates | ate
> *<Adverb> ::=* vigorously | ravenously |
> *<Object> ::=* steak |
> *<Determiner> ::=* The |

Even if one uses alternatives such as YACC, the de facto standard for around 20 years, or ANTLR, or many other extended BNF (EBNF) forms, the highest level of semantic rule based specification one might achieve would be by introducing grammatical categories such as DETERMINER, NOUN, VERB, ADVERB and by having words such as *The, dog, ate, ravenously*, respectively, belonging to one of these categories. The intended grammar may take the following form,

```
sentence : subject verbPhrase (object)?;
subject : (DETERMINER)? NOUN;
verbPhrase : VERB (ADVERB)?;
object : NOUN;
```

which still leaves plenty of space for construction of meaningless sentences such as *The dog gestured ravenously*. It is also worth mentioning that even with alternatives for CFG such as *regular expressions*, which were meant to simplify things by working with characters only and no rules referencing other ones on the right-hand side, things did not improve towards embedding of semantic rules in a language specification. In fact, things turned out to be more complex with *regular expressions,* since without recursion (no stack), one cannot specify repeated structures.

In short, one needs to think about the difference between sequence of words in a sentence and what really dictates the validity of sentences. Even with the programming expression, if one is about to design state machinery capable of recognizing semantically sensitive sentences, the key idea must be that a sentence is not merely a cleverly combined sequence of words, but rather groups of words and groups of groups of words. Even with the programming expression (a[i+3]), humans can immediately recognize that there is something wrong with the expression, whereas it is notoriously difficult to design state machinery recognizing the faulty expression, since the number of left parentheses and brackets matches the number of one on the right.

In other words, sentences have a *structure* like this book. This book is organized into a series of chapters each containing sections, which, in turn, contain subsections and so on. Nested structures abound in computer science too. For example, in an object-oriented class library, classes group all elements beneath them in the hierarchy into categories (any kind of cat might be a subclass of feline etc...). The first hint of a solution to the underpowered state machinery is now apparent. Just as a class library is not an unstructured category of classes, a sentence is not just a flat list of words. Can one, therefore, argue that the role one gave each word, plays an equally large part in one's understanding of a sentence? Certainly, it does, but it is not enough. The examples used earlier, highlight the fact that structure imparts meaning very clearly. It is not purely the words though, nor the sequence that impart meaning. Can it also be argued that if state machines can generate invalid sentences, they must be having trouble with structure? These questions will be left unanswered for the time being, or perhaps in the near future. It turns out that even if we manage to define state machinery to cope with structure in a sentence, claiming that once semantic rules are perfectly defined, it is far reaching, since there are more significant issues to consider, as we will see in the following sections.

1.2 Meaning and Context

The difficulty of defining semantic rules to cope with meaningful states, operations or statements is exacerbated by the conclusions drawn from the study of "Meaning" as a key concept for understanding a variety of processes in living systems. It turns out that "Meaning" has an elusive nature and "subjective" appearance, which is, perhaps the reason why it has been ignored by information science. Attempts have been made to circumscribe a theory of meaning in order to determine the meaning of an indeterminate sign. Meaning-making, however, has been considered as the procedure for extracting the information conveyed by a message, in which the former is considered to be the set of values one might assign to an indeterminate signal. In this context, meaning-making is described in terms of a constraint-satisfaction problem that relies heavily on contextual cues and inferences.

The lack of any formalization of the concepts "meaning" and "context", for the working scientist, is probably due to the theoretical obscurity of concepts associated with the axis of semiotics in information processing and science. Even with regard to information and information flow, it has been argued that "the formulation of a precise, qualitative conception of information and a theory of the transmission of information has proved elusive, despite the many other successes of computer science" (Barwise and Seligman 1993). Since Barwise's publication, little has changed. Researchers in various fields still find it convenient to conceptualize the data in terms of information theory. By doing so, they are excluding the more problematic concept of meaning from the analysis. It is clear, however, that the meaning of a message cannot be reduced to the information content.

In a certain sense, the failure to reduce meaning to information content is like the failure to measure organization through information content. Moreover, the relevance of information theory is criticized by those who argue that when we study a living system, as opposed to artificial devices, our focus should be on meaning-making rather than information processing per se. In the context of artificial devices, the probabilistic sense of information prevails. Meaning, however, is a key concept for understanding a variety of processes in living systems, from recognition capacity of the immune system to the neurology of the perception. Take, for instance, the use of information theory in biology, as stated by (Emmeche and Hoffmeyer 1991). They argue that unpredictable events are an essential part of life and it is impossible to assign distinct probabilities to any event and conceptualize the behavior of living systems in terms of information theory. Therefore, biological information must embrace the "semantic openness" that is evident, for example, in human communication.

In a nutshell, it is worth mentioning that meaning has taken both main views, divorced from information and non-reducible to each other. The first is due to the fact that the concept of information relies heavily on "information theory" like Shannon's statistical definition of information, whereas the latter is due to the conception that information can broadly be considered as something conveyed by a message in order to provoke a response (Bateson 2000). Hence, the message can be considered as a portion of the world that comes to the attention of a cogitative system, human or non-human. Simply stated, information is a differentiated portion of reality (i.e., a message), a bit of information as a difference, which makes a difference, a piece of the world that comes to notice and results in some response (i.e., meaning). In this sense, information is interactive. It is something that exists in between the responding system and the differentiated environment, external or internal. An example, if one leaves one's house to take a walk, notices that the sky is getting cloudy, one is likely to change one's plans in order to avoid the rain. In this care the cloudy sky may be considered the message (i.e., the difference) and one's avoidance will be the information conveyed by the message (i.e., a difference that makes a difference). In this context, information and meaning are considered synonymous and without any clear difference between them. Though they are intimately related, they cannot be reduced to each other.

In the same spirit, Bateson presents the idea that a differentiated unit, e.g., *a word*, has meaning only on a higher level of logical organization, e.g., *the sentence*, only in context and as a result of interaction between the organism and the environment. In this sense, the internal structure of the message is of no use in understanding the meaning of the message.

The pattern(s) into which the sign is woven and the interaction in which it is located is what turns a differentiated portion of the world into a response by the organism. This idea implies that turning a signal (i.e., a difference) into a meaningful event (i.e., a difference that makes a difference) involves an active extraction of information from the message. Based on the suggestions, the following ideas have been suggested:

a) Meaning-making is a procedure for extracting the information conveyed by a message.
b) Information is the value one may assign to an indeterminate signal (i.e., a sign).

These ideas are very much in line with conceptions that see meaning-making as an active process that is a condition for information-processing rather than the product of information-processing per se. The most interesting things in the conception of meaning-making as an active process, are the three organizing concepts of a) indeterminacy of the signal, b) contextualization, c) transgradience. The indeterminacy (or variability) of the signal is an important aspect of any meaning-making process. *It answers the question what is the indeterminacy of the signal and why is it important for a theory of meaning-making?* The main idea is that in itself every sign/unit is devoid of meaning until it is contextualized in a higher-order form of organization such as a *sentence*. It can be assigned a range of values and interpretations.

For instance, in natural language the sign "shoot" can be used in one context to express an order to a soldier to fire his gun and in a different context as a synonym for "speak". In immunology, the meaning of a molecule's being an antigen is not encapsulated in the molecule itself. That is, at the most basic level of analysis, a sign has the potential to mean different things (i.e., to trigger different responses) in different contexts, a property that is known in linguistics as polysemy and endows language with enormous flexibility and cognitive economy. In the field of linguistics it is called pragmatics, which deals with meaning in context, the single most obvious way in which the relation between language and context is reflected in the structure of languages themselves is often called *deixis* (pointing or indicating in Greek). To this extent, linguistic variables (e.g., this, he, that) are used to indicate something in a particular context. They are indeterminate signals.

Nevertheless, the indeterminacy of a signal or word can be conceived as a constraint satisfaction problem. This, in turn, is defined as a triple {V, D, C}, where: (a) V is a set of variables, (b) D is a domain of values, and (c) C is a set of constraints {C1,C2, . . . Cq}. In the context of semiotics, V is considered to be the set of indeterminate signals and D the finite set of interpretations/values one assigns to them. Based on the above definition, a sign is indeterminate if assigning it a value is a constraint-satisfaction problem. One should note that solving the constraint-satisfaction problem is a meaning-making process, since it involves the extraction of the information conveyed by a message (e.g., to whom does the "he" refer?). However, rather than a simple mapping from V to D, this process also involves contextualization and inference.

The problematic notion of *context*, in the conception of meaning-making as an active process, can be introduced better as an *environmental setting composed of communicating units and their relation in time and space*. The general idea and situation theory (Seligman and Moss 1997) is one possible way of looking into these aspects. In situation theory, a situation is *"individuals in relations having properties and standing in relations to various spatiotemporal relations"*. In a more general way, we can define a situation as a pattern, a meaning, an ordered array of objects that have an identified and repeatable form. In an abstract sense, a contextualization process can be conceived as a functor or a structure-preserving mapping of the particularities or the token of the specific occurrence onto the generalities of the pattern.

Regarding the interpretation of things as a constraints satisfaction problem, a context forms the constraints for the possible values (i.e., interpretations) that one may attribute to a sign. According to this logic, a situation type is a structure or pattern of situations. In other words, a situation is defined as a set of objects organized both spatially and temporally in a given relation. If this relation characterizes a set of situations, one can consider it a structure or a situation type. For example, the structure of hierarchical relations is the same no matter who the boss is. The situation type is one of hierarchical relations. Based on this type of a situation, we can make inferences about other situations. For example, violations of a rigid hierarchical relationship by a subordinate are usually responded to with another situation of penalties imposed by the superiors.

Although a sign, like the meaning of a word in a sentence, is indeterminate, in a given context one would like to use it to communicate only one meaning (i.e., to invite only one specific response) and not others. Therefore, the word disambiguation problem arises. In a sense, inferences via contextualization work as a zoom-in, zoom-out function. Contextualization offers the ability to zoom out and reflexively zoom back, in a way that constrains the possible values we may assign to the indeterminate signal. In other words, in order to determine the meaning of a microelement and extract the information it conveys, one has to situate it on a level of higher order of organization.

Let us consider the following example: "I hate turkeys". The vertical zooming-out from the point representing "I" to the point representing "human" captures the most basic denotation of "I," given that *denotation* is the initial meaning captured by a sign. As such, it could be considered the most reasonable starting point for contextualization. It is also a reasonable starting point because both evolutionary and ontological denotation is the most basic semiotic category. According to the *Oxford English Dictionary*, a turkey can be zoomed out to its closest ontological category, a "bird". This ontological category commonly describes any feathered vertebrate animal. Therefore, if we are looking for a function from the indeterminate sign "love" to a possible value/interpretation, the first constraint is that it is a relation between a *human being* and an *animal*.

There is, however, one more contextual cue. This is dictated by the denotation of dinner as a token of a *meal*. In that situation, there is a relationship of eating between *human beings* and *food*. Given that the zoomed-out concept of human beings for the sign "I" does participate in this relationship as well, another candidate value for the interpretation of "love" will arise, which, apparently, makes more sense, since it may be much closer to the meaning of the sentence "I hate turkeys". The situation where humans consume turkeys as food is the one giving meaning to this sentence.

Contextualization, however, is not sufficient to meaning-making processes. *Transgradience*, as the third dimension of the meaning-making process, refers to the need for interpretation, inference and integration as a process in which *inferences are applied to a signal-in-context in order to achieve a global, integrated view of a situation*. In general terms, transgradience refers to the ability of the system to achieve a global view of a situation by a variety of means. An interesting parallel can be found by the immune system deciding whether a molecule is an antigen by means of a complex network of immune agents that communicate, make inferences, and integrate the information they receive. Further sub-dimensions may arise though, which could potentially complicate things: (1) the spatiotemporal array in which the situation takes place; (2) our background knowledge of the situation and (3) our beliefs concerning the situation.

In brief, our ability to extract the information that a word may convey as a signal, is a meaning-making process that relies heavily on contextual cues and inferences. Now the challenge is to pick the right situation, which constrains the interpretation values to be allocated to indeterminate signals, i.e., ambiguous words. Although one's understanding of semiotic systems has advanced (Sebeok and Danesi 2000) and computational tools have reached a high level of sophistication, one still does not have a satisfactory answer to the question of how the meaning emerges in a particular context.

1.3 The Symbol Grounding Problem

The whole discussion, so far, is underpinned by the assumption that the adherence of meaning to symbols and signals is a result of a meaning-making process rather than something intrinsic to the symbols and the chosen symbolic system itself. It is this innate feature of any symbolic system, which poses limitations to any symbol manipulator, to the extent to which one can interpret symbols as having meaning systematically. This turns interpretation of any symbol such as letters or words in a book parasitic, since they derive their meaning from us similarly, none of the symbolic systems can be used as a cognitive model and therefore, cognition cannot just be a manipulation of a symbol. Spreading this limitation would mean grounding every symbol in a symbolic system with its meaning and not leaving interpretation merely to its shape. This has been referred to in the 90s as the famous 'symbol grounding problem' (Harnad 1990) by raising the following questions :

• "How can the semantic interpretation of a formal symbol system be made intrinsic to the system, rather than remain parasitic, depending solely on the meanings in our heads?"

• "How can the meanings of the meaningless symbol tokens, manipulated solely on the basis of their (arbitrary) shapes, be grounded in anything but other meaningless symbols?"

The problem of constructing a symbol manipulator able to understand the extrinsic meaning of symbols, has been brought into analogy with another famous problem of trying to learn Chinese from a Chinese/Chinese dictionary. It also sparked off the discussion about symbolists, symbolic Artificial Intelligence and the symbolic theory of mind, which has been challenged by Searle's "Chinese Room Argument". According to these trends, it has been assumed that if a system of symbols can generate indistinguishable behavior in a person, this system must have a mind. More specifically, according to the symbolic theory of mind, if a computer could pass the Turing Test in Chinese, i.e., if it could respond to all Chinese symbol strings it receives as input from Chinese symbol strings that are indistinguishable from the replies a real Chinese speaker would make (even if we keep on testing infinitely), the computer would understand the meaning of Chinese symbols in the same sense that one understands the meaning of English symbols. Like Searle's demonstration, this turns out to be impossible, for both humans and computers, since the meaning of the Chinese symbols is not intrinsic and depends on the shape of the chosen symbols. In other words, imagine that you try to learn Chinese with a Chinese/Chinese dictionary only. The trip through the dictionary would amount to a merry-go-round, passing endlessly from one meaningless symbol or symbol-string (the definientes) to another (the definienda), never stopping to explicate what anything meant.

The standard reply and approach of the symbolists and the symbolic theory of mind, which prevails in the views of semantic aspects in natural language processing within this book as well, is that the meaning of the symbols comes from connecting the symbol system to the world "in the right way." Though this view trivializes the symbol grounding problem and the meaning making process in a symbolic system, it also highlights the fact that if each definiens in a Chinese/Chinese dictionary were somehow connected to the world in the right way, we would hardly need the definienda. Therefore, this would alleviate the difficulty of picking out the objects, events and states of affairs in the world that symbols refer to.

With respect to these views, hybrid non-symbolic / symbolic systems have been proposed in which the elementary symbols are grounded in some kind of non-symbolic representations that pick out, from their proximal sensory projections, the distal object categories to which the elementary symbols refer. These groundings are driven by the insights of how humans can (1) discriminate, (2) manipulate, (3) identify and (4) describe the objects, events and states of affairs in the world they live in and they can also (5) "produce descriptions" and (6) "respond to descriptions" of those objects, events and states of affairs.

The attempted groundings are also based on discrimination and identification, as two complementary human skills. To be able to discriminate one has to judge whether two inputs are the same or different, and, if different, to what degree. Discrimination is a relative judgment, based on our capacity to tell things apart and discern the degree of similarity.

Identification is based on our capacity to tell whether a certain input is a member of a category or not. Identification is also connected with the capacity to assign a unique response, e.g., a name, to a class of inputs. Therefore, the attempted groundings must rely on the answers to the question asking what kind or internal representation would be needed in order to be able to discriminate and identify. In this context, iconic representations have been proposed (Harnad 1990). For instance, in order to be able to discriminate and identify horses, we need horse icons. Discrimination is also made independent of identification in that one might be able to discriminate things without knowing what they are. According to the same theorists, icons alone are not sufficient to identify and categorize things in an underdetermined world full with infinity of potentially confusable categories. In order to identify, one must selectively reduce those to "invariant features" of the sensory projection that will reliably distinguish a member of a category from any non-members with which it could be confused. Hence, the output is named "categorical representation". In some cases these representations may be innate, but since evolution could hardly anticipate all the categories one may ever need or choose to identify, most of these features must be learned from experience. In a sense, the categorical representation of a horse is probably a learned one.

It must be noted, however, that both representations are still sensory and non- symbolic. The former are analogous copies of the sensory projection, preserving its "shape" faithfully. The latter are supposed to be icons that have been filtered selectively to preserve only some of the features of the shape of the sensory projection, which distinguish members of a category from non-members reliably. This sort of non-symbolic representation seems to differ from the symbolic theory of mind and currently known symbol manipulators such as conventional computers trying to cope with natural language processing and their semantic aspects.

Despite the interesting views emerging from the solution approaches to the symbol grounding problem, the symbol grounding scheme, as introduced above has one prominent gap: no mechanism has been suggested to explain how the all-important categorical representations can be formed. How does one find the invariant features of the sensory projection that make it possible to categorize and identify objects correctly? To this extent, connectionism, with its general pattern learning capability, seems to be one natural candidate to complement identification. In effect, the "connection" between the names and objects that give rise to their sensory projections and icons would be provided by connectionist networks. Icons, paired with feedback indicating their names, could be processed by a connectionist network that learns to identify icons correctly from the sample of confusable alternatives it has encountered. This can be done by adjusting the weights of the features and feature combinations that are reliably associated with the names in a way that may resolve the confusion. Nevertheless, the choice of names to categorize things is not free from extrinsic interpretation of things, since some symbols are still selected to describe categories.

PART II

Mathematics

CHAPTER 2

Relations

The concept of a relation is fundamental in order to understand a broad range of mathematical phenomena. In natural language, *relation* is understood as *correspondence, connection*. We say that two objects are related if there is a common property linking them.

Definition 2.0.1. Consider the sets A and B. We call *(binary) relation* between the elements of A and B any subset $R \subseteq A \times B$. An element $a \in A$ is in relation R with an element $b \in B$ if and only if $(a, b) \in R$. An element $(a, b) \in R$ will be denoted by aRb.

Definition 2.0.2. If $A_1, \ldots, A_n$, $n \geq 2$ are sets, we call an *n*-ary relation any subset $R \subseteq A_1 \times \ldots \times A_n$. If $n = 2$, the relation R is called *binary*, if $n = 3$ it is called *ternary*. If $A_1 = A_2 = \ldots = A_n = A$, the relation R is called *homogenous*.

In the following, the presentation will be restricted only to binary relations.

Remark 1 *The direct product $A \times B$ is defined as the set of all ordered pairs of elements of A and B, respectively:*

$$A \times B := \{(a, b) \mid a \in A, b \in B\}.$$

Remark 2 *If A and B are finite sets, we can represent relations as cross tables. The rows correspond to the elements of A, while the columns correspond to the elements of B. We represent the elements of R, i.e., $(a, b) \in R$, by a cross (X) in this table. Hence, the relation R is represented by a series of entries (crosses) in this table. If at the intersection of row a with column b there is no entry, it means that a and b are not related by R.*

Example 2.0.1. Relations represented as cross-tables

(1) Let $A = \{a\}$. There are only two relations on A, the empty relation, $R = \varnothing$, and the total relation, $R = A \times A$.

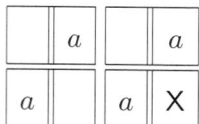

(2) $A := \{1, 2\}$, $B := \{a, b\}$. Then, all relations $R \subseteq A \times B$ are described by

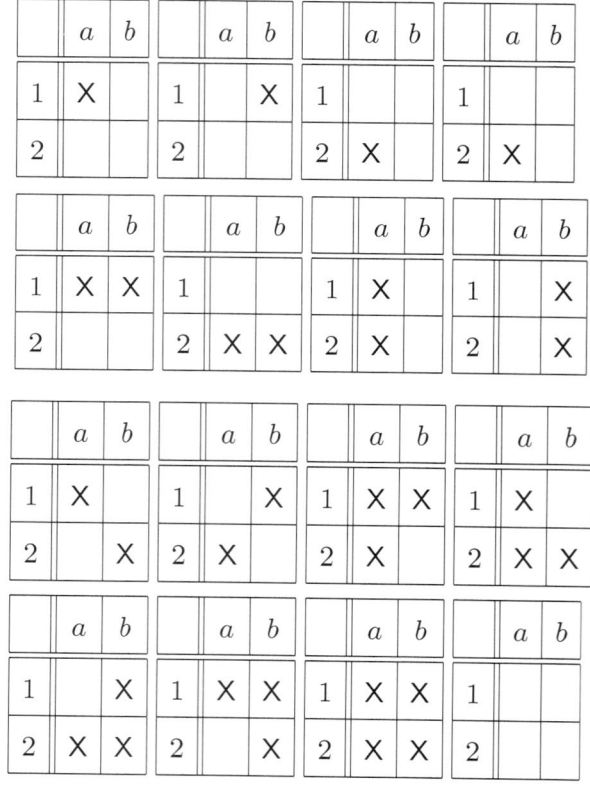

Example 2.0.2. In applications, A, B, and $R \subseteq A \times B$ are no longer abstract sets, they have a precise semantics, while the relation R represents certain correspondences between the elements of A and B. The following example describes some arithmetic properties of the first ten natural numbers:

	even	odd	div.by 3	div.by 5	div.by 7	prime	$x^2 + y^2$	$x^2 - y^2$
1		X						
2	X					X	X	
3		X	X			X		X
4	X							
5		X		X		X		X
6	X		X					
7		X			X	X		
8	X						X	X
9		X	X					
10	X			X				

Example 2.0.3. Other relations

(1) The divisibility relation in $\mathbb{Z}$: $R := \{(m, n) \in \mathbb{Z}^2 \mid \exists k \in \mathbb{Z}.\ n = km\}$.

(2) $R := \{(x, y) \in \mathbb{R}^2 \mid x^2 + y^2 = 1\}$. This relation consists of all points located on the circle centered in the origin and radius 1.

(3) The equality relation on set A:
$$\Delta_A := \{(x, x) \mid x \in A\}.$$

(4) The equality relation in $\mathbb{R}$ consists of all points located on the first bisecting line.

(5) The universal relation on a set A expresses the fact that all elements of that set are related to each other:
$$\nabla_A := \{(x, y) \mid x, y \in A\}.$$

(6) The empty relation means that none of the elements of A and B are related:
$$R = \varnothing \subseteq A \times B.$$

(7) Let $A = B = \mathbb{Z}$ and R the divisibility relation on $\mathbb{Z}$:

$$R := \{(x, y) \in \mathbb{Z} \times \mathbb{Z} \mid \exists k \in \mathbb{Z}.\ y = kx\}.$$

2.1 Operations with Relations

Let (A_1, B_1, R_1) and (A_2, B_2, R_2) be two binary relations. They are equal if and only if $A_1 = A_2$, $B_1 = B_2$, $R_1 = R_2$.

Definition 2.1.1. Let A and B be two sets, R and S relations on $A \times B$. Then R is *included* in S if $R \subseteq S$.

Definition 2.1.2. Let $R, S \subseteq A \times B$ be two relations on $A \times B$. The *intersection* of R and S is defined as the relation $R \cap S$ on $A \times B$.

Definition 2.1.3. Let $R, S \subseteq A \times B$ be two relations on $A \times B$. The *union* of R and S is defined as the relation $R \cup S$ on $A \times B$.

Definition 2.1.4. Let $R \subseteq A \times B$ be a relation on $A \times B$. The *complement* of R is defined as the relation $\complement R$ on $A \times B$, where

$$\complement R := \{(a, b) \in A \times B \mid (a, b) \notin A \times B\}.$$

Remark 3 *If R and S are relations on $A \times B$ then*

(1) $a(R \cap S)b \Leftrightarrow aRb$ and aSb.

(2) $a(R \cup S)b \Leftrightarrow aRb$ or aSb.

(3) $a(\complement R)b \Leftrightarrow (a, b) \in A \times B$ and $(a, b) \notin R$.

Definition 2.1.5. Let $R \subseteq A \times B$ and $S \subseteq C \times D$ be two relations. The *product* or *composition* of R and S is defined as the relation $S \circ R \subseteq A \times D$ by

$$S \circ R := \{(a, d) \in A \times D \mid \exists b \in B \cap C. \ (a, b) \in R \text{ and } (b, d) \in S\}.$$

If $B \cap C = \varnothing$, then $S \circ R = \varnothing$.

Definition 2.1.6. *Let $R \subseteq A \times B$ be a relation. The *inverse* of R is a relation $R^{-1} \subseteq B \times A$ defined by*

$$R^{-1} := \{(b, a) \in B \times A \mid (a, b) \in R\}.$$

Theorem 2.1.1. *Let $R \subseteq A \times B$, $S \subseteq C \times D$, and $T \subseteq E \times F$ be relations.*

Then the composition of relations is associative:

$$(T \circ S) \circ R = T \circ (S \circ R).$$

Proof. Let $(a, f) \in (T \circ S) \circ R$. By the definition of the relational product, there exists $b \in B \cap C$ with $(a, b) \in R$ and $(b, f) \in T \circ S$. By the same definition, there exists $d \in D \cap E$ with $(b, d) \in S$ and $(d, f) \in T$. Now, $(a, b) \in R$ and $(b, d) \in S$ imply $(a, d) \in S \circ R$. Together with $(d, f) \in T$, this implies that $(a, f) \in T \circ (S \circ R)$. Hence,

$$(T \circ S) \circ R \subseteq T \circ (S \circ R).$$

The converse implication is proved analogously.

Theorem 2.1.2. *Let* $R_1 \subseteq A \times B$, $R_2 \subseteq A \times B$, $S_1 \subseteq C \times D$, $S_2 \subseteq C \times D$ *be relations. The following hold true:*

(1) $R_1 \circ (S_1 \cup S_2) = (R_1 \circ S_1) \cup (R_1 \circ S_2)$.

(2) $(R_1 \cup R_2) \circ S_1 = (R_1 \circ S_1) \cup (R_2 \circ S_1)$.

(3) $R_1 \circ (S_1 \cap S_2) \subseteq (R_1 \circ S_1) \cap (R_1 \circ S_2)$.

(4) $(R_1 \cap R_2) \circ S_1 \subseteq (R_1 \circ S_1) \cap (R_2 \circ S_1)$.

Proof.

(1) Let $(c, b) \in R_1 \circ (S_1 \cup S_2)$. Then, there exists an element $d \in D \cap A$ with $(c, d) \in S_1 \cup S_2$ and $(d, b) \in R$, i.e., $(c, d) \in S_1$ or $(c, d) \in S_2$ and $(d, b) \in R$. This means that $(c, b) \in R_1 \circ S_1$ or $(c, b) \in R_1 \circ S_2$, i.e., $(c, b) \in (R_1 \circ S_1) \cup (R_1 \circ S_2)$. We have proved that $R_1 \circ (S_1 \cup S_2) \subseteq (R_1 \circ S_1) \cup (R_1 \circ S_2)$.

The converse inclusion follows analogously.

(2) By a similar argument, one obtains $(R_1 \cup R_2) \circ S_1 = (R_1 \circ S_1) \cup (R_2 \circ S_1)$.

(3) Let $(c, b) \in R_1 \circ (S_1 \cap S_2)$. Then, there exists an element $d \in D \cap A$ with $(c, d) \in S_1 \cap S_2$ and $(d, b) \in R$. Hence $(c, d) \in S_1$, $(c, d) \in S_2$ and $(d, b) \in R$. By the definition of the relational product, $(c, b) \in R_1 \circ S_1$ and $(c, b) \in R_1 \circ S_2$, hence $(c, b) \in (R_1 \circ S_1) \cap (R_1 \circ S_2)$.

(4) Left to the reader.

Theorem 2.1.3. *Let* R_1, $R_2 \subseteq A \times B$ *and* $S \subseteq C \times D$ *be binary relations.*

Then

(1) $(R_1 \cup R_2)^{-1} = R_1^{-1} \cup R_2^{-1}$.

(2) $(R_1 \cap R_2)^{-1} = R_1^{-1} \cap R_2^{-1}$.

(3) $(R_1 \circ S)^{-1} = S^{-1} R_1^{-1}$.

(4) $(\complement R_1)^{-1} = \complement (R_1)^{-1}$.

(5) $((R_1)^{-1})^{-1} = R_1$.

The proof is left as an exercise to the reader.

Corollary 2.1.4.

(1) *If* $S_1 \subseteq S_2$, *then* $R \circ S_1 \subseteq R \circ S_2$.

(2) *If* $R_1 \subseteq R_2$, *then* $R_1 \circ S \subseteq R_2 \circ S$.

(3) $R_1 \subseteq R_2 \Leftrightarrow R_1^{-1} \subseteq R_2^{-1}$.

Definition 2.1.7. Let $R \subseteq A \times B$ be a binary relation, $X \subseteq A$ a subset of A and $a \in A$ an arbitrary element. The *cut* of R after X is defined by

$$R(X) := \{b \in B \mid \exists x \in X.\ (x, b) \in R\}.$$

The set $R(A)$ is called *image* of R and $R^{-1}(B)$ is called *preimage* of R. The cut after a is denoted by $R(a)$.

Theorem 2.1.5. *Let $R \subseteq A \times B$ be a binary relation and $X_1, X_2 \subseteq A$ subsets. Then*

(1) $R(X_1 \cup X_2) = R(X_1) \cup R(X_2)$.

(2) $R(X_1 \cap X_2) \subseteq R(X_1) \cap R(X_2)$.

(3) *If $X_1 \subseteq X_2$ then $R(X_1) \subseteq R(X_2)$.*

Proof.

(1) Suppose $y \in R(X_1 \cup X_2)$. Then there is an $x \in X_1 \cup X_2$ with $(x, y) \in R$, i.e., $\exists x \in X_1.(x, y) \in R$ or $\exists x \in X_2.(x, y) \in R$, hence $y \in R(X_1)$ or $y \in R(X_2)$, i.e., $y \in R(X_1) \cup R(X_2)$. The converse inclusion is proved by a similar argument.

(2) If $y \in R(X_1 \cap X_2)$, then there is an $x \in X_1 \cap X_2$ with $(x, y) \in R$, i.e., $\exists x \in X_1.(x, y) \in R$ and $\exists x \in X_2.(x, y) \in R$, hence $y \in R(X_1)$ and $y \in R(X_2)$, which implies $y \in R(X_1) \cap R(X_2)$. The converse inclusion does not hold in general.

(3) Follows directly from $R(X_1 \cup X_2) = R(X_1) \cup R(X_2)$.

Theorem 2.1.6. *Let $R_1, R_2 \subseteq A \times B$ and $S \subseteq B \times C$ be relations and $X \subseteq A$. Then*

(1) $(R_1 \cup R_2)(X) = R_1(X) \cup R_2(X)$.

(2) $(R_1 \cap R_2)(X) \subseteq R_1(X) \cap R_2(X)$.

(3) $(S \circ R_1)(X) = S(R_1(X))$.

(4) *If $R_1 \subseteq R_2$ then $R_1(X) \subseteq R_2(X)$.*

Proof.

(1) Suppose $y \in (R_1 \cup R_2)(X)$. Then there is an $x \in X$ with $(x, y) \in R_1 \cup R_2$, i.e., $(x, y) \in R_1$ or $(x, y) \in R_2$, hence $y \in R(X_1)$ or $y \in R(X_2)$ which implies $y \in R_1(X) \cup R_2(X)$. The converse inclusion follows similarly.

(2) Let $y \in (R_1 \cap R_2)(X)$. Then there is an $x \in X$ with $(x, y) \in R_1 \cap R_2$, i.e., $(x, y) \in R_1$ and $(x, y) \in R_2$, hence $y \in R_1(X)$ and $y \in R_2(X)$

which implies $y \in R_1(X) \cap R_2(X)$. The converse inclusion is not generally true.

(3) Let $z \in (S \circ R_1)(X)$. Then there is an $x \in X$ with $(x, z) \in S \circ R_1$, hence there is also a $y \in B$ with $(x, y) \in R_1$ and $(y, z) \in S$, i.e., $\exists y \in B.y \in R_1(X)$ and $(y, z) \in S$ wherefrom follows $z \in S(R_1(X))$. The converse inclusion follows similarly.

(4) Follows directly from $(R_1 \cup R_2)(X) = R_1(X) \cup R_2(X)$.

2.2 Homogenous Relations

Definition 2.2.1. A relation $R \subseteq A \times A$ on set A is called

(1) *reflexive* if for every $x \in A$, we have $(x, x) \in R$;
(2) *transitive* if for every $x, y, z \in A$, from $(x, y) \in R$ and $(y, z) \in R$ follows $(x, z) \in R$;
(3) *symmetric* if for every $x, y \in A$, from $(x, y) \in R$ follows $(y, x) \in R$;
(4) *antisymmetric* if for every $x, y \in A$, from $(x, y) \in R$ and $(y, x) \in R$ follows $x = y$.

Remark 4 *As we have seen before, if A is finite, a relation $R \subseteq A \times A$ can be represented as a cross table. For homogenous relations, this cross table is a square.*

(1) *A finite relation is reflexive if all elements from the main diagonal are marked:*

Example 2.2.1. The equality relation on set A is always reflexive.

For $A := \{1, 2, 3, 4, 5\}$, the equality on A is given by

	1	2	3	4	5
1	X				
2		X			
3			X		
4				X	
5					X

Another example of a reflexive relation on a five element set $A := \{1, 2, 3, 4, 5\}$:

	1	2	3	4	5
1	X		X	X	
2		X		X	
3	X		X		
4		X		X	
5	X		X		X

(2) R is symmetric if and only if the distribution of crosses in the table is symmetric with respect to transposition:

	1	2	3	4	5
1			X	X	X
2				X	X
3	X				
4	X	X			
5	X	X			

Classification is one of the major quests in mathematics. For this, we need to group elements according to some analogies, similarities or rules.

The following definition introduces the concept of equivalence between elements of a set.

Definition 2.2.2. A relation (A, A, R) is called *equivalence relation* on the set A, if R is reflexive, transitive and symmetric.

Example 2.2.2

(1) Let $n \in \mathbb{N}$. On the set $\mathbb{Z}$ of integers, we define the relation

$$aRb \Leftrightarrow \exists k \in \mathbb{Z}.\ b - a = kn \Leftrightarrow b = a + kn.$$

This relation is an equivalence relation.

Proof.

(a) *(Reflexivity:)* Let $a \in \mathbb{Z}$. Then $a - a = 0$ and we choose $k = 0$.

(b) *(Transitivity:)* Let a, b, $c \in \mathbb{Z}$ and k_1, $k_2 \in \mathbb{Z}$ with $b = a + k_1 n$, $c = b + k_2 n$. Then $c = (a + k_1 n) + k_2 n = a + (k_1 + k_2)n$, hence aRc.

(c) *(Symmetry:)* Let a, $b \in \mathbb{Z}$ with aRb and bRa. Then we can find k_1, $k_2 \in \mathbb{Z}$ with $b = a + k_1 n$, $a = b + k_2 n$. Then $b - a = k_1 n$ and $a - b = k_2 n$, hence $k_1 = -k_2$, i.e, $a = b$.

This relation is denoted by $\equiv_n$ and called **congruence modulo n relation**. If $a \equiv b \pmod n$, we say that a is congruent with b modulo n.

(2) The equality relation on A is an equivalence relation. It is the only relation on A being simultaneously symmetric and antisymmetric.

Definition 2.2.3. Let M be a set. A *partition* of M is a collection $\mathcal{P}$ of subsets of M satisfying

(1) $\forall P \in \mathcal{P}.\ P \neq \varnothing$.

(2) $\forall P, Q \in \mathcal{P},\ P \cap Q \neq \varnothing$ or $P = Q$.

(3) $M = \bigcup_{P \in \mathcal{P}} P$.

Remark 5 *A partition of M is a cover of M with disjoint, non-empty sets, every element of M lies in exactly one cover subset.*

Example 2.2.3.

(1) In the set of integers, $2\mathbb{Z}$ denotes the set of even numbers and $2\mathbb{Z} + 1$ the set of odd numbers. Then $\{2\mathbb{Z}, 2\mathbb{Z} + 1\}$ is a partition of $\mathbb{Z}$.

(2) If $f : M \to N$ is a mapping, the set $\{f^{-1}(n) \mid n \in Im(f)\}$ is a partition of M.

(3) The set of all quadratic surfaces in $\mathbb{R}^3$ can be partitioned into ellipsoids, one-sheeted hyperboloids and two-sheeted hyperboloids (if the degenerated ones are ignored).

In the following, Part(A) denotes the set of all partitions of A, and EqRel(A) the set of all equivalence relations on A.

Definition 2.2.4. Let R be an equivalence relation on A. For every $a \in A$, define *the equivalence class* of a by

$$\bar{a} = [a]_R := \{b \in A \mid aRb\}.$$

Proposition 2.2.1. *Let R be an equivalence relation on A. Then, for every a ∈ A, the equivalence classes* $[a]_R$ *are partitioning A.*

Proof.

(1) The relation R being reflexive, for every $a \in A$, we have aRa, hence $a \in [a]_R$. This proves that every equivalence class $[a]_R$ is not empty.

(2) Suppose two classes $[a]_R$ and $[b]_R$ have a common element $x \in [a]_R \cap [b]_R$. Then aRx and bRx. Using symmetry and transitivity of R, we get aRb, hence $[a]_R = [b]_R$.

(3) $\forall a \in A. \ a \in [a]_R$. It follows that $M = \bigcup_{a \in A} [a]_R$.

Example 2.2.4. In the set $\mathbb{Z}$ consider the congruence relation modulo n. The equivalence classes are:

$$\begin{aligned}
\overline{0} &= \{kn \mid k \in \mathbb{Z}\} & &= n\mathbb{Z} \\
\overline{1} &= \{1 + kn \mid k \in \mathbb{Z}\} & &= 1 + n\mathbb{Z} \\
\overline{2} &= \{2 + kn \mid k \in \mathbb{Z}\} & &= 2 + n\mathbb{Z} \\
&\ \vdots \\
\overline{n-1} &= \{n - 1 + kn \mid k \in \mathbb{Z}\} &&= (n-1) + n\mathbb{Z}.
\end{aligned}$$

We obtain exactly n equivalence classes, since
$$\overline{n} = \overline{0}, \ \overline{n+1} = \overline{1}, \ \overline{n+2} = \overline{2}, \ \ldots$$

Definition 2.2.5. Let M be a set and R an equivalence relation on M. The *quotient set* of M with respect to R is the set of all equivalence classes of R

$$M/R := \{[a]_R | a \in M\}.$$

Example 2.2.5. Consider on the set of all integers, $\mathbb{Z}$, the congruence relation modulo $n \equiv_n$. Then $\mathbb{Z}_n := \mathbb{Z}/\equiv_n = \{\overline{0}, \overline{1}, \ldots, \overline{n-1}\}$.

Definition 2.2.6. Let $n \in \mathbb{N}$. On $\mathbb{Z}_n$ define the operations $+_n$ *addition* modulo n, and *multiplication* modulo n as follows:

$\bar{a} +_n \bar{b} := \bar{r}$ if the rest of the division of $a + b$ to n is r.

$\bar{a} \cdot_n \bar{b} := \bar{s}$ if the rest of the division of $a \cdot b$ to n is s.

Example 2.2.6.

(1) $n = 2$:

$+_2$	$\bar{0}$	$\bar{1}$
$\bar{0}$	$\bar{0}$	$\bar{1}$
$\bar{1}$	$\bar{1}$	$\bar{0}$

$\cdot_2$	$\bar{0}$	$\bar{1}$
$\bar{0}$	$\bar{0}$	$\bar{0}$
$\bar{1}$	$\bar{0}$	$\bar{1}$

(2) $n = 3$:

$+_3$	$\bar{0}$	$\bar{1}$	$\bar{2}$
$\bar{0}$	$\bar{0}$	$\bar{1}$	$\bar{2}$
$\bar{1}$	$\bar{1}$	$\bar{2}$	$\bar{0}$
$\bar{2}$	$\bar{2}$	$\bar{0}$	$\bar{1}$

$\cdot_3$	$\bar{0}$	$\bar{1}$	$\bar{2}$
$\bar{0}$	$\bar{0}$	$\bar{0}$	$\bar{0}$
$\bar{1}$	$\bar{0}$	$\bar{1}$	$\bar{2}$
$\bar{2}$	$\bar{0}$	$\bar{2}$	$\bar{1}$

(3) $n = 4$:

$+_4$	$\bar{0}$	$\bar{1}$	$\bar{2}$	$\bar{3}$
$\bar{0}$	$\bar{0}$	$\bar{1}$	$\bar{2}$	$\bar{3}$
$\bar{1}$	$\bar{1}$	$\bar{2}$	$\bar{3}$	$\bar{0}$
$\bar{2}$	$\bar{2}$	$\bar{3}$	$\bar{0}$	$\bar{1}$
$\bar{3}$	$\bar{3}$	$\bar{0}$	$\bar{1}$	$\bar{2}$

$\cdot_4$	$\bar{0}$	$\bar{1}$	$\bar{2}$	$\bar{3}$
$\bar{0}$	$\bar{0}$	$\bar{0}$	$\bar{0}$	$\bar{0}$
$\bar{1}$	$\bar{0}$	$\bar{1}$	$\bar{2}$	$\bar{3}$
$\bar{2}$	$\bar{0}$	$\bar{2}$	$\bar{0}$	$\bar{2}$
$\bar{3}$	$\bar{0}$	$\bar{3}$	$\bar{2}$	$\bar{1}$

(4) $n = 5$:

$+_5$	$\bar{0}$	$\bar{1}$	$\bar{2}$	$\bar{3}$	$\bar{4}$
$\bar{0}$	$\bar{0}$	$\bar{1}$	$\bar{2}$	$\bar{3}$	$\bar{4}$
$\bar{1}$	$\bar{1}$	$\bar{2}$	$\bar{3}$	$\bar{4}$	$\bar{0}$
$\bar{2}$	$\bar{2}$	$\bar{3}$	$\bar{4}$	$\bar{0}$	$\bar{1}$
$\bar{3}$	$\bar{3}$	$\bar{4}$	$\bar{0}$	$\bar{1}$	$\bar{2}$
$\bar{4}$	$\bar{4}$	$\bar{0}$	$\bar{1}$	$\bar{2}$	$\bar{3}$

$\cdot_5$	$\bar{0}$	$\bar{1}$	$\bar{2}$	$\bar{3}$	$\bar{4}$
$\bar{0}$	$\bar{0}$	$\bar{0}$	$\bar{0}$	$\bar{0}$	$\bar{0}$
$\bar{1}$	$\bar{0}$	$\bar{1}$	$\bar{2}$	$\bar{3}$	$\bar{4}$
$\bar{2}$	$\bar{0}$	$\bar{2}$	$\bar{4}$	$\bar{1}$	$\bar{3}$
$\bar{3}$	$\bar{0}$	$\bar{3}$	$\bar{1}$	$\bar{4}$	$\bar{2}$
$\bar{4}$	$\bar{0}$	$\bar{4}$	$\bar{3}$	$\bar{2}$	$\bar{1}$

(5) $n = 6$:

$+_6$	$\bar{0}$	$\bar{1}$	$\bar{2}$	$\bar{3}$	$\bar{4}$	$\bar{5}$
$\bar{0}$	$\bar{0}$	$\bar{1}$	$\bar{2}$	$\bar{3}$	$\bar{4}$	$\bar{5}$
$\bar{1}$	$\bar{1}$	$\bar{2}$	$\bar{3}$	$\bar{4}$	$\bar{5}$	$\bar{0}$
$\bar{2}$	$\bar{2}$	$\bar{3}$	$\bar{4}$	$\bar{5}$	$\bar{0}$	$\bar{1}$
$\bar{3}$	$\bar{3}$	$\bar{4}$	$\bar{5}$	$\bar{0}$	$\bar{1}$	$\bar{2}$
$\bar{4}$	$\bar{4}$	$\bar{5}$	$\bar{0}$	$\bar{1}$	$\bar{2}$	$\bar{3}$
$\bar{5}$	$\bar{5}$	$\bar{0}$	$\bar{1}$	$\bar{2}$	$\bar{3}$	$\bar{4}$

$\cdot_6$	$\bar{0}$	$\bar{1}$	$\bar{2}$	$\bar{3}$	$\bar{4}$	$\bar{5}$
$\bar{0}$	$\bar{0}$	$\bar{0}$	$\bar{0}$	$\bar{0}$	$\bar{0}$	$\bar{0}$
$\bar{1}$	$\bar{0}$	$\bar{1}$	$\bar{2}$	$\bar{3}$	$\bar{4}$	$\bar{5}$
$\bar{2}$	$\bar{0}$	$\bar{2}$	$\bar{4}$	$\bar{0}$	$\bar{2}$	$\bar{4}$
$\bar{3}$	$\bar{0}$	$\bar{3}$	$\bar{0}$	$\bar{3}$	$\bar{0}$	$\bar{3}$
$\bar{4}$	$\bar{0}$	$\bar{4}$	$\bar{2}$	$\bar{0}$	$\bar{4}$	$\bar{2}$
$\bar{5}$	$\bar{0}$	$\bar{5}$	$\bar{4}$	$\bar{3}$	$\bar{2}$	$\bar{1}$

If n is prime, every element occurs just once on every row and column of the addition and multiplication mod n tables (except 0 for multiplication). This is no longer true if n is not prime. If n is composed but a is relatively prime to n (i.e., they do not share any common divisor), then again on the line and column of a, every element occurs just once.

In the following, we are going to prove that there exists a bijection between Part(A) and EqRel(A), i.e., partitions and equivalence relations describe the same phenomenon.

Theorem 2.2.2. *Let A be a set.*

(1) *Let $\mathcal{P}$ be a partition of A. Then there is an equivalence relation $E(\mathcal{P})$, called the equivalence relation induced by $\mathcal{P}$.*
(2) *Let R be an equivalence relation on A. The equivalence class $[x]_R$ of an element $x \in A$ induces a map $[\cdot]_R : A \to \mathbb{P}(A)$. The quotient set*

$$A/R := \{[x]_R \mid x \in A\} \subseteq \mathbb{P}(A)$$

is a partition of A.
(3) *If $\mathcal{P}$ is a partition of A, $\mathcal{P} = A/E(\mathcal{P})$.*
(4) *If R is an equivalence relation on A, then $R = E(A/R)$.*

Proof.

(1) Let $\mathcal{P}$ be a partition of A. Then there exist a surjective mapping

$$\pi_p : A \to \mathcal{P}$$

with $x \in \pi_p(x)$ for every $x \in A$.

If $\mathcal{P}$ is a partition of A, it follows that for every $x \in A$ there is a unique class A_x with $x \in A_x$. Define $\pi_p : A \to \mathcal{P}$ by $\pi_p(x) := A_x$. Because of the definition of a partition, we have that π_p is well defined and surjective, since the partition $\mathcal{P}$ covers A with disjoint, non-empty classes.

The relation $\ker(\pi_p) \subseteq A \times A$ defined by

$$(x, y) \in \ker(\pi_p) :\Leftrightarrow \pi_p(x) = \pi_p(y)$$

is an equivalence relation on A, denoted by E_p, and is called *the equivalence relation induced by $\mathcal{P}$.*

To prove this, we are going to prove a more general fact, namely that for every map $f : A \to B$, its *kernel*, given by the relation $\ker(f) \subseteq A \times A$ defined by

$$x \ker(f) y :\Leftrightarrow f(x) = f(y)$$

is an equivalence relation on A.

Reflexivity: For all $x \in A$, $f(x) = f(x)$, hence $x \ker(f) x$.

Transitivity: For all $x, y, z \in A$, from $x \ker(f) y$ and $y \ker(f) z$ follows $f(x) = f(y) = f(z)$ and so $f(x) = f(z)$, i.e., $x \ker(f) z$.

Antisymmetry: For all $x, y \in A$, if $x \ker(f) y$ then $f(x) = f(y)$, i.e., $f(y) = f(x)$, and so $y \ker(f) x$.

(2) Let R be an equivalence relation on A. It has been proved in Proposition 2.2.1 that A/R is a partition of A.

(3) Let now $\mathcal{P}$ be a partition of A. Let $P \in \mathcal{P}$. We prove that for $x \in P$, $P = [x]_{E(\mathcal{P})}$. Let $y \in P$, then because of the first part of the theorem, we have $(x, y) \in E(\mathcal{P})$ and so $y \in [x]_{E(\mathcal{P})}$.

If $y \in [x]_{E(\mathcal{P})}$ then $(x, y) \in E(\mathcal{P})$, hence $y \in \pi_p(y) = \pi_p(x) = P$. This proves that $P = [x]_{E(\mathcal{P})}$.

For every $x, y \in P$, $\pi_p(x) = \pi_p(y)$, i.e., $(x, y) \in E(\mathcal{P})$, hence $P = [x]_{E(\mathcal{P})} = [y]_{E(\mathcal{P})}$, wherefrom follows that $P \in A/R$ and so $P \subseteq A/E(\mathcal{P})$.

For the inverse inclusion, take $[x]_{E(\mathcal{P})}$. Then, for every $y \in [x]_{E(\mathcal{P})}$ there is a unique $P \in \mathcal{P}$ with $x, y \in P$. Hence $A/E(\mathcal{P}) \subseteq P$, which proves the equality.

(4) Let R be an equivalence relation on A. We prove that $R = E(A/R)$.

For this, we first prove the inclusion $R \subseteq E(A/R)$. Let $(x, y) \in R$ arbitrarily chosen. Then $[x]_R = [y]_R$, hence $(x, y) \in E(A/R)$. This proves the above inclusion.

For the converse inclusion, $E(A/R) \subseteq R$, take an arbitrary pair of elements $(x, y) \in E(A/R)$. Then, there exists an element $z \in A$ with $x, y \in [z]_R$, since A/R is a partition of A. We deduce that $(x, z) \in R$ and $(y, z) \in R$. Using the symmetry of R, we have $(x, z) \in R$ and $(z, y) \in R$. By transitivity, $(x, y) \in R$, i.e., $E(A/R) \subseteq R$.

2.3 Order Relations

Definition 2.3.1. A reflexive, transitive and antisymmetric relation $R \subseteq A \times A$ is called *order relation* on the set A. The tuple (A, R) is called *ordered set.*

Remark 6 *An order relation on a A is often denoted by $\leq$. If $x, y \in A$ and $x \leq y$ then x and y are comparable. Else, they are called incomparable.*

Definition 2.3.2. The ordered set $(A, \leq)$ is called *totally ordered* or *chain*, if for every $x, y \in A$ we have $x \leq y$ or $y \leq x$.

Example 2.3.1

(1) The relation $\leq$ on $\mathbb{R}$ is an order, hence $(\mathbb{R}, \leq)$ is an ordered set. Moreover, it is totally ordered.
(2) The inclusion on the power set $\mathfrak{P}(M)$ is an order. It is not a chain, since we can find subsets $A, B \in \mathfrak{P}(M)$ which are not comparable, neither $A \subseteq B$, nor $B \subseteq A$.
(3) The divisibility relation on $\mathbb{N}$ is an order.
(4) The equality relation on a set A is an order, the smallest order on A. Moreover, it is the only relation on A which is simultaneously an equivalence and an order relation.

Definition 2.3.3. Let $(A, \leq)$ an ordered set and $a \in A$. Then

(1) a is a *minimal element* of A, if

$$\forall x \in A. \ x \leq a \Leftrightarrow x = a.$$

(2) a is a *maximal element* if

$$\forall x \in A.\ a \le x \Leftrightarrow x = a.$$

(3) a is *the smallest element* of A, if

$$\forall x \in A.\ a \le x.$$

(4) a is *the greatest element* of A, if

$$\forall x \in A.\ x \le a.$$

(5) a is *covered* by $b \in A$ if

$$a < b \text{ and there is no } c \in A \text{ such that } a < c < b.$$

In this case, we say that a is a *lower neighbor* of b and b is an *upper neighbor* of a.

Remark 7 *An ordered set may have more than one minimal element or maximal element. If $(A, \le)$ has a smallest element or a greatest element, respectively, this is uniquely determined. The smallest element of an ordered set is called* the minimum *of that set. Similarly, the greatest element of an ordered set is called the* maximum *of that set.*

Example 2.3.2

(1) 0 is the smallest element of $(\mathbb{N}, \le)$ but there are no maximal elements nor a maximum.
(2) The ordered set $(\mathbb{N}, |)$ has 1 as a minimum and 0 as a maximum.
(3) The ordered set $(\mathfrak{P}(M)\backslash\{\varnothing\}), \subseteq)$ has no minimum; every set $\{x\}$, $x \in M$ is minimal.

Definition 2.3.4. An ordered set $(A, \le)$ is called *well ordered* if every non empty subset B of A has a minimum.

Example 2.3.3. The set $(\mathbb{N}, \le)$ is well ordered. The sets $(\mathbb{Z}, \le)$, $(\mathbb{Q}, \le)$, $(\mathbb{R}, \le)$ are totally ordered but not well ordered.

Theorem 2.3.1. *If $(A, \le)$ is an ordered set, the following are equivalent:*

(1) **The minimality condition:** *Every non-empty subset $B \subseteq A$ has at least one minimal element.*
(2) **The inductivity condition:** *If $B \subseteq A$ is a subset satisfying*

   (a) *B contains all minimal elements of A;*
   (b) *From $a \in A$ and $\{x \in A \,|\, x < a\} \subseteq B$ follows $a \in B$, then $A = B$.*

(3) **The decreasing chains condition:** *Every strictly decreasing sequence of elements from A is finite.*

Definition 2.3.5. Let $(A, \leq)$ be an ordered set and $B \subseteq A$. An element $a \in A$ is called *upper bound* of B if all elements of B are smaller than a: $\forall x \in B.\ x \leq a$. The element $a \in A$ is called *lower bound* of B if all elements of B are greater than a: $\forall x \in B.\ a \leq x$.

Lemma 2.3.2. *(Zorn's Lemma). If $(A, \leq)$ is a non-empty ordered set and every chain in A has an upper bound in A, then A has maximal elements.*

2.4 Lattices. Complete Lattices

Let $(L, \leq)$ be an ordered set and $x, y \in L$.

Definition 2.4.1. We call *infimum* of x, y an element $z \in L$ denoted by $z = \inf(x, y) = x \wedge y$ with

(1) $z \leq x, z \leq y$,
(2) $\forall a \in L.\ a \leq x, a \leq y \Rightarrow a \leq z$.

Definition 2.4.2. If $X \subseteq L$, we call *infimum* of the set X the element $z \in L$ denoted by $z = \inf X = \wedge X$ with

(1) $\forall x \in L.\ z \leq x$.
(2) $\forall a \in L.\forall x \in X, a \leq x \Rightarrow a \leq z$.

Definition 2.4.3. We call *supremum* of x, y an element $z \in L$ denoted by $z = \sup(x, y) = x \vee y$ with

(1) $x \leq z, y \leq z$,
(2) $\forall a \in L.\ x \leq a, y \leq a \Rightarrow z \leq a$.

Definition 2.4.4. If $X \subseteq L$, we call *supremum* of the set X the element $z \in L$ denoted by $z = \sup X = \vee X$ with

(1) $\forall x \in L.x \leq z$.
(2) $\forall a \in L.\forall x \in X, x \leq a \Rightarrow z \leq a$.

Remark 8 (1) *The infimum of a subset $X \subseteq L$ is the greatest lower bound of the set X.*
(2) *The supremum of a subset $X \subseteq L$ is the greatest upper bound of the set X.*

Definition 2.4.5. An ordered set $(L, \leq)$ is called *lattice* if for every $x, y \in L$ exist $x \wedge y$ and $x \vee y$.

Definition 2.4.6. An ordered set $(L, \leq)$ is called *complete lattice* if every subset of L has an infimum and a supremum.

Theorem 2.4.1. $(L, \leq)$ *is a complete lattice if and only if* $\forall X \subseteq L$. $\exists$ inf $X \in L$.

Example 2.4.1

(1) $(\mathbb{N}, \mid)$ is a lattice. If $m, n \in \mathbb{N}$ then $m \wedge n = \gcd(m, n)$ and $m \vee n = \operatorname{lcm}(m, n)$.
(2) $(\mathfrak{P}(M), \subseteq)$ is a complete lattice, $\inf(X_i)_{i \in I} = \bigcap_{i \in I} X_i$ and $\sup(X_i)_{i \in I} = \bigcup_{i \in I} X_i$.

Definition 2.4.7. Let $(P, \leq)$ and $(Q, \leq)$ be ordered sets. A map $f : P \to Q$ is called *order-preserving* or *monotone* if for every $x, y \in P$ from $x \leq y$, follows $f(x) \leq f(y)$ in Q. An *order-isomorphism* is a bijective map $f : P \to Q$ such that both f and f^{-1} are order-preserving.

Definition 2.4.8. Let $(A, \leq)$ be an ordered set and $Q \subseteq A$ be a subset.

The set Q is called *downward closed set* if $x \in Q$ and $y \leq x$ always imply $y \in Q$. Dually, Q is an *upward closed set* if $x \in Q$ and $x \leq y$ always imply $y \in Q$.

For $P \subseteq A$, we define

$$\downarrow P := \{y \in A \mid \exists x \in P. \ y \leq x\}$$

$$\uparrow P := \{y \in A \mid \exists x \in P: \ x \leq y\}.$$

The set $\downarrow P$ is the smallest downward closed set containing P. The set P is downward closed if and only if $P = \downarrow P$. Dually, $\uparrow P$ is the smallest upward closed set containing P. The set P is upward closed if and only if $P = \uparrow P$.

Definition 2.4.9. Let $(L, \leq)$ be a lattice. An element $a \in L$ is called *join-irreducible* or $\vee$-*irreducible* if

$$\forall x, y \in L. \ x < a, y < a \Rightarrow x \vee y < \mathrm{a}.$$

A *meet-irreducible* or $\wedge$-*irreducible* element is defined dually.

Remark 9 *In a finite lattice, an element is join-irreducible if and only if it has exactly one lower neighbor. An element is meet-irreducible if and only if it has exactly one upper neighbor.*

The upper neighbors of 0 are called **atoms** *if they exist, they are always V-irreducible, while the lower neighbors of 1 (called* **coatoms** *if they exist), are Λ-irreducible.*

Definition 2.4.10. Let L be a lattice. A subset $X \subseteq L$ is called *supremum-dense* in L if

$$\forall a \in L. \exists Y \subseteq X. \, a = \vee Y.$$

Dually, it is called *infimum-dense* if

$$\forall a \in L. \exists Y \subseteq X. \, a = \wedge Y.$$

Remark 10 *In a finite lattice L, every supremum-dense subset of L contains all join-irreducible elements and every infimum-dense subset of L contains all meet-irreducible elements.*

2.5 Graphical Representation of Ordered Sets

Every finite ordered set $(A, \leq)$ can be represented graphically by an order diagram. Every element is represented by a circle, the order relation is represented by ascending or descending lines. An element x is placed directly below an element y if x is lower neighbor of y. The order is then represented (because of the transitivity) by an ascending path connecting two elements $a \leq b$. If there is no strict ascending (or descending path) between two elements $a, b \in A$, a and b are incomparable with respect to the order of A.

Example 2.5.1. Here we can see some examples of ordered sets and their graphical representation. The practical use of this graphical representation lies at hand.

(1) The ordered set (A, R) where $A := \{a, b, c, d, e, f\}$, and

$$R := \{(a, d), (a, e), (b, d), (b, f) \, (c, e), (c, f),$$

$$(a, a), (b, b,), (c, c), (d, d,), (e, e), (f, f)\}$$

is graphically represented by the following order diagram:

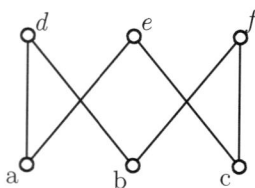

(2) The following order diagram represents an ordered set which proves to be a lattice: $A := \{0, a, b, c, d, e, f, 1\}$, where the smallest element of A is denoted by 0 and the greatest element of A is denoted by 1.

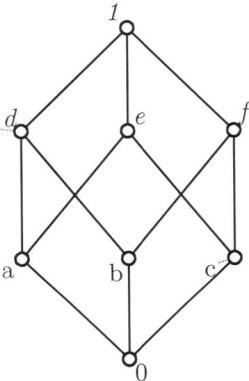

As one can easily check, for every pair (x, y) of elements in A, we can compute the infimum and the supremum of (x, y), $x \wedge y$ is the greatest lower neighbor of x and y, while $x \vee y$ is the smallest upper neighbor of x and y.

(3) The following diagram, slightly different from the previous one, displays just an ordered set and not a lattice, since there is no infimum for the elements d and e, for instance.

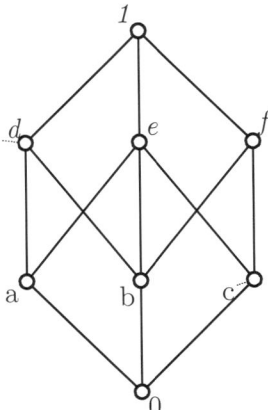

Indeed, d and e have three lower neighbors, $\{0, a, b\}$ but there is no greatest lower neighbor, since a and b are not comparable.

2.6 Closure Systems. Galois Connections

Definition 2.6.1. Let M be a set and $P \subseteq \mathfrak{P}(M)$ a family of subsets of M. The family P is a **closure system** on M if

(1) For every non-empty family $(A_i)_{I \in I} \subseteq P$, we have $\bigcap_{i \in I} A_i \in P$;
(2) $M \in P$.

The elements of a closure system are called **closed sets**.

Example 2.6.1

(1) The powerset of set M is a closure system.
(2) The set $E(M)$ of all equivalence relations on M is a closure system on M.

Proposition 2.6.1. *Let P be a closure system on a set M. Then $(P, \subseteq)$ is a complete lattice.*

Proof.

Let $(A_i)_{i \in I}$ be an arbitrary family of sets of P. Then

$$\bigwedge_{i \in I}^{Ai} = \bigcap_{i \in I}^{Ai}$$

is the infimum of $(A_i)_{i \in I}$. Hence, by Theorem 2.4.1, $(P, \subseteq)$ is a complete lattice. The supremum of $(A_i)_{i \in I}$ is given by

$$\bigvee_{i \in I}^{Ai} = \bigcap\{X \in P \mid \bigcup_{i \in I}^{Ai} \subseteq X\}.$$

Definition 2.6.2. A map $\varphi: \mathfrak{P}(M) \to \mathfrak{P}(N)$ is a *closure operator* if, for all $A, B \subseteq M$ the following holds true:

(1) *Extensivity:* $A \subseteq \varphi(A)$,
(2) *Monotony:* If $A \subseteq B$, then $\varphi(A) \subseteq \varphi(B)$,
(3) *Idempotency:* $\varphi(\varphi(A)) = \varphi(A)$.

For every $A \subseteq M$, the set $\varphi(A)$ is called *closure* of A. The set A is called *closed* if $\varphi(A) = A$.

Remark 11 (1) *The idempotency condition is equivalent to*

$$\forall A \in \mathfrak{P}(M).(\varphi \circ \varphi)(A) = \varphi(A).$$

(2) *A subset $A \in \mathfrak{P}(A)$ is closed if and only if $A \in \varphi(\mathfrak{P}(M))$.*

Proof. Suppose A is closed, then $A = \varphi(A)$, hence $A \in \varphi(\mathfrak{P}(M))$. Conversely, if $A \in \varphi(\mathfrak{P}(M))$, then $A = \varphi(X)$, with $X \subseteq A$. Hence $\varphi(A) = (\varphi \circ \varphi)(X) = \varphi(X) = A$, i.e., A is closed.

(3) *If we take instead of* $\mathfrak{P}(M)$ *an arbitrary ordered set, we can define a closure operator over an ordered set as well.*

Example 2.6.2. Let $(P, \leq)$ be an ordered set. A closure operator is given by

$$\downarrow [-]: \mathfrak{P}(P) \to \mathfrak{P}(P), X \to \downarrow X.$$

The corresponding closure system is the set of all downward closed sets of P.

Theorem 2.6.2. *Let P be a closure system on A. The map* $\varphi: \mathfrak{S}(A) \to \mathfrak{P}(A)$ *defined by*

$$\varphi(X) := \cap \{Y \in P \mid X \subseteq Y\}$$

is a closure operator.

Proof. We have to check the three properties of a closure operator, extensivity, monotony and idempotency. From $\varphi(X) := \cap\{Y \in P \mid X \subseteq Y\}$ follows that extensivity and monotony are true for φ. In order to prove the idempotency, we have to prove that

$$\varphi(X) = X \Leftrightarrow X \in P.$$

If $\varphi(X) = X$, then X is the intersection of some sets of P. Hence, $X \in P$. Conversely, if $X \in P$, then X is a term of the intersection

$$\varphi(X) := \cap\{Y \in P \mid X \subseteq Y\},$$

hence $\varphi(X) \subseteq X$. By extensivity, we have $\varphi(X) = X$.

From $\varphi(X) \in P$ and the monotony property, one has $\varphi(\varphi(X)) = \varphi(X)$. Hence, φ is a closure operator. Moreover, $\varphi(A)$ is the smallest subset of P which contains X.

Theorem 2.6.3. *If* $\varphi: \mathfrak{P}(A) \to \mathfrak{P}(A)$ *is a closure operator on A, then*

$$P := \{X \subseteq A \mid \varphi(X) = X\}$$

is a closure system on A.

Proof. If $R \subseteq P$ and $X = \cap_{Y \in R} Y$, then $X \subseteq Y$, hence $\varphi(X) \subseteq \varphi(Y) = Y$ for every $Y \in R$.

We have

$$\varphi(X) \subseteq \cap_{Y \in R} Y = X,$$

and by extensivity, we get $\varphi(X) = X$, wherefrom follows $X \in F$.

Remark 12 *It has just been proved that the closed sets of a closure operator are a closure system. Every closure system is the system of all closed sets of a closure operator.*

If φ is a closure operator on A, denote by P_φ the corresponding closure system. If P is a closure system, denote by φ_P the induced closure operator.

Theorem 2.6.4. $P_{\varphi P} = P$ *and* $\varphi_{P\varphi} = \varphi$.

Example 2.6.3

(1) The set of all closed subsets of a topological space is a closure system. Moreover, a special property holds true: the join of many finite closed sets is again closed.
(2) Let (X, d) be a metric space. A closed set is defined to be a subset $A \subseteq X$ which contains all limit points. All closed sets in a metric space are a closure system. The corresponding closure operator joins to a subset A all its limit points.

The definitions of closure system and closure operators can be easily generalized on ordered sets.

Definition 2.6.3. A *closure operator* on an ordered set $(P, \leq)$ is a map $\varphi: P \to P$, satisfying for every $a, b \in P$ the following conditions:

(1) *Extensivity:* $a \leq \varphi(a)$;
(2) *Monotony:* If $a \leq b$, then $\varphi(a) \leq \varphi(b)$;
(3) *Idempotency:* $\varphi(a) = \varphi(\varphi(a))$.

If $a \in P$, then $\varphi(a)$ is called the *closure* of a. An element $a \in P$ is called *closed* if $a = \varphi(a)$.

Remark 13 *The subset of P consisting of all closed elements with respect to φ is $\varphi(P)$.*

Definition 2.6.4. A *closure system* on a complete lattice $(L, \leq)$ is a subset $P \subseteq L$, closed under arbitrary infima:

$$H \subseteq P \Rightarrow \wedge H \in P.$$

Definition 2.6.5. Let $(P, \leq)$ and $(Q, \leq)$ be ordered sets. A pair of maps $\varphi: P \to Q$ and $\Psi: Q \to P$ is called a *Galois-connection* between P and Q if

(1) For every $x, y \in P$ from $x \leq y$, follows $\varphi(y) \leq \varphi(x)$;
(2) For every $a, b \in Q$ from $a \leq b$, follows $\Psi(y) \leq \Psi(x)$;

(3) For every $p \in P$ and every $q \in Q$, we have $p \leq \Psi \circ \varphi(p)$ and $q \leq \varphi \circ \Psi(q)$.

Proposition 2.6.5. *Let $(P, \leq)$ and $(Q, \leq)$ be ordered sets. A pair of maps $\varphi: P \to Q$ and $\Psi: Q \to P$ is a Galois-connection if and only if*

$$(4) \forall p \in P. \ \forall q \in Q. \ p \leq \Psi(q) \Leftrightarrow q \leq \varphi(p).$$

Proof. Suppose (φ, Ψ) is a Galois-connection and let $p \in P$ and $q \in Q$ be arbitrary but fixed. Suppose $p \leq \varphi(q)$. Then, by (1), we have $\varphi(p) \geq \varphi(\Psi(q))$. Condition (3) yields $\varphi(p) \geq q$. The converse follows by a similar argument.

Suppose now

$$\forall p \in P. \ \forall q \in Q. \ p \leq \Psi(q) \Leftrightarrow q \leq \varphi(p).$$

Choose in (4) $q := \varphi(p) \leq \varphi(p)$, then $p \leq \Psi(\varphi(p))$, i.e., condition (3). If $x, y \in P$ with $x \leq y$ we deduce that

$$x \leq \Psi(\varphi(y))$$

and using (4) we conclude that $\varphi(y) \ \varphi(x)$.

Proposition 2.6.6. *For every Galois-connection (φ, Ψ) we have*

$$\varphi = \varphi \circ \Psi \circ \varphi \ and \ \Psi = \Psi \circ \varphi \circ \Psi.$$

Proof. Choose $q := \varphi(p)$. Then, by (3), we obtain $\varphi(p) \leq \varphi \circ \Psi \circ \varphi(p)$. From $p \leq \Psi \circ \varphi(p)$ using (1), we have that $\varphi(p) \geq \varphi \circ \Psi \circ \varphi(p)$.

Theorem 2.6.7. *If (φ, Ψ) is a Galois connection between $(P, \leq)$ and $(Q, \leq)$, then*

(1) *The map $\Psi \circ \varphi$ is a closure operator on P and $\varphi \circ \Psi$ is a closure operator on P and the subset of all closed elements in P is $\Psi(Q)$.*
(2) *The map $\varphi \circ \Psi$ is a closure operator on Q and the subset of all closed elements in Q is $\varphi(P)$.*

Proof.
(1) Using (2) from 2.6.5 we obtain the extensivity of $\Psi \circ \varphi$ and from (1) of 2.6.5, we obtain its monotony. Suppose $p \in P$. Then $\varphi(p) \in Q$ and by (2) of 2.6.5, we have

$$\varphi(\mathrm{p})(\varphi \circ \Psi)(\varphi(p))$$

and $p \leq \Psi(\varphi(p))$. Using the monotony of φ, we have

$$(\varphi \circ \Psi) \ (\varphi(p)) \leq \varphi(a).$$

Hence $(\varphi \circ \Psi)(\varphi(p)) = \varphi(p)$. Applying Ψ, one gets

$$(\Psi \circ \varphi) \circ (\Psi \circ \varphi)(p) = \Psi \circ \varphi(p),$$

i.e., the idempotency of $\Psi \circ \varphi$. This proves that $\Psi \circ \varphi$ is a closure operator.

The subset P_0 of all closed elements in P is $\Psi(\varphi(p))$. Using $\varphi(P)$ $\subseteq Q$, we have $P_0 \subseteq \Psi(Q)$. If in the equality $(\varphi \circ \Psi)(\varphi(p)) = \varphi(p)$ the roles of φ and Ψ are changed, we get

$$(\Psi \circ \varphi)(\varphi(q)) = \varphi(q), \forall q \in Q.$$

Hence $\Psi(Q) \subseteq P_0$, wherefrom follows that $P_0 = \Psi(Q)$.

(2) We proceed similarly, changing roles of φ and Ψ.

Definition 2.6.6. Let A and B be sets and $R \subseteq A \times B$ a relation. We define the following derivations by

$$X^R := \{b \in B \,|\, \forall a \in X.\ aRb\}, X \subseteq A$$

and

$$Y^R := \{a \in A \,|\, \forall b \in Y.\ aRb\}, Y \subseteq B.$$

Proposition 2.6.8. *If $R \subseteq A \times B$ is a binary relation, the derivations defined in 2.6.6 define a Galois-connection $\varphi_R \colon \mathfrak{P}(A) \to \mathfrak{P}(B)$ and $\Psi_R \colon \mathfrak{P}(B) \to \mathfrak{P}(A)$ by*

$$\varphi_R(X) := X^R, X \in \mathfrak{P}(A)$$

and

$$\Psi_R(Y) := Y^R, Y \in \mathfrak{P}(B).$$

Conversely, if (φ, Ψ) is a Galois-connection between $\mathfrak{P}(A)$ and $\mathfrak{P}(B)$, then

$$R(\varphi, \Psi) := \{(a, b) \in A \times B \mid a \in \varphi(\{a\})\} = \{(a, b) \in A \times B \,|\, b \in \varphi(\{a\})\}$$

is a binary relation between A and B. Moreover,

$$\varphi_{R(\varphi, \Psi)} = \varphi,$$
$$\Psi_{R(\varphi, \Psi)} = \Psi,$$

and

$$R_{(\varphi R, \Psi R)} = R.$$

CHAPTER 3

Algebraic Structures

Algebraic structures play a major role in mathematics and its applications. Being a natural generalization of arithmetic operations with numbers and their properties, different algebraic structures are used to describe general patterns and behavior of elements and their operations. This chapter briefly introduces basic algebraic structures.

3.1 Functions

Definition 3.1.1. A function or map is a triple $f = (A, B, F)$, where A and B are sets and $F \subseteq A \times B$ is a relation satisfying

$(xFy \wedge xFy') \Rightarrow y = y'$ (Uniqueness of F)
$\forall x \in A.\exists y \in B.xFy$ (Totality of F)

The set A is called **domain** of f, B is called **codomain** of f, and F is called the **graph** of f. We write $f : A \to B$ if there is a graph F, such that $f = (A, B, F)$ is a map. We define

$$B^A := \{f \mid f : A \to B\}$$

the set of all functions from A to B. If $f = (A, B, F)$ and $g = (B, C, G)$ are functions, the **composition** of f and g, denoted by $g \circ f$, is defined by (A, C, R), where

$$xRz :\Leftrightarrow \exists y \in B.xFy \wedge yGz.$$

The function $\mathrm{id}_A := (A, A, \{(x, x) \mid x \in A\})$ is called the **identity** of A.

Remark 14 (1) *In practice, we often identify the function f with its graph F.*

(2) *The graph of the identity of A is defined by*

$$\Delta_A := \{(a, a) \mid a \in A\}$$

and is called the diagonal of A.

Definition 3.1.2. A function $f : A \to B$ is called

- *injective* or *one-to-one* if for every $a_1, a_2 \in A$, from $f(a_1) = f(a_2)$ always follows that $a_1 = a_2$.
- *surjective* or *onto*, if for every $b \in B$ there exist $a \in A$ with $f(a) = b$.
- *bijective*, if f is both injective and surjective, i.e., for every $b \in B$ there exist a **unique** $a \in A$ with $f(a) = b$.

Remark 15 (1) *A function $f : A \to B$ is injective if and only if for every $a_1, a_2 \in A$, from $a_1 \neq a_2$ follows $f(a_1) \neq f(a_2)$.*

(2) *A function $f : A \to B$ is surjective if and only if $f(A) = B$.*
(3) *A function $f = (A, B, F)$ is*
   (a) *injective if and only if $F^{-1} \circ F = \Delta_A$.*
   (b) *surjective if and only if $F \circ F^{-1} = \Delta_B$.*
   (c) *bijective if and only if $F^{-1} \circ F = \Delta_A$ and $F \circ F^{-1} = \Delta_B$. In this case, F^{-1} is the graph of the function $f^{-1} : B \to A$ with*

$$f \circ f^{-1} = \mathrm{id}_B \text{ and } f^{-1} \circ f = \mathrm{id}_A.$$

(4) *If f^{-1} is a map such that*

$$f \circ f^{-1} = \mathrm{id}_B \text{ and } f^{-1} \circ f = \mathrm{id}_A,$$

*then f is called **invertible** and f^{-1} is called the **inverse map** of f.*

Definition 3.1.3. Let $f : A \to B$ be a function. Then $f(A) := \{f(a) \mid a \in A\}$ is called the **image** of f. If $X \subset B$, the set $f^{-1}(X) := \{a \in A \mid f(a) \in X\}$ is called the **preimage** of X.

Theorem 3.1.1. *A function $f : A \to B$ is invertible if and only if f is bijective.*

Proof. Let $f : A \to B$ be an invertible map. Then, the inverse function $f^{-1} : B \to A$ exists with

$$f \circ f^{-1} = \mathrm{id}_B \text{ and } f^{-1} \circ f = \mathrm{id}_A.$$

Let now $a_1, a_2 \in A$ with $f(a_1) = f(a_2)$. Using f^{-1}, we get $a_1 = f^{-1}(f(a_1)) = f^{-1}(f(a_2)) = a_2$, hence f is injective.

Let $b \in B$. Then $f(f^{-1}(b)) = b$. Define $a: = f^{-1}(b)$. Hence there exists an $a \in A$ with $f(a) = b$, i.e., f is surjective. Since f is both injective and surjective, it is bijective.

Suppose now f is bijective. Let $b \in B$ arbitrarily chosen. Then there exists an $a \in A$ with $f(a) = b$. Because of the injectivity of f, the element a is uniquely determined. Hence

$$\{(b, a) \in B \times A \mid b = f(a)\}$$

is the graph of a function $g : B \to A$ with $g(b) = a$ for every $b \in B$. Since $b = f(a)$ and $a = g(b)$, we have

$$\forall b \in B. \ b = f(g(y)),$$

i.e., $f \circ g = \mathrm{id}_B$. Similarly, we can prove that $g \circ f = \mathrm{id}_A$, hence f is invertible.

Theorem 3.1.2. *Let $f : A \to B$ and $g : B \to C$ be bijective maps. Then*
(1) $g \circ f : A \to C$ *is bijective;*
(2) $(g \circ f)^{-1} = f^{-1} \circ g^{-1}$.

Proof.

(1) Let a_1, $a_2 \in A$ with $a_1 \neq a_2$. Then $f(a_1) \neq f(a_2)$ and $g(f(a_1)) \neq g(f(a_2))$, since both f and g are injective. It follows that $g \circ f$ is injective.
Let now $c \in C$. Since g is surjective, there exists $b \in B$ with $g(b) = c$. Using the surjectivity of f, we can find an $a \in A$ with $f(a) = b$, hence $g(f(a)) = c$, i.e., $g \circ f$ is surjective.
The function $g \circ f$ is injective and surjective, hence bijective.
(2) $g \circ f$ is bijective, hence invertible. We denote its inverse by $(g \circ f)^{-1}$.

From $c = (g \circ f)(a)$ follows that $a = (g \circ f)^{-1}(c)$. Then there exists a $b \in B$ with

$$b = g^{-1}(c) \text{ and } a = f^{-1}(b).$$

We conclude $a = f^{-1}(b) = f^{-1}(g^{-1}(c)) = (f^{-1} \circ g^{-1})(c)$, i.e.,

$$\forall c \in C. \ (g \circ f)^{-1}(c) = (f^{-1} \circ g^{-1})(c),$$

which means $(g \circ f)^{-1} = f^{-1} \circ g^{-1}$.

Theorem 3.1.3. *Let A be a finite set and $f: A \to A$ a function. The following are equivalent:*

(1) *f is injective*;
(2) *f is surjective;*
(3) *f is bijective.*

Proof. Since A is finite, we can write $A := \{a_1, a_2, \ldots, a_n\}$. Suppose f is injective but not surjective. Then there exists $a_k \in A$ with $a_k \neq f(a_i)$ for $i \in \{1, 2, \ldots, n\}$. Then,

$$f(A) = \{f(a_1), f(a_2), \ldots, f(a_n)\} \subset \{a_1, a_2, \ldots a_n\}.$$

Because of the strict inclusion, we conclude that the existence of different elements a_i, $a_j \in f(A)$ with $f(a_i) = f(a_j)$, wherefrom follows $a_i = a_j$, which is a contradiction.

Suppose now f is surjective, but not injective. Then there exist a_i, $a_j \in A$ with $a_i \neq a_j$ and $f(a_i) = f(a_j)$, i.e., $f(A)$ would have at most $n - 1$ elements. Since A has n elements, it follows that there exists $a_k \in A$ with $a_k \notin f(A)$, hence f is not surjective.

3.2 Binary Operations

The concept of a binary operation is a natural generalization of real number arithmetic operations.

Definition 3.2.1. A **binary operation** on a set M is a map $*: M \times M \rightarrow M$, which assigns to every pair $(a, b) \in M \times M$ an element $a * b$ of M. A set M together with a binary operation $*$ is sometimes called a **groupoid**.

Many other symbols might be used instead of $*$, like $\perp$, $\circ$, $+$, $\cdot$, $\vee$, $\wedge$, $\top$. Usually, the usage of $+$ is called **additive notation:** $a + b$, often means that $a + b = b + a$ for every $a, b \in M$. The element $a + b$ is called the **sum** of a and b.

The symbol $\cdot$ is called **multiplicative notation:** $a \cdot b$ or ab which is called the **product** of a and b.

Example 3.2.1. Examples of operations

(1) The usual addition on $\mathbb{N}, \mathbb{Z}, \mathbb{Q}, \mathbb{R}, \mathbb{C}, M(m \times n, \mathbb{R}), \mathbb{R}^n$;
(2) The usual multiplication on $\mathbb{N}, \mathbb{Z}, \mathbb{Q}, \mathbb{R}, \mathbb{C}, M_n(\mathbb{R})$;
(3) The cross product on $\mathbb{R}^3$;
(4) Let A be a set. The composition of functions is an operation on the set A^A of all maps of a set A into itself.
(5) Let M be a set. Denote by $\mathfrak{P}(M)$ the power set of M. The union and intersection of subsets of M are binary operations on $\mathfrak{P}(M)$.

The following are not operations in the sense defined above:

(1) Scalar multiplication on $\mathbb{R}^n$:

$$\mathbb{R} \times \mathbb{R}^n \to \mathbb{R}^n,$$

$$(\lambda, \mathbf{v}) \to \lambda\mathbf{v}.$$

We combine objects of different types, scalars and vectors, which is not included in the definition of an operation.

(2) The scalar product on $\mathbb{R}^n$:

$$\mathbb{R}^n \times \mathbb{R}^n,$$

$$(\mathbf{v}, \mathbf{w}) \to \langle \mathbf{v}, \mathbf{w} \rangle$$

Here we combine objects of the same type, i.e., vectors in $\mathbb{R}^n$ but the result is an object of a different type (a scalar), which is not included in the definition of an operation.

(3) Subtraction $(a, b) \to a - b$ is not an operation on $\mathbb{N}$, but it is an operation on $\mathbb{Z}, \mathbb{Q}, \mathbb{R}, \mathbb{C}, \mathbb{R}^n$.

Definition 3.2.2. Let M be a set and $*$ an operation on M. If

$$a * b = b * a \text{ for every } a,\, b \in M,$$

the operation $*$ is called **commutative.**

3.3 Associative Operations. Semigroups

Definition 3.3.1. The operation $*$ defined on the set S is called **associative** if

$$a * (b * c) = (a * b) * c \text{ for every } a, b, c \in M.$$

A pair $(S, *)$ where S is a set and $*$ is on associative operation on S is called **semigroup.** A semigroup with a commutative operation is called a **commutative semigroup.**

Remark 16 *There are five ways to form a product of four factors:*

$((a * b) * c) *d,\ (a * b) * (c * d),\ a * (b * (c * d))$

$(a * (b * c)) * d, \qquad\qquad\quad a * ((b * c) * d)$

If the operation $*$ is associative, all these five products are equal.

Example 3.3.1

(1) Addition and multiplication on $\mathbb{N}, \mathbb{Z}, \mathbb{Q}, \mathbb{R}, \mathbb{C}, M_{m \times n}(\mathbb{R}), M_n(\mathbb{R})$ are commutative.

(2) Subtraction on $\mathbb{N}, \mathbb{Z}, \mathbb{Q}, \mathbb{R}, \mathbb{C}$ is not associative.
$$2 - (3 - 4) = 3 \neq -5 = (2 - 3) - 4.$$

(3) Multiplication on $\mathbb{N}, \mathbb{Z}, \mathbb{Q}, \mathbb{R}, \mathbb{C}$ is commutative but it is no longer commutative on $M_n(\mathbb{R})$ if $n \geq 2$:

$$\begin{pmatrix} 0 & 1 \\ 0 & 0 \end{pmatrix} \begin{pmatrix} 0 & 0 \\ 1 & 0 \end{pmatrix} = \begin{pmatrix} 1 & 0 \\ 0 & 0 \end{pmatrix} \neq \begin{pmatrix} 0 & 0 \\ 0 & 1 \end{pmatrix} = \begin{pmatrix} 0 & 0 \\ 1 & 0 \end{pmatrix} \begin{pmatrix} 0 & 1 \\ 0 & 0 \end{pmatrix}.$$

(4) The composition of permutations, i.e., of bijective functions is associative but in general not commutative.

(5) The composition of functions is not commutative: Let $f, g : \mathbb{R} \to \mathbb{R}$ defined by $f(x) := x^3$ and $g(x) := x + 1$. Then

$$(f \circ g)(x) = f(g(x)) = (x + 1)^3$$

and

$$(g \circ f)(x) = g(f(x)) = x^3 + 1.$$

Hence $f \circ g \neq g \circ f$.

Example 3.3.2

(1) Let A be a singleton, i.e., an one-element set. Then there exists a unique binary operation on A, which is associative, hence A forms a semigroup.

(2) The set $2\mathbb{N}$ of even natural numbers is a semigroup with respect to addition and multiplication.

Definition 3.3.2. Let M be a set with two binary operations $\top, \bot$. We say that $\top$ is **distributive at right** with respect to $\bot$ if for every $x, y, z \in m$, we have

$$(x \bot y)\top z = (x\top z) \bot (y\top z)$$

and **distributive at left** if

$$z\bot(x \bot y) = (z\top x) \bot (z\top y).$$

We say that $\top$ is distributive with respect to $\bot$ if it is left and right distributive.

Definition 3.3.3. Let $(S,*)$ be a semigroup, $a \in S$ and $n \in \mathbb{N}^*$. We define the nth **power** of a to be

$$a^n := \underbrace{a * a * \ldots * a.}_{n \text{ times}}$$

The following computation rules hold true:

$$a^n * a^m = a^{(n+m)}, \ (a^n)^m = a^{n \times m} \text{ for every } n, m \in \mathbb{N}.$$

In the additive notation, we speak about **multiples** of a and write

$$n \times a := \underbrace{a + a + \ldots + a}_{n}$$

3.4 Neutral Elements. Monoids

Definition 3.4.1. Let S be a set and $*$ an operation on S. An element $e \in S$ is called **neutral element** of the operation $*$ if

$$a * e = a = e * a, \text{ for every } a \in S.$$

Remark 17 *Not every operation admits a neutral element. If a neutral element exists, then it is uniquely determined. If e_1 and e_2 would be both neutral elements, then*

$$e_1 = e_1 * e_2 = e_2.$$

Definition 3.4.2. The triple $(M, *, e)$ is called **monoid** if M is a set, $*$ is an associative operation having e as its neutral element. If the monoid operation $*$ is commutative, then $(M, *, e)$ is called **commutative monoid.**

In the additive notation, the neutral element is denoted by 0, while in the multiplicative notation it is denoted by 1. This does not mean that we refer to the *numbers* 0 and 1. This notation only suggests that these abstract elements behave with respect to some operations in a way similar to the role played by 0 and 1 for arithmetic operations.

Example 3.4.1

(1) $(\mathbb{N}, \times, 1)$ and $(\mathbb{Z}, \times, 1)$ are commutative monoids. Their neutral element is 1.

(2) $\left(M_2(\mathbb{R}), \cdot, \begin{pmatrix} 1 & 0 \\ 0 & 1 \end{pmatrix} \right)$ is a non-commutative monoid. The neutral element is the unit matrix $\begin{pmatrix} 1 & 0 \\ 0 & 1 \end{pmatrix}$.

(3) Let M be a set. Consider $\mathfrak{P}(M)$ with the operations of set union and set intersection. Then $(\mathfrak{P}(M), \cup, \varnothing)$ and $(\mathfrak{P}(M), \cap, M)$ are commutative monoids.

(4) **The word monoid:** Let A be a set of symbols called *alphabet.* A *word* over the alphabet A is a finite sequence of symbols from A: $w = a_1 a_2 \ldots a_n$, $a_i \in A$, $i = 1, \ldots, n$. The *length* of w is defined to be n. We denote by A^* the set of words defined over the alphabet A. The empty sequence is denoted by ε. On the set A^*, we define a binary operation called *juxtaposition:* for every $v = a_1 a_2 \ldots a_n, w = b_1 b_2 \ldots b_m \in A^*$, define

$$v \circ w := a_1 a_2 \ldots a_n b_1 b_2 \ldots b_m \in A^*.$$

An easy computation shows that juxtaposition is associative and has the empty sequence ε as neutral element. Hence, $(A^*, \circ, e)$ is a monoid, called the *word monoid over A.*

Lemma 3.4.1. *If A is a set, then $(A^A, \circ, id_A)$ is a monoid.*

Proof. Let $f, g, h: A \to A$ be maps in A^A. We want to prove that $\circ$ is associative, i.e.,

$$f \circ (g \circ h) = (f \circ g) \circ h.$$

Let F, G, H be the corresponding graphs for the maps f, g, h, respectively. Then, for every $a_1, a_2 \in A$, we have

$$a_1(F \circ (G \circ H))a_2 \qquad \Leftrightarrow \exists x \in A . a_1 Fx \wedge x(G \circ H)a_2$$

$$\Leftrightarrow \exists x \in A . a_1 Fx \wedge \exists y \in A . xGy \wedge yHa_2$$

$$\Leftrightarrow \exists x \in A . \exists y \in A . a_1 Fx \wedge xGy \wedge yHa_2$$

$$\Leftrightarrow \exists x \in A . (a_1 Fx \wedge xGy), \exists y \in A . yHa_2$$

$$\Leftrightarrow a_1((F \circ G) \circ H)a_2.$$

We have just proved that maps composition is associative in A^A. The identity map always acts as a neutral element with respect to maps composition, hence $(A^A, \circ, id_A)$ is a monoid.

3.5 Morphisms

Every algebraic structure defines a natural class of maps called *morphisms.* These maps allow us to compare processes that take place in a structure with those from a similar one. In the definition of morphisms the so-called *compatibility* condition with the operations of an algebraic structure occurs.

Definition 3.5.1

- Let $(S_1, *_1)$ and $(S_2, *_2)$ be semigroups. A map $f : S_1 \to S_2$ is a **semigroup morphism** if

$$f(a *_1 b) = f(a) *_2 f(b) \text{ for every } a, b \in S_1.$$

The above relation is called *compatibility of the semigroup structure with the map f.*

- Let $(M_1, \circ_1, e_1)$ and $(M_2, \circ_2, e_2)$ be monoids. A map $f : M_1 \to M_2$ is a **monoid morphism** if f is a semigroup morphism and

$$f(e_1) = e_2.$$

- A semigroup **endomorphism** is a semigroup morphism $f : S \to S$. We denote the set of all endomorphisms of S with $\text{End}(S)$. A semigroup **isomorphism** is a bijective morphism. A semigroup **automorphism** is a bijective endomorphism. The set of all automorphisms of a semigroup S is denoted by $\text{Aut}(S)$. These notions are defined analogously for monoids.

Example 3.5.1

(1) Let $(S, *)$ be a semigroup and $s \in S$. The map

$$p_s : \mathbb{N}\backslash\{0\} \to S, \, p_s(n) := s \times n$$

is a semigroup morphism from $(\mathbb{N}\backslash \{0\}, +)$ to $(S, *)$.

Proof. Remember that $p_s(n) := s \times n = \underbrace{s * s * \ldots * s}_{n}$. Let $m, n \in \mathbb{N}\backslash \{0\}$ be two arbitrary natural numbers. The semigroup morphism condition would require that $p_s(n + m) = p_s(n) * p_s(m)$. We have $p_s(n + m) = \underbrace{s * s * \ldots * s}_{n+m} = \underbrace{(s * s * \ldots * s)}_{n} * \underbrace{(s * s * \ldots * s)}_{m} = p_s(n) * p_s(m)$ which proves that p_s is indeed a semigroup morphism.

(2) Let $(M, \circ, e)$ be a monoid. For $m \in M$, we define $m0 := e$. The map

$$p_m : \mathbb{N} \to M, p_m(n) := mn$$

is a monoid morphism.

Proof. Remember that a monoid morphism is a semigroup morphism with the neutral elements correspondence condition. In order to check the semigroup morphism condition, proceed similar to the previous example. The neutral elements correspondence condition means $p_m(0)$ = e. By definition, $p_m(0) = m0 = e$, hence p_m is a monoid morphism.

(3) Let $A := \{a\}$ be a singleton and A^* the set of words over A. The map

$$p_a : \mathrm{N} \to A^* \quad p_a(n) := \underbrace{aa \ldots a}_{n \text{ times}}$$

is a monoid isomorphism.

Proof. For the monoid $(M, \circ, e)$ in (2) take the word monoid A^* with juxtaposition as monoid operation and ε as neutral element. Then p_a is a monoid morphism as proved above.
Let $m, n \in \mathrm{N}$ be arbitrary natural numbers. If $p_a(m) = p_a(n)$, then

$$\underbrace{aa \ldots a}_{m \text{ times}} = \underbrace{aa \ldots a}_{n \text{ times}},$$

i.e., a word of length m equals a word of length n. Hence $m = n$ and p_a is one-to-one.
Let now $w \in A^*$ be a word. Because of the definition of the word monoid A^*, there is an $n \in \mathrm{N}$ with $w = \underbrace{aa \ldots a}_{n \text{ times}}$, i.e., $w = p_a(n)$, hence p_a is onto.

(4) Consider the semigroups $(\mathbb{R}, +)$ and $(\mathbb{R}_+, \times)$. Then the map $f : \mathbb{R} \to \mathbb{R}_+$ given by $f(x) := a^x$ for an $a \in \mathbb{R}_+$, $a \neq 1$ is a semigroup morphism.
Proof. For every $x, y \in \mathbb{R}$, we have

$$f(x + y) = a^{x+y} = a^x \times a^y = f(x) \times f(y).$$

(5) Let $x \in \mathbb{R}^n$. The map

$$f_x : \mathbb{R} \to \mathbb{R}^n, \ f_x(\lambda) := \lambda x$$

is a monoid morphism from $(\mathbb{R}, +, 0)$ to $(\mathbb{R}^n, +, 0_n)$.
The proof is left to the reader.

(6) The morphism $f : \mathbb{R} \to \mathbb{R}_+$, defined by $f(x) := a^x$, $a \in \mathbb{R}_+$, $a \neq 1$ is an isomorphism.

Proof.
Define $g : \mathbb{R}_+ \to \mathbb{R}$ by $g(x) := \log_a(x)$. Then
  $g(x \times y) = \log_a(xy) = \log_a x + \log_a y = g(x) + g(y)$
for every $x, y \in \mathbb{R}_+$. Hence g is a semigroup morphism.
Let $x \in \mathbb{R}$ be arbitrarily chosen. Then

$$(g \circ f)(x) = g(f(x)) = g(a^x) = \log_a a^x = x,$$

i.e., $g \circ f = \text{id}_{\mathbb{R}}$.

If $x \in \mathbb{R}_+$ is arbitrarily chosen, then

$$(f \circ g)(x) = f(g(x)) = f(\log_a x) = a^{\log_a x} = x,$$

i.e., $f \circ g = \text{id}_{\mathbb{R}_+}$.

Theorem 3.5.1. *Let $(M, *, e)$ be a monoid. We define $h: M \to M^M$ by $h(x)(y) := x * y$. The map $h: (M, *, e) \to (M^M, \circ, \text{id}_M)$ is an injective morphism.*

Proof. Let $x, y \in M$ with $h(x) = h(y)$. Then

$$x = x * e = h(x)(e) = h(y)(e) = y * e = y,$$

hence h is injective.

We are going to check the morphism conditions for h: Since $h(e)(x) = e * x = x = \text{id}_M(x)$, we have $h(e) = \text{id}_M$.

For every $x, y, z \in M$, we have $h(x * y) = h(x) \circ h(y)$, because $h(x * y)(z) = (x * y) * z = x * (y * z) = x * h(y)(z) = h(x)(h(y)(z)) = (h(x) \circ h(y))(z)$.

Theorem 3.5.2. *Let A be a set, $i_A : A \to A^*$ a function mapping every element $a \in A$ to the one element sequence $a \in A^*$ and M a monoid. Then, for every map $f : A \to M$, there exists exactly one monoid morphism $h: A^* \to M$ with $h \circ i_A = f$. This unique map h is called the continuation of f to a monoid morphism on A^*.*

Proof. Uniqueness: Let $h_1, h_2 : A^* \to M$ be monoid morphisms satisfying $h_1 \circ i_A = h_2 \circ i_A$. We have to prove that $h_1 = h_2$. This will be done by induction over the length n of a sequence $s \in A^*$, $n = |s|$.

Suppose $n = 0$. Then $s = \varepsilon$ and $h_1(\varepsilon) = e = h_2(\varepsilon)$.

Suppose the assertion is true for n and prove it for $n + 1$: Let s be a sequence of length $n + 1$. Then, $s = a \cdot s'$ for an $a \in A$ and $s' \in A^*$, with $|s'| = n$. Because of the induction hypothesis, we have $h_1(s') = h_2(s')$. Moreover,

$$h_1(a) = h_1(i_A(a)) = h_2(i_A(a)) = h_1(a).$$

Then

$$h_1(s) = h_1(a \cdot s') = h_1(a) \circ h_1(s') = h_2(a) \circ h_2(s') = h_2(a \cdot s') = h_2(s).$$

We have proved that if such a continuation exists, then it is uniquely determined.

Existence: Define $h(a_1 \ldots a_n) := h(a_1) \ldots h(a_n)$. We can easily prove the morphism property for h, hence such a continuation always exists.

Consider S an arbitrary set, and $M = (S^S, \circ, \mathrm{id})$. We obtain a 1-1-correspondence between mappings $f: A \to S^S$ and morphisms from A^* to $(S^S, \circ, \mathrm{id})$.

The previous result can be used in order to interpret a mapping $f: A \to S^S$ as an **automaton:**

- S is the set of internal states of the automaton;
- A is the input alphabet;
- For every $a \in A$, f describes the state transition $f(a): S \to S$.

The continuation of $f: A \to S^S$ to a morphism $h: A^* \to S^S$ describes the state transition induced by a sequence from A^*.

Definition 3.5.2. Let M be a monoid and S a set of states. An M-automaton is a monoid morphism

$$h: M \to (S^S, \circ, id_S).$$

Remark 18 *If we choose the monoid* $(\mathbb{N}, +, 0)$ *for M, then* $h: M \to S^S$ *is a so-called* discrete system. *Choosing* $(\mathbb{R}_0, +, 0)$ *for M, we obtain a so-called* continuous system.

3.6 Invertible Elements. Groups

Definition 3.6.1. Let $(M, *, e)$ be a monoid with neutral element e. An element $a \in M$ is called **invertible,** if there is an element $b \in M$ such that

$$a * b = e = b * a.$$

Such an element is called an **inverse** of a.

Remark 19 *If* $m \in M$ *has an inverse, then this inverse is unique: Let* m_1 *and* m_2 *be inverse to* $m \in M$. *Then*

$$m_1 = m_1 * e = m_1 * (m * m_2) = (m_1 * m) * m_2 = e * m_2 = m_2.$$

This justies the notation m^{-1} *for the inverse of m.*

Remark 20 *In the additive notation, an element* a *is invertible if there is an element b such that* $a + b = 0 = b + a$. *The inverse of a is denoted by* $-a$ *instead of* a^{-1}.

Example 3.6.1

(1) In every monoid, the neutral element e is invertible, its inverse is again e.
(2) In $(\mathbb{N},+, 0)$ the only invertible element is 0.
(3) In $(\mathbb{Z},+, 0)$ all elements are invertible.
(4) In $(\mathbb{Z}, \times, 1)$ the only invertible elements are 1 and -1.
(5) If A is a set, then the invertible elements of $(A^A, \circ, \mathrm{id}_A)$ are exactly the bijective functions on A.

Proposition 3.6.1. *Let $(M, *, e)$ be a monoid. We denote the set of inverse elements with $M^\times$. The following hold true:*

(1) $e \in M^\times$ *and* $e^{-1} = e$.
(2) *If* $a \in M^\times$, *then* $a^{-1} \in M^\times$ *and* $(a^{-1})^{-1} = a$.
(3) *If* $a, b \in M^\times$, *then* $a * b \in M^\times$ *and* $(a * b)^{-1} = b^{-1} * a^{-1}$.

Proof.

1. As $e * e = e$, the neutral element e is invertible and equal to its inverse.
2. If a is invertible, then $a * a^{-1} = e = a^{-1} * a$, hence a^{-1} is invertible and the inverse of it is a.
3. Let a and b be invertible. Then
$(a * b)(b^{-1} * a^{-1}) = a * (b * b^{-1}) * a^{-1} = a * e * a^{-1} = a * a^{-1} = e$
$(b^{-1} * a^{-1}) * (a * b) = b^{-1} * (a^{-1} * a) * b = b^{-1} * e * b = b^{-1} * b = e$.

Thus $a * b$ is invertible and its inverse is $b^{-1} * a^{-1}$.

Definition 3.6.2. A monoid $(G, *, e)$ in which every element is invertible is called a **group**. If the operation $*$ is commutative, the group is called *abelian*.

Remark 21 (1) *A group is a set G with a binary operation $*$: $G \times G \to G$ such that the following axioms are satisfied:*

(G1) $a * (b * c) = (a * b) * c$ *for every* $a, b, c \in G$.
(G2) $\exists e \in G.a * e = e * a = a$ *for every* $a \in G$.
(G3) $\forall a \in G.\exists a^{-1} \in G.a * a^{-1} = a^{-1} * a = e$.
(2) *If G is a group, then $G = G^\times$.*
(3) *For abelian groups, we generally use the additive notation.*

Example 3.6.2

(1) $(\mathbb{Z},+, 0), (\mathbb{Q}, +, 0), (\mathbb{R}, +, 0), (\mathbb{C}, +, 0)$ are abelian groups.
(2) $(\{-1, 1\}, \times, 1)$ is a group.
(3) $(\mathbb{Q}\setminus\{0\}, \times, 1), (\mathbb{R}\setminus\{0\}, \times, 1), (\mathbb{C}\setminus\{0\}, \times, 1)$ are abelian groups.

(4) $(S_n, \circ, id)$, where S_n is the set of all permutations of an n elements set and id is the identity is a non-commutative group.

(5) The set of all $n \times n$ invertible matrices $\mathrm{GL}_n(\mathbb{R}) := \{A \in M_n(\mathbb{R}) \mid \det A \neq 0\}$ together with the matrix multiplication is a group. The neutral element is the matrix I_n. For $n = 1$, this is an abelian group, isomorphical to $(\mathbb{R} \setminus \{0\}, \times, 0)$. For $n \geq 2$, this group is generally not abelian:

$$\begin{pmatrix} 1 & 1 \\ 0 & 1 \end{pmatrix}\begin{pmatrix} 1 & 0 \\ 1 & 1 \end{pmatrix} = \begin{pmatrix} 2 & 1 \\ 1 & 1 \end{pmatrix} \neq \begin{pmatrix} 1 & 1 \\ 1 & 2 \end{pmatrix} = \begin{pmatrix} 1 & 0 \\ 1 & 1 \end{pmatrix}\begin{pmatrix} 1 & 1 \\ 0 & 1 \end{pmatrix}$$

This group is called *general linear group of dimension n.*

(6) $\mathrm{SL}_n(\mathbb{R}) := \{A \in GL_n(\mathbb{R}) \mid \det A = 1\}$ together with matrix multiplication is a group, called *special linear group of dimension n.*

Proposition 3.6.2. *A triple $(G, *, e)$, where G is a set, $*$ an associative operation on G and $e \in G$ is a group if and only if for every a, b $\in G$ the equations*

$$a * x = b, \, y * a = b$$

have unique solutions in G.

Proof. Let $(G, *, e)$ be a group and the equations $a * x = b$, $y * a = b$. From condition (G3) follows that every element $a \in G$ is invertible, its inverse being denoted by a^{-1}.

From $a * x = b$, we get $x = a^{-1} * b$, hence
$$a * (a^{-1} * b) = (a * a^{-1}) * b = e * b = b,$$

i.e., $x = a^{-1} * b$ is a solution.

Suppose there are two different solutions x_1 and x_2, i.e.,

$$a * x_1 = b \text{ and } a * x_2 = b.$$

We have $a*x_1 = a*x_2$. Multiplying to the left with a^{-1}, we obtain $x_1 = x_2$, i.e., the solution of the equation is uniquely determined.

Similarly, we get the existence of a unique solution of the equation

$$y * a = b.$$

Suppose now that for a triple $(G, *, e)$, the equations

$$a * x = b, \, y * a = b$$

have unique solutions in G for every $a, b \in G$. Choose $b = a$, hence the equation $a * x = a$ has a unique solution for every $a \in G$. We denote this unique solution by e_a. We shall now prove that all e_a are equal.

Let $b \in G$ arbitrary chosen and z the unique solution of $y * a = b$, i.e., $z * a = b$.

Then

$$b * e_a = (z * a) * e_a = z * (a * e_a) = z * a = b,$$

i.e., e_a does not depend on a. We denote $e_a =: e$, then $x * e = x$, for every $x \in G$.

Similarly, starting with $y * a = b$, we conclude the existence of a unique element $e' \in G$ with $e' * x = x$ for every $x \in G$.

From $x * e = x$ and $e' * x = x$, we get $e' * e = e'$ and $e' * e = e$, i.e., $e = e'$.

We conclude that in G, the binary operation $*$ admits a neutral element.

From the uniqueness of the solutions of $a * x = b$, $y * a = b$, we obtain that there exist unique $a', a'' \in G$ with

$$a * a' = e, \; a'' \, a = e.$$

Since $*$ is associative and admits a neutral element, we have

$$a' = e * a' = (a'' * a) * a' = a'' * (a * a') = a'' * e = a''.$$

We have just proved that in G every element is invertible hence G is a group.

Theorem 3.6.3. *Let $(M, *, e)$ be a monoid with neutral element e. Then $M^\times$ is a group with neutral element e with respect to the restriction of the operation $*$ to $M^\times$.*

Proof. For $a, b \in M^\times$, the product $a * b$ is also in $M^\times$, since the product of invertible elements is again invertible. Thus, we have obtained an operation on $M^\times$ by restricting $*$ to the set $M^\times$.

The operation $*$ on M is associative, hence its restriction to $M^\times$ is associative too. Also, $e \in M^\times$ and clearly e is a neutral element for $M^\times$ too. Every $a \in M^\times$ is invertible in M by definition, and its inverse a^{-1} also belongs to $M^\times$.

Lemma 3.6.4. *For a group $(G, \times, e)$ the following hold true:*

(1) $x \times y = e \Rightarrow y \times x = e$;

(2) $x \times y = x \times z \Rightarrow y = z$.

Proof. Let $x \times y = e$. Then, there is a $z \in G$ with $y \times z = e$. Hence

$$x = x \times e = x \times y \times z = e \times z = z$$

and so $y \times x = e$.

Suppose now that $x \times y = x \times z$. Then there is an element $u \in G$ with $u \times x = e$. Hence

$$y = e \times y = u \times x \times y = u \times x \times z = e \times z = z.$$

Definition 3.6.3

(1) A group G is called **finite** if G is finite, and **infinite** at contrary.
(2) If G is a group, the **order** of G is the cardinality of G. We write $\mathrm{ord}(G) = |G|$.

Proposition 3.6.5. *Let $(G, \times, 1)$ be a group and $x \in G$. Then, for every $m, n \in \mathbb{Z}$*

(1) $x^n x^m = x^{n+m}$;
(2) $(x^n)^m = x^{nm}$.

3.7 Subgroups

Definition 3.7.1. Let G be a group. A subset $H \subseteq G$ is called **subgroup** if

(1) $e \in H$,
(2) $g, h \in H$ then $gh \in H$,
(3) $g \in H$ then $g^{-1} \in H$.

Remark 22 (1) *The last two properties can be merged into one single property:*

$$g, h \in H \Rightarrow gh^{-1} \in H.$$

(2) *Every subgroup is a group.*

Example 3.7.1

(1) All subgroups of $(\mathbb{Z}, +)$ are of the form $H = n\mathbb{Z}$.

Proof. Let H be a subgroup of $\mathbb{Z}$. If $H = \{0\}$, then let $n = 0$. Suppose $H \neq \{0\}$. Then there is at least one element $a \in H$ with a $\neq 0$. If $a \in H$, then $-a \in H$, so there is at least one element $a \in H$ with $a > 0$. Let $A := \{a \in H \mid a > 0\}$. Then $A \neq \varnothing$. The set A has a minimum, denoted by n. We are going to prove that $H = n\mathbb{Z}$.

Since $n \in A \subseteq H$, then $n\mathbb{Z} \subseteq \mathbb{Z}$. Let $a \in H$. Then there are $q, r \in \mathbb{Z}$ with

$$a = nq + r, 0 \leq r < d.$$

Since $a \in H$ and $nq \in n\mathbb{Z} \subseteq H$, then $r = a - nq \in H$. Because n is the least positive element of H and $r < d$, we have that $r = 0$, i.e., $a = nq \in n\mathbb{Z}$.

(2) $(\mathbb{Z}, +)$ *is a subgroup of* $(\mathbb{Q}, +)$, $(\mathbb{Q}, +)$ *is a subgroup of* $(\mathbb{R}, +)$, *and* $(\mathbb{R}, +)$ *is a subgroup of* $(\mathbb{C}, +)$.

(3) The orthogonal group, i.e., the group of all orthogonal $n \times n$ matrices and the special orthogonal group, i.e., the group of all orthogonal $n \times n$ matrices with determinant 1 are subgroups of $GLn(K)$.

Proposition 3.7.1. *Let H_i, $i \in I$ be a family of subgroups of a group G.*

Then $H := \cap_{i \in I} H_i$ is a subgroup of G.

Proposition 3.7.2. External characterization. *Let G be a group and M a subset of G. Then there is a least subgroup of G containing M. This semigroup is called* **the subgroup generated by** *M and is denoted by $\langle M \rangle$.*

Proof. Let H_i, $i \in I$, be the family of all subgroups of G containing M. Then their intersection is again a subgroup of G containing M. Hence $\cap_{i \in I} H_i$ is the smallest subgroup of G containing M.

Proposition 3.7.3. Internal characterization. *Let M be a subset of a group G. Then $\langle M \rangle$ is the set of all finite products of elements of M and their inverses:*

$$(*) \ g = x_1^{\varepsilon_1} x_2^{\varepsilon_2} \ldots x_n^{\varepsilon_n}, x_i \in M, \varepsilon_i \in \{-1, 1\}, i = 1, \ldots, n.$$

Proof. We are going to prove that the elements of form (*) form a subgroup H of G. The neutral element ε is the empty product with $n = 0$ factors. Hence it is an element of H. The product of two elements of the form (*) is again an element of this form, hence in H. The inverse of g is

$$g^{-1} = x_n^{-\varepsilon n} \ldots x_1^{-\varepsilon 1}$$

having again the form *(*).*

Moreover, M is a subset of H, since every element x of M has the form (*) with just one factor x.

Every subgroup U containing M contains all inverses of $x \in M$, hence all elements of form (*), i.e., $H \subseteq U$.

Remark 23 *If the group operation is commutative then the elements of the subgroup generated by $\{a_1, \ldots, a_n\}$ are of the form*

$$a_1^{k_1} a_2^{k_2} \ldots a_n^{k_n}, \ k_1, \ldots, k_n \in \mathbb{Z}.$$

Definition 3.7.2. A group is called **cyclic** if it is generated by one of its elements.

Remark 24 *Let G be a group and $a \in G$ an element. The subgroup generated by the element a consists of all powers of a:*

$$\langle a \rangle = \{a^k \mid k \in \mathbb{Z}\}.$$

Definition 3.7.3. Let G be a group and H a subgroup of G. For every $g \in G$, the set

$$gH := \{gh \mid h \in H\}$$

is called the **left coset** of g with respect to H.

Proposition 3.7.4. *The cosets of H form a partition of G with disjoint sets. The equivalence relation generated by this partition is given by*

$$x \equiv y :\Leftrightarrow x^{-1}y \in H \Leftrightarrow y \in xH.$$

Proof. Since $e \in H$, it follows that $ge = g \in gH$ for every $g \in G$, hence the cosets are covering G. Suppose $z \in xH \cap yH$. Then $z = xh_1 = yh_2$. It follows, $x = yh_2 h_1^{-1} \in yH$ and so $xH \subseteq yH$. Dually, $y = xh_1 h_2^{-1} \in xH$ and so $yH \subseteq xH$. We have proved that if $xH \cap yH \neq \emptyset$ then $xH = yH$ which concludes the proof.

Example 3.7.2

(1) In the group $(\mathbb{Z}, +)$ consider the subgroup $H = n\mathbb{Z}$. The cosets are exactly the equivalence classes modulo n:

$\overline{0} = 0 + n\mathbb{Z}$, $\overline{1} = 1 + n\mathbb{Z}$, . . ., $\overline{n-1} = (n-1) + n\mathbb{Z}$.

(2) Consider the permutation group of a three element set $\{1, 2, 3\}$:

$$S_3 = \left\{ \mathrm{id} = \begin{pmatrix} 1 & 2 & 3 \\ 1 & 2 & 3 \end{pmatrix}, p_1 = \begin{pmatrix} 1 & 2 & 3 \\ 2 & 3 & 1 \end{pmatrix}, p_2 = \begin{pmatrix} 1 & 2 & 3 \\ 3 & 1 & 2 \end{pmatrix}, \right.$$

$$\left. t_{12} = \begin{pmatrix} 1 & 2 & 3 \\ 2 & 1 & 3 \end{pmatrix}, t_{13} = \begin{pmatrix} 1 & 2 & 3 \\ 3 & 2 & 1 \end{pmatrix}, t_{23} = \begin{pmatrix} 1 & 2 & 3 \\ 1 & 3 & 2 \end{pmatrix} \right\}.$$

Then $H = \{\mathrm{id}, t_{12}\}$ is a subgroup. The cosets are given by

$$H, \ t_{23}H = \{t_{23}, p_2\}, \ t_{13}H = \{t_{13}, p_1\}.$$

Definition 3.7.4. If G is a group, the right cosets are defined dually:

$$Hx := \{hx \,|\, h \in H\}.$$

Remark 25 *If the group G is commutative, then the left and right cosets are the same. In the non-commutative case, this is no longer the case. In the above example for the permutation group S_3 and the subgroup H, we have*

$$t_{13}H = \{t_{13}, p_1\} \neq \{t_{13}, p_2\} = Ht_{13}.$$

3.8 Group Morphisms

Definition 3.8.1. Let $(G_1, \circ_1, e_1)$ and $(G_2, \circ_2, e_2)$ be groups. We call **group morphism** a monoid morphism $f: G_1 \to G_2$ with

$$f(a^{-1}) = (f(a))^{-1} \text{ for every } a \in G.$$

The morphism f is an **isomorphism**, if f is bijective. An isomorphism $f: G \to G$ is called an **automorphism.**

Example 3.8.1

(1) Let $(\mathbb{Z}, +, 0)$ be the additive group of integers and $(G, +, 0)$ an arbitrary group. Let $a \in G$ be fixed. The map $f: \mathbb{Z} \to G$ defined by $f(n) := na$ is a group morphism.

(2) Let $(\mathbb{R}\backslash\{0\}, \times, 1)$ be the multiplicative group of reals and $GL_2(\mathbb{R}) := \{A \in M_2(\mathbb{R}) \,|\, \det A \neq 0\}$ be the multiplicative group of invertible matrices over $\mathbb{R}$. The map $f: \mathbb{R}\backslash\{0\} \to G$ defined by

$$f(x) := \begin{pmatrix} x & 0 \\ 0 & x \end{pmatrix}, x \in \mathbb{R}$$

is a group morphism.

Proof. Indeed, for every $x, y \in \mathbb{R} \setminus \{0\}$,

$$f(x) \times f(y) = \begin{pmatrix} x & 0 \\ 0 & x \end{pmatrix} \cdot \begin{pmatrix} y & 0 \\ 0 & y \end{pmatrix} = \begin{pmatrix} xy & 0 \\ 0 & xy \end{pmatrix} = f(xy).$$

(3) Let $(G, \times, 1)$ be a commutative group and $f : G \to G$ defined by $f(x) := x^{-1}$. The function f is a group morphism.

Proof. For every $x, y \in G$,

$$f(xy) = (xy)^{-1} = y^{-1}x^{-1} = x^{-1}y^{-1} = f(x)f(y).$$

(4) The identity function $\mathrm{id}_G : G \to G$ is a group isomorphism, hence an automorphism of G.

(5) The map $f : (\mathbb{R}, +, 0) \to (\mathbb{R}_+, \times, 1)$ defined by $f(x) := e^x$ is a group isomorphism. Its inverse is $g : (\mathbb{R}_+, \times, 1) \to (\mathbb{R}, +, 0)$, given by $g(x) := \ln x$.

(6) Let $(G, \times, 1)$ be a group and $x \in G$. The function $\iota_x : G \to G$, defined by $\iota_x(g) := xgx^{-1}$ for every $g \in G$ is an automorphism of G.

Proof. For $g, g' \in G$,

$$\iota_x(gg') = xgg'x^{-1} = (xgx^{-1})(xg'x^{-1}) = \iota_x(g)\iota_x(g'),$$

i.e., ι_x is a group morphism. We will now prove the bijectivity of ι_x. Suppose $\iota_x(g) = \iota_x(g')$. Then $xgx^{-1} = xg'x^{-1}$, hence $g = g'$, so ι_x is injective.

For every $h \in G$, take $g := x^{-1}hx \in G$. Then $\iota_x(g) = h$, so ι_x is surjective, hence bijective.

The automorphisms ι_x are called *inner automorphisms* of G. If G is abelian, all inner automorphisms are equal to id_G, the identity on G.

(7) The set of all inner automorphisms of a group G with map composition and the identity is a group.

Proposition 3.8.1. *If $f : G_1 \to G_2$ and $g : G_2 \to G_3$ are group morphisms, then $g \circ f$ is a group morphism.*

Proof. For every $x, y \in G$,

$$(g \circ f)(xy) = g(f(xy)) \qquad = g(f(x)f(y)) = g(f(x))g(f(y))$$
$$= (g \circ f)(x)(g \circ f)(y).$$

Proposition 3.8.2. *Let $f: G \to H$ be a group morphism. Then*

(1) $f(e) = e$;
(2) $f(x^{-1}) = f(x)^{-1}$ *for all* $x \in G$;
(3) *If* G_1 *is a subgroup of* G, *then* $f(G_1)$ *is a subgroup of* H;
(4) *If* H_1 *is a subgroup of* H, *then* $f^{-1}(H_1)$ *is a subgroup of* G.

Proof.

(1) We have $f(e \times e) = f(e) = f(e)f(e)$. Multiplying to the left with $f(e)^{-1}$, we obtain $e = f(e)$.
(2) $e = x \times x^{-1}$ for every $x \in G$. Then $e = f(e) = f(x \times x^{-1}) = f(x)f(x^{-1})$ and so $f(x^{-1}) = f(x)^{-1}$.
(3) For every $x, y \in G_1$, we have $x \times y^{-1} \in G_1$ from the subgroup definition. Let now $s, t \in f(G_1)$. Then there exist $x, y \in G_1$ with $s = f(x)$, $t = f(y)$. We want to prove that $st^{-1} \in f(G_1)$. Indeed

$$st^{-1} = f(x)f(y)^{-1} = f(x \times y^{-1}) \in f(G_1).$$

(4) Let $x, y \in f^{-1}(H_1)$. Then there exist $s, t \in H_1$ such that $f(x) = s$, $f(y) = t$. Then

$$x \times y^{-1} \in f^{-1}(s) \times f^{-1}(t^{-1}) = f^{-1}(st^{-1}) \in f^{-1}(H_1).$$

Hence $f^{-1}(H_1)$ is a subgroup.

Definition 3.8.2. Let $f: G \to H$ be a group morphism. The **kernel** of f is defined as

$$\ker(f) = \{x \in G \mid f(x) = e\}.$$

The **image** of f is defined as

$$Im(f) = f(G).$$

Remark 26 $\ker(f) = f^{-1}(e)$, *hence the kernel of a group morphism is a subgroup of* G. *The image of* f *is a subgroup of* H.

Theorem 3.8.3. *A group morphism* $f: G \to H$ *is injective if and only if* $\ker(f) = \{e\}$.

Proof. Since $f(e) = e$, we have $e \in \ker(f)$. Let $x \in \ker(f)$. Then $f(x) = e = f(e)$. If f is injective, then $x = e$, i.e., $\ker(f) = \{e\}$.

Suppose now $\ker(f) = \{e\}$ and $f(x) = f(y)$. It follows that $e = f(x)^{-1}f(y) = f(x^{-1}y)$, i.e., $x^{-1}y \in \ker(f) = \{e\}$. We have $x^{-1}y = e$, wherefrom follows $x = y$, hence f is injective.

3.9 Congruence Relations

Let G be a set and $*$ be a binary operation on G.

Definition 3.9.1. An equivalence relation R on G is called **congruence relation** if for all $a, b, c, d \in G$

$$(C) \ aRc, bRd \Rightarrow a * bRc * d.$$

The equivalence classes are called **congruence classes.** The congruence class of an element $a \in G$ is denoted by $\bar{a} = \{g \in G \mid gRa\}$. The set of all congruence classes is denoted by G/R and is called **quotient** of G.

Remark 27 *(1) The notion of a congruence is a natural generalization of congruence modulo n for integers. For every natural $n \in \mathbb{N}$, the relation $\equiv_n$ is not only an equivalence relation but also a congruence relation on $\mathbb{Z}$.*

(2) We can define the following operation on the quotient G/R:

$$\bar{a} * \bar{b} := \overline{a * b}.$$

This operation is well defined, being independent on the representative choice. The quotient map

$$\pi: G \rightarrow G/R, \ \pi(a) := \bar{a}$$

is a surjective morphism. This can be easily seen from

$$\pi(a * b) = \overline{a * b} = \bar{a} * \bar{b} = \pi(a) * \pi(b).$$

The quotient set with this induced operation is called the **quotient groupoid** *of G and π is called the* **quotient morphism.**

(3) If G is a semigroup, monoid or group, respectively, then so is the quotient of G modulo the congruence R.

Proposition 3.9.1. *Let R be a congruence on a group G. Then*

(1) $xRy \Leftrightarrow xy^{-1}Re$;
(2) $xRy \Rightarrow x^{-1}Ry^{-1}$;
(3) $\bar{e}$ *is a subgroup of G;*
(4) $eRx \Leftrightarrow eRyxy^{-1}$.

Proof.

(1) $xRy \Rightarrow xy^{-1}Ryy^{-1} \Leftrightarrow xy^{-1}Re$ and $xy^{-1}Re \Rightarrow xy^{-1}yRe \Leftrightarrow xRy$.
(2) $xRy \Rightarrow xy^{-1}Re \Rightarrow x^{-1}xy^{-1}Rx^{-1} \Leftrightarrow y^{-1}Rx^{-1} \Leftrightarrow x^{-1}Ry^{-1}$.

(3) Because of the reflexivity of R, we have eRe, and so $e \in \bar{e}$. If $x, y \in \bar{e}$, then eRx and eRy. It follows $e = eeRxy$, hence $xy \in \bar{e}$. Let now $x \in \bar{e}$. Then eRx and so $e = e^{-1}Rx^{-1}$, hence $x^{-1} \in \bar{e}$.

(4) $eRx \Rightarrow yey^{-1}Ryxy^{-1} \Rightarrow eRyxy^{-1} \Rightarrow y^{-1}eyRy^{-1}yxy^{-1}y \Rightarrow eRx$

Definition 3.9.2. Let G be a group, H a subgroup is called **normal** if for all $g \in G$

$$gHg^{-1} = H.$$

Proposition 3.9.2. *Let N be a normal subgroup of a group G. The relation $R_H \subseteq G \times G$ with*

$$xR_Hy :\Leftrightarrow xy^{-1} \in H$$

is a congruence on G.

Proof. Let $x_1, x_2, y_1, y_2 \in G$ with $x_1R_Hx_2$ and $y_1R_Hy_2$, i.e., $x_1x_2^{-1} \cdot y_1y_2^{-1} \in H$. Then

$$x_1y_1(x_2y_2)^{-1} = x_1y_1y_2^{-1}x_2^{-1} = x_1y_1y_2^{-1}x_1^{-1}x_1x_2^{-1} \in H$$

since $x_1y_1y_2^{-1}x_1^{-1} \in H$, because $y_1y_2^{-1} \in H$ and H is a normal subgroup, and $x_1x_2^{-1} \in H$.

It follows $x_1y_1R_Hx_2y_2$.

The following theorem states that there is a one to one correspondence between congruences on a group G and some subgroups of G.

Theorem 3.9.3. *Let G be a group. Then*

(1) $\bar{e}_{R_H} = H$ *for all normal subgroups H of G;*

(2) $R_{\bar{e}_R} = R$ *for all congruences R on G.*

Proof.

(1) $x \in \bar{e}_{R_H} \Leftrightarrow xR_He \Leftrightarrow xe^{-1} \in H \Leftrightarrow x \in H$.

(2) $xR_{\bar{e}_R}y \Leftrightarrow xy^{-1} \in \bar{e}_R \Leftrightarrow xy^{-1}Re \Leftrightarrow xRy$.

3.10 Rings and Fields

The usual arithmetic operations on number sets are addition and multiplication. The same situation occurs in the case of matrix sets. We investigate now the general situation:

Definition 3.10.1. A set R with two operations + and ×, called *addition* and *multiplication,* and a fixed element $0 \in R$ is called a **ring** if

(R1) $(R, +, 0)$ is a commutative monoid;
(R2) $(R, \times)$ is a semigroup;
(R3) The following *distributivity* condition holds true

$$a \times (b + c) = a \times b + a \times c \text{ for every } a, b, c \in R,$$
$$(b + c) \times a = b \times a + c \times a \text{ for every } a, b, c \in R.$$

If the ring R has an element $1 \in R$ such that $(R, \times, 1)$ is a monoid, then R is called **unit ring.** If the multiplication commutes, the ring is called **commutative ring.**

Example 3.10.1

(1) $(\mathbb{Z}, +, \times, 0, 1)$ is a commutative unit ring.
(2) $(M_n(\mathbb{R}), +, \times, 0_n, I_n)$ is a non-commutative unit ring for $n \geq 2$.
(3) $\mathbb{Z}[i] := \{z = a + bi \in \mathbb{C} \mid a, b \in \mathbb{Z}\}$ is a commutative ring.

Definition 3.10.2. A commutative ring K with $0 \neq 1$ and $K^{\times} := K \setminus \{0\}$ is called **field.**

Example 3.10.2

(1) $\mathbb{Q}, \mathbb{R}$ are fields.
(2) **Quaternions field:** Consider the set of matrices over $\mathbb{C}$:

$$H := \left\{ \begin{pmatrix} z & -\overline{w} \\ w & \overline{z} \end{pmatrix} \in M_2(\mathbb{C}) \mid z, w \in \mathbb{C} \right\}.$$

Remember that for a complex number $z = x + iy \in \mathbb{C}$, we have denoted $\overline{x} := x - iy \in \mathbb{C}$ *the complex conjugate* of z. The set H with matrix addition and multiplication form a non-commutative field called *the quaternions field.*

The elements of H

$$e := \begin{pmatrix} 1 & 0 \\ 0 & 1 \end{pmatrix}, u := \begin{pmatrix} i & 0 \\ 0 & -i \end{pmatrix}, v := \begin{pmatrix} 0 & - \\ 1 & 1 \end{pmatrix} 0, w := \begin{pmatrix} 0 & i \\ i & 0 \end{pmatrix}$$

satisfy

$$u^2 = v^2 = w^2 = -e$$

and

$$uv = -vu = w, \; vw = -wv = u, \; uw = -wu = v.$$

(3) We consider the set of quotients modulo 2, $\mathbb{Z}_2 = \{\bar{0}, \bar{1}\}$, and define the following operations:

+	$\bar{0}$	$\bar{1}$
$\bar{0}$	$\bar{0}$	$\bar{0}$
$\bar{1}$	$\bar{1}$	$\bar{0}$

.	$\bar{0}$	$\bar{1}$
$\bar{0}$	$\bar{0}$	$\bar{0}$
$\bar{1}$	$\bar{0}$	$\bar{1}$

$\mathbb{Z}_2$ with these operations is a commutative field.

Remark 28 *A field is a set K with two operations $+, \times$ and two elements $0 \neq 1$, satisfying the following axioms:*

(K1) $\forall a, b, c \in K.\ a + (b + c) = (a + b) + c,\ a \times (b \times c) = (a \times b) \times c;$

(K2) $\forall a \in K\ .\ a + 0 = 0 + a = a,\ a \times 1 = 1 \times a = 1;$

(K3) $\forall a \in K\ .\ \exists b \in K. a + b = b + a = 0;$

(K4) $\forall a \in K\backslash\{0\}.\ \exists b \in K. a \times b = b \times a = 1;$

(K5) $\forall a, b \in K\ .\ a + b = b + a;$

(K6) $\forall a, b; c \in K.\ a \times (b + c) = a \times b + a \times c.$

CHAPTER 4

Linear Algebra

4.1 Vectors

In geometry, a vector is an oriented segment, with a precise direction and length. Real numbers are called **scalars**. Two oriented segments represent sent the same vector if they have the same length and direction. In other words, no difference is made between parallel vectors.

Let **u** and **v** be two vectors. The **addition** of **u** and **v** is described by the *paralellogram rule*.

The **substraction** of two vectors is defined by the *triangle rule*, adding the opposite vector.

The **zero vector** is denoted by 0, being a vector of length 0 and no direction.

Let **v** be a vector and α a scalar. The **scalar multiplication** of v with the scalar α is denoted by $\alpha\mathbf{v}$ and is defined as follows:

- If $\alpha > 0$ then $\alpha\mathbf{v}$ has the same direction with **v** and its length equals α-times the length of **v**.
- If $\alpha = 0$ then $0\mathbf{v} = 0$.
- If $\alpha < 0$ then $\alpha\mathbf{v}$ has the opposite direction as **v** and its length equals $|\alpha|$-times the length of **v**.

Remark 29 *The previous operations satisfy the following rules:*

(1) $\mathbf{u} + (\mathbf{v} + \mathbf{w}) = (\mathbf{u} + \mathbf{v}) + \mathbf{w}$, *for every vector* **u, v,w**.

(2) $\mathbf{v} + 0 = 0 + \mathbf{v} = \mathbf{v}$, *for every vector* **v**.

(3) $\mathbf{v} + (-\mathbf{v}) = -\mathbf{v} + \mathbf{v} = 0$, *for every vector* **v**.

(4) $\mathbf{v} + \mathbf{w} = \mathbf{w} + \mathbf{v}$, *for every vector* **v, w**.

(5) $\lambda(\mu\mathbf{v}) = \lambda(\mu)\mathbf{v}$, *for every scalar* λ, μ *and every vector* **v**.

(6) $1\mathbf{v} = \mathbf{v}$, *for every vector* $\mathbf{v}$.

(7) $\lambda\,(\mathbf{v} + \mathbf{w}) = \lambda\mathbf{v} + \lambda\mathbf{w}$, *for every scalar* λ *and every vector* $\mathbf{v}$ *and* $\mathbf{w}$.

(8) $(\lambda + \mu)\mathbf{v} = \lambda\mathbf{v} + \mu\mathbf{v}$, *for every scalar* λ, μ *and every vector* $\mathbf{v}$.

Remark 30 *The following* derived rules *hold true*:

   (1) $\mathbf{u} + \mathbf{w} = \mathbf{v} + \mathbf{w}$ *implies* $\mathbf{u} = \mathbf{v}$.
   (2) $0\mathbf{v} = 0$ and $\lambda 0 = 0$.
   (3) $\lambda\mathbf{v} = 0$ implies $\lambda = 0$ *or* $\mathbf{v} = 0$.

Proof.

   (1) Suppose $\mathbf{u} + \mathbf{w} = \mathbf{v} + \mathbf{w}$. Adding $-\mathbf{w}$, we obtain $(\mathbf{u} + \mathbf{w})+(-\mathbf{w})$ $= (\mathbf{v} + \mathbf{w})+(-\mathbf{w})$. Applying rules (1), (3), and (2), we obtain $(\mathbf{u} + \mathbf{w})+(-\mathbf{w}) = \mathbf{u}+(\mathbf{w}+(-\mathbf{w})) = \mathbf{u} + 0 = \mathbf{u}$. Similarly, $(\mathbf{v} + \mathbf{w}) + (-\mathbf{w}) = \mathbf{v}$, hence $\mathbf{u} = \mathbf{v}$.
   (2) We have $0\mathbf{v} = (1 + (-1))\mathbf{v} = \mathbf{v} + (-\mathbf{v}) = 0$ by (8) and (3). We proceed for the second part similarly.
   (3) Suppose $\lambda\mathbf{v} = 0$. If $\lambda = 0$ the proof has been finished. If $\lambda \neq 0$, then, by multiplicating equation $\lambda\mathbf{v} = 0$ with λ^{-1}, right hand is still 0, the left hand becomes $(\lambda^{-1})(\lambda\mathbf{v}) = (\lambda^{-1}\lambda)\mathbf{v} = 1\mathbf{v} = \mathbf{v}$, using (5) and (6). We obtain $\mathbf{v} = 0$.

Definition 4.1.1. Two vectors $\mathbf{v}$ and $\mathbf{w}$ in the plane are called **linearly independent** if they are not parallel. If they are parallel they are called **linearly dependent**.

Remark 31 *The vectors* $\mathbf{v}$ *and* $\mathbf{w}$ *are linearly independent if and only if there is no scalar* λ, *such that* $\mathbf{v} = \lambda\mathbf{w}$ *or* $\mathbf{w} = \lambda\mathbf{v}$. *If there is a scalar* λ, *such that* $\mathbf{v} = \lambda\mathbf{w}$ *or* $\mathbf{w} = \lambda\mathbf{v}$, *the vectors* $\mathbf{v}$ *and* $\mathbf{w}$ *are linearly dependent*.

Proposition 4.1.1. *Let* $\mathbf{v}$ *and* $\mathbf{w}$ *be two vectors in plane. The following conditions are equivalent*

   (1) $\mathbf{v}$ *and* $\mathbf{w}$ *are linearly independent*.
   (2) *If* $\lambda\mathbf{v} + \mu\mathbf{w} = 0$, *then* $\lambda = 0$ *and* $\mu = 0$.
   (3) *If* $\lambda\mathbf{v} + \mu\mathbf{v} = \lambda'\,\mathbf{v} + \mu'\mathbf{w}$, *then* $\lambda = \lambda'$ *and* $\mu = \mu'$.

Proof. $(1 \Rightarrow 2)$ Suppose $\lambda\mathbf{v} + \mu\mathbf{w} = 0$ and $\lambda \neq 0$. Then $\lambda\mathbf{v} = -\mu\mathbf{w}$ and, by multiplying with λ^{-1}, one gets $\mathbf{v} = (-\mu/\lambda^{-1})\mathbf{w}$, a contradiction with the linearly independence of the two vectors.

$(2 \Rightarrow 3)$ Suppose $\lambda\mathbf{v} + \mu\mathbf{v} = \lambda'\mathbf{v} + \mu'\mathbf{w}$, then $(\lambda - \lambda')\mathbf{v} +(\mu -\mu^{-1})\mathbf{w} = 0$, implying $\lambda = \lambda'$ and $\mu = \mu'$.

$(3 \Rightarrow 1)$ Suppose $\mathbf{v}$ and $\mathbf{w}$ are linearly independent. Then, there exists a scalar $\lambda \neq 0$ such that $\mathbf{v} = \lambda\mathbf{w}$ or $\mathbf{w} = \lambda\mathbf{v}$. In the first case, $\mathbf{v} = \lambda\mathbf{w}$ leads to $1\mathbf{v} + 0\mathbf{w} = 0\mathbf{v} + \lambda\mathbf{w}$. From the hypothesis, one gets $1 = 0$, contradiction!

In the following, we will describe the conditions that three vectors are not coplanar.

Definition 4.1.2. Let $\mathbf{v}$ and $\mathbf{w}$ be two vectors and λ, $\mu \in \mathbb{R}$ scalars. A vector $\lambda\mathbf{v} + \mu\mathbf{w}$ is called **linear combination** of $\mathbf{v}$ and $\mathbf{w}$.

Remark 32 *If $\mathbf{v}$ and $\mathbf{w}$ are fixed and the scalars λ and μ run over the set of all reals, the set of all linear combinations of $\mathbf{v}$ and $\mathbf{w}$ is the set of vectors in the plane defined by the support lines of the two vectors $\mathbf{v}$ and $\mathbf{w}$.*

Definition 4.1.3. Three vectors $\mathbf{u}$, $\mathbf{v}$ and $\mathbf{w}$ in space are called **linearly independent** if none of them is a linear combination of the other two. Otherwise, they are called **linearly dependent.**

Remark 33 *Three vectors $\mathbf{u}$, $\mathbf{v}$ and $\mathbf{w}$ are linearly independent if and only if they are not coplanar.*

Proposition 4.1.2. *For every three vectors in space $\mathbf{u}$, $\mathbf{v}$ and $\mathbf{w}$, the following are equivalent:*

   (1) $\mathbf{u}$, $\mathbf{v}$ *and* $\mathbf{w}$ *are linearly independent.*
   (2) $\lambda\mathbf{u} + \mu\mathbf{w} + \nu\mathbf{w} = 0$ *implies* $\lambda = \mu = \nu = 0$.
   (3) $\lambda\mathbf{u} + \mu\mathbf{v} + \nu\mathbf{w} = \lambda'\mathbf{u} + \mu'\mathbf{v} + \nu'\mathbf{w}$ *implies* $\lambda = \lambda'$, $\mu = \mu'$, $\nu = \nu'$.

4.2 The space $\mathbb{R}^n$

Let n be a non-zero natural number. The set $\mathbb{R}^n$ consists of all n-tuples

$$\mathbf{v} = \begin{pmatrix} \alpha_1 \\ \alpha_2 \\ \vdots \\ \alpha_n \end{pmatrix},$$

where $\alpha_1, \alpha_2 \ldots, \alpha_n \in \mathbb{R}$. The elements of $\mathbb{R}^n$ are called **vectors** and they describe the coordinates of a point in $\mathbb{R}^n$ with respect to the origin of the chosen coordinates system

$$0 = \begin{pmatrix} 0 \\ 0 \\ \vdots \\ 0 \end{pmatrix}.$$

Addition, multiplication and scalar multiplication are defined as follows:

$$\mathbf{v}+\mathbf{w} = \begin{pmatrix} a_1 \\ a_2 \\ \vdots \\ a_n \end{pmatrix} + \begin{pmatrix} b_1 \\ b_2 \\ \vdots \\ b_n \end{pmatrix} = \begin{pmatrix} a_1 + b_1 \\ a_2 + b_2 \\ \vdots \\ a_n + b_n \end{pmatrix}, \lambda\mathbf{v} = \lambda \begin{pmatrix} a_1 \\ a_2 \\ \vdots \\ a_n \end{pmatrix} = \begin{pmatrix} \lambda a_1 \\ \lambda a_2 \\ \vdots \\ \lambda a_n \end{pmatrix}.$$

The properties (1)-(8) of addition and multiplication of vectors hold true in $\mathbb{R}^n$.

Remark 34 (Special cases) (1) $n = 0$. Define $\mathbb{R}^0 := \{0\}$.

(2) $n = 1$. *Define* $\mathbb{R}^1 := \mathbb{R}$.

(3) $n = 2$. *In order to identify the set of all vectors in the plane with $\mathbb{R}^2$, a coordonatization is necessary. For this, we choose an origin O and two linearly independent vectors $\mathbf{u}_1$, $\mathbf{u}_2$ in plane. Any other vector $\mathbf{v}$ in the plane can be represented as a linear combination of $\mathbf{u}_1$ and $\mathbf{u}_2$:*

$$\mathbf{v} = \alpha_1\mathbf{u}_1 + \alpha_2\mathbf{u}_2.$$

The scalars α_1 and α_2 are uniquely determined by the linearly independent vectors $\mathbf{u}_1$ and $\mathbf{u}_2$. Suppose there are other scalars b_1 and b_2 with $\mathbf{v} = \alpha_1\mathbf{u}_1 + \alpha_2\mathbf{u}_2 = \beta_1\mathbf{u}_1 + \beta_2\mathbf{u}_2$, then from the linear independence condition we obtain that $\alpha_1 = \beta_1$, $\alpha_2 = \beta_2$. The system $B : (O, \mathbf{u}_1, \mathbf{u}_2)$ is called coordinate system and the scalars α_1, α_2 are called coordinates of the vector $\mathbf{v}$ with respect to the system B. The $\mathbb{R}^2$ vector

$$[\mathbf{v}]_B := \begin{pmatrix} \alpha_1 \\ \alpha_2 \end{pmatrix}$$

consisting of the coordinates of **v** *with respect to the system B is called* **coordinates vector** *of* **v** *with respect to B.*

   This coordinatization gives a one-to-one correspondence between the set of all vectors in the plane and the elements of $\mathbb{R}^2$. *Indeed, every vector in the plane has a unique set of coordinates in* $\mathbb{R}^2$. *Conversely, to every element* $\begin{pmatrix} a_1 \\ a_2 \end{pmatrix} \in \mathbb{R}^2$ *can be associated a vector* **v** *in the plane, the linear combination of the vectors from the coordinates system B with the scalars* a_1 *and* a_2, *namely* $\mathbf{v} = a_1\mathbf{u}_1 + a_2\mathbf{u}_2$.

Remark 35 *The coordinates system B must not be the same as the cartesian one, which consists of two perpendicular axes intersecting in the origin. Any two non parallel vectors can be used as a coordinate system, together with a common point as origin.*

(4) $n = 3$. *Consider now the set of all vectors in space. A reference point O is chosen along with three noncoplanar vectors* $\mathbf{u}_1$, $\mathbf{u}_2$ *and* $\mathbf{u}_3$. *Denote* $B := (O, \mathbf{u}_1, \mathbf{u}_2, \mathbf{u}_3)$. *This system will allow one to coordinate the space as follows:*

   Every vector **v** *in space is a linear combination of the three vectors in the chosen system:* $\mathbf{v} = a_1\mathbf{u}_1 + a_2\mathbf{u}_2 + a_3\mathbf{u}_3$. *Similarly as in the planar case, the scalars* a_1, a_2, a_3 *are uniquely determined by the vector* **v** *and the system B. The* $\mathbb{R}^3$

$$[\mathbf{v}]_B := \begin{pmatrix} a_1 \\ a_2 \\ a_3 \end{pmatrix}$$

is called **the coordinate vector** *of* **v** *with respect to the system B. With the help of this construction, every vector from the space is associated by a one-to-one correspondence with an element from* $\mathbb{R}^3$, *allowing us to identify the set of all vectors in space with the set* $\mathbb{R}^3$.

4.3 Vector Spaces Over Arbitrary Fields

Let K be a field, called *scalar field*.

Definition 4.3.1. A **vector space** over K (for short a K-vector space) is defined to be a set V with

- A fixed element $0 \in V$,
- A binary operation $+: V \times V \to V$ called *addition,*
- A binary operation $\cdot: K \times V \to V$ called *scalar multiplication*
  satisfying

(1) $\forall \mathbf{u}, \mathbf{v}, \mathbf{w} \in V.\ \mathbf{u} + (\mathbf{v} + \mathbf{w}) = (\mathbf{u} + \mathbf{v}) + \mathbf{w}$.

(2) $\forall \mathbf{v} \in V.\ \mathbf{v} + 0 = 0 + \mathbf{v} = \mathbf{v}$.

(3) $\forall \mathbf{v} \in V.\ \mathbf{v} + (-1)\mathbf{v} = 0$.

(4) $\forall \mathbf{v}, \mathbf{w} \in V.\ \mathbf{v} + \mathbf{w} = \mathbf{w} + \mathbf{v}$.

(5) $\forall \mathbf{v} \in V, \forall \lambda, \mu \in K.\ \lambda(\mu \mathbf{v}) = (\lambda\mu)\mathbf{v}$.

(6) $\forall \mathbf{v} \in V.\ 1\mathbf{v} = \mathbf{v}$.

(7) $\forall \mathbf{v}, \mathbf{w} \in V, \forall \lambda \in K.\ \lambda(\mathbf{v} + \mathbf{w}) = \lambda\mathbf{v} + \lambda\mathbf{w}$.

(8) $\forall \mathbf{v}, \in V, \forall \lambda, \mu \in K.\ (\lambda = \mu)\mathbf{v} = \lambda\mathbf{v} + \mu\mathbf{v}$.

Remark 36 $(V, +, 0)$ *is a commutative group.*

Remark 37 *Let V be a* K*-vector space. Then*

(1) $\forall \mathbf{u}, \mathbf{v}, \mathbf{w} \in V.\ \mathbf{u} + \mathbf{w} = \mathbf{v} + \mathbf{w}$ *implies* $\mathbf{u} = \mathbf{v}$.

(2) $\forall \mathbf{v} \in V.\ 0\mathbf{v} = 0$.

(3) $\forall \lambda \in K.\ \lambda 0 = 0$.

(4) $\forall \mathbf{v} \in V, \forall \lambda \in K.\ \lambda\mathbf{v} = 0$ *implies* $\lambda = 0$ *or* $\mathbf{v} = 0$.

Remark 38 *Even if the same symbol is used, it is important to distinguish between the scalar $0 \in K$ and the vector $0 \in V$.*

Example 4.3.1

(1) $\mathbb{R}^n$ is an $\mathbb{R}$-vector space with vector addition and scalar multiplication defined above.

(2) The set of all oriented segments in the plane is a vector space identified with $\mathbb{R}^2$.

(3) The set of all oriented segments in space is a vector space identified with $\mathbb{R}^3$.

(4) Let $n \in \mathbb{N}$. We denote by $\mathbb{C}^n$ the elements of type

$$\mathbf{v} := \begin{pmatrix} \alpha_1 \\ \alpha_2 \\ \vdots \\ \alpha_n \end{pmatrix},$$

where $\alpha_1, \alpha_2, \ldots, \alpha_n \in \mathbb{C}$. Vector addition and scalar multiplication are defined as follows:

$$\mathbf{v} + \mathbf{w} = \begin{pmatrix} \alpha_1 \\ \alpha_2 \\ \vdots \\ \alpha_n \end{pmatrix} + \begin{pmatrix} \beta_1 \\ \beta_2 \\ \vdots \\ \beta_n \end{pmatrix}, \lambda \mathbf{v} = \lambda \begin{pmatrix} \alpha_1 \\ \alpha_2 \\ \vdots \\ \alpha_n \end{pmatrix} = \begin{pmatrix} \lambda \alpha_1 \\ \lambda \alpha_2 \\ \vdots \\ \lambda \alpha_n \end{pmatrix}.$$

The set $\mathbb{C}^n$ enhanced with these operations is a $\mathbb{C}$-vector space.

(5) Let $m, n \in \mathbb{N}$. Consider $M_{m \times n}(\mathbb{C})$ the set of all matrices with m rows and n columns over the set of all complex numbers $\mathbb{C}$. This forms a complex vector space together with the operations of matrix addition and complex scalar multiplication.

(6) Let X be a set. We denote by $F(X, \mathrm{K}) := \{f \mid f : X \to \mathrm{K}\}$. This set is enhanced with a K-vector space structure as follows: Define $0 \in F(X, \mathrm{K})$ as the map whose values are always $0 \in \mathrm{K}$. For $f, g \in F(X, \mathrm{K})$ and $\lambda \in \mathrm{K}$ define

$$\forall x \in X.(f + g)(x) := f(x) + g(x)$$

$$\forall x \in X.(\lambda f)(x) := \lambda f(x).$$

4.4 Linear and Affine Subspaces

Let K be a field and V a K-vector space.

Definition 4.4.1. A subset $U \subseteq V$ is called **linear subspace** (or **vector subspace**) if

(1) $0 \in U$;

(2) $\forall \mathbf{u}, \mathbf{v} \in U. \; \mathbf{u} + \mathbf{v} \in U$;

(3) $\forall \lambda \in \mathrm{K}, \forall \mathbf{u} \in U. \; \lambda \mathbf{u} \in U$.

Remark 39 *(1) Conditions 2 and 3 can be synthesized into an equivalent condition*

$$\forall \lambda, \mu \in \mathrm{K}, \forall \mathbf{u}, \mathbf{v} \in U. \; \lambda \mathbf{u} + \mu \mathbf{v} \in U.$$

(2) By induction, if U is a linear subspace of V, then for every $\mathbf{u}_1, \mathbf{u}_2, \ldots, \mathbf{u}_n \in U$ and every $\lambda_1, \lambda_2, \ldots, \lambda_n \in \mathrm{K}$ one has $\lambda_1 \mathbf{u}_1 + \lambda_2 \mathbf{u}_2 + \ldots + \lambda_n \mathbf{u}_n \in U$.

Example 4.4.1.

(1) Every vector space V has two *trivial subspaces* $\{0\}$ and V. All other subspaces of V are called *proper subspaces*.
(2) The linear subspaces of $\mathbb{R}^2$ are $\{0\}$, lines through 0 and $\mathbb{R}^2$.
(3) The linear subspaces of $\mathbb{R}^3$ are $\{0\}$, lines through 0, planes through 0 and $\mathbb{R}^3$.
(4) Let $P(\mathbb{R})$ be the set of all polynomial mappings $p\colon \mathbb{R} \to \mathbb{R}$ with real coefficients and $P_n(\mathbb{R})$ the set of polynomial maps of degree at most n with real coefficients.

$P(\mathbb{R})$ and $P_n(\mathbb{R})$ are linear subspaces of the $\mathbb{R}$-vector space $F(\mathbb{R}, \mathbb{R})$.

Remark 40 (1) *Let U be a linear subspace of V. Then U is itself a K- vector space.*

Proof. By definition, $0 \in U$. The restriction of vector addition and scalar multiplication to U define two similar operations on U. The vector space axioms are obviously fulfilled.

(2) *If U is a linear subspace of V, we denote this by $U \leq V$. Let $S(V)$ be the set of all linear subspaces of V. The relation $\leq$ defined on $S(V)$ by $U_1 \leq U_2$ if and only if U_1 is a linear subspace of U_2, is an order relation.*

Proof. If U is a subspace of V, then U is a subspace of itself, hence reflexivity is proven. Let now U_1, U_2, U_3 be subspaces of V with $U_1 \leq U_2$, $U_2 \leq U_3$. Herefrom follows that U_1 is a subset of U_3. Since vector addition and scalar multiplication are closed in U_1 and U_3, it follows that U_1 is a subspace of U_3, hence the relation is transitive. The antisymmetry follows immediately.

Proposition 4.4.1. *Let U_1 and U_2 be linear subspaces of V. Then $U_1 \cap U_2$ is a linear subspace of V.*

Proof. We are checking the subspace axioms:

1. Since $U_1, U_2 \leq V$, we have $0 \in U_1$ and $0 \in U_2$, so $0 \in U_1 \cap U_2$.
2. Let $\mathbf{u}, \mathbf{v} \in U_1 \cap U_2$. Then $\mathbf{u}, \mathbf{v} \in U_1$ and $\mathbf{u}, \mathbf{v} \in U_2$. It follows $\mathbf{u} + \mathbf{v} \in U_1$, $\mathbf{u} + \mathbf{v} \in U_2$, hence $\mathbf{u} + \mathbf{v} \in U_1 \cap U_2$.
3. Let $\lambda \in K$ and $\mathbf{u} \in U_1 \cap U_2$. Then $\mathbf{u} \in U_1$ and $\mathbf{u} \in U_2$. It follows $\lambda \mathbf{u} \in U_1$ and $\lambda \mathbf{u} \in U_2$, hence $\lambda \mathbf{u} \in U_1 \cap U_2$.

Corollary 4.4.2. *Let $U_i \in S(V)$, $i \in I$. Then $\bigcap_{i \in I} U_i \in S(V)$.*

Remark 41 (1) *The set $S(V)$ is closed under arbitrary intersections.*

(2) *If U_1, $U_2 \leq V$, then $U_1 \cap U_2 = \inf(U_1, U_2)$ in the ordered set $(S(V),$ $\leq)$, i.e., the infimum of two subspaces always exists and is given by their intersection.*

(3) *For an arbitrary index set I, if $U_i \leq V$, $i \in I$, then $\cap_{i \in I} U_i = \inf(U_i) I$ in the ordered set $(S(V), \leq)$, i.e., the infimum of an arbitrary family of subspaces always exists and is given by their intersection.*

(4) *$(S(V), \leq)$ is a complete lattice, the infimum being described by arbitrary intersections of linear subspaces.*

(5) *In general, the union of two subspaces is no longer a subspace. Consider $\mathbb{R}^2$ and the subspaces U_1 and U_2 given by the vectors $\mathbf{v}_1 :=$*

$$\begin{pmatrix} 1 \\ 1 \end{pmatrix} and \ \mathbf{v}_2 = \begin{pmatrix} 1 \\ -1 \end{pmatrix} respectively, i.e., the lines through zero having$$

as support $\mathbf{v}_1$ and $\mathbf{v}_2$, respectively. If $U_1 \cup U_2$ is a subspace, then

$$\forall \mathbf{u}, \mathbf{v} \in U_1 \cup U_2. \ \mathbf{u} + \mathbf{v} \in U_1 \cup U_2. \ But \ \mathbf{v}_1 + \mathbf{v}_2 = \begin{pmatrix} 2 \\ 0 \end{pmatrix} \notin U_1 \cup U_2.$$

(6) *The supremum in the complete lattice $S(V)$, $\leq)$ is not the union of subspaces. Another construction is needed for this.*

Definition 4.4.2. Let U_1, U_2, ..., U_n be linear subspaces of V. **The sum** of these subspaces is defined as

$$U_1 + U_2 + \cdots + U_n := \{\mathbf{u}_1 + \mathbf{u}_2 + \cdots + \mathbf{u}_n \mid \mathbf{u}_1 \in U_1, \mathbf{u}_2 \in U_2, \ldots, \mathbf{u}_n \in U_n\}.$$

For an infinite family of subspaces $(U_i)_{i \in I}$, their sum is defined by the set of all finite sums of elements from different U_i,

$$\sum_{i \in I} U_i := \{\mathbf{u}_{i1} + \mathbf{u}_{i2} + \cdots + \mathbf{u}_{ik} \mid \mathbf{u}_{ij} \in U_{ij}, i_j \in I, j = 1, \ldots k, k \in \mathbb{N}\}.$$

Proposition 4.4.3. *The sum of the linear subspaces U_1, U_2, ..., U_n is a linear subspace and*

$$U_1 + U_2 + \cdots + U_n = \sup(U_1, U_2, \ldots, U_n).$$

Proof. One has $0 \in U_i$ for every $i = 1, \ldots, n$, hence $0 \in U_1 + U_2 + \cdots + U_n$.

Let $\mathbf{v}_1, \mathbf{v}_2 \in U_1 + U_2 + \cdots + U_n$. Then $\mathbf{v}_1 = \mathbf{u}_1 + \mathbf{u}_2 + \cdots + \mathbf{u}_n$, with $\mathbf{u}_i \in U_i$, $i = 1, \ldots, n$ and $\mathbf{v}_2 = \mathbf{w}_1 + \mathbf{w}_2 + \cdots + \mathbf{w}_n$, with $\mathbf{w}_i \in U_i$, $i = 1, \ldots n$. Then $\mathbf{v}_1 + \mathbf{v}_2 = (\mathbf{u}_1 + \mathbf{w}_1) + (\mathbf{u}_2 + \mathbf{w}_2) + \cdots + (\mathbf{u}_n + \mathbf{w}_n)$. Since $\mathbf{u}_i + \mathbf{w}_i \in U_i$, $i = 1, \ldots, n$, we have $\mathbf{v}_1 + \mathbf{v}_2 \in U_1 + U_2 + \cdots + U_n$.

The third condition is verified in a similar way.

We have to prove that $U_1 + U_2 + \cdots + U_n = \sup(U_1, U_2, \ldots, U_n)$. On the one hand, $U_i \subseteq U_1 + U_2 + \cdots + U_n$ since one can choose all $u_j = 0$ in the definition of the sum, except the elements of U_i. Suppose $W \le V$ and $U_i \le W$, $i = 1, \ldots, n$, hence all elements of all subspaces U_i are in W, in particular all sums $u_1 + u_2 + \cdots + u_n$ with $u_1 \in U_1$, $u_2 \in U_2, \ldots, u_n \in U_n$ are in W. Hence $U_1 + U_2 + \cdots + U_n \le W$ and so $U_1 + U_2 + \cdots + U_n = \sup(U_1, U_2, \ldots, U_n)$.

Remark 42 *Similarly, the supremum of an arbitrary family of subspaces* $(U_i)_{i \in I}$ *is defined by their sum.*

Remark 43 *$(S(V), \le)$ is a complete lattice, the infimum is given by subspace intersection and the supremum by their sum.*

Example 4.4.2. Consider in $\mathbb{R}^3$ the planes $U_1 := x0y$ and $U_2 := x0z$. Their intersection is $0x$, a line through 0. Their sum is the entire space $\mathbb{R}^3$.

Consider the vector $\mathbf{v} := \begin{pmatrix} 1 \\ 1 \\ 1 \end{pmatrix} \in \mathbb{R}^3$. This vector can be described as the sum of vectors from U_1 and U_2 as follows:

$$\mathbf{v} = \begin{pmatrix} 1 \\ 1 \\ 1 \end{pmatrix} = \begin{pmatrix} 1 \\ 1 \\ 0 \end{pmatrix} + \begin{pmatrix} 0 \\ 0 \\ 1 \end{pmatrix} = \begin{pmatrix} 0 \\ 1 \\ 0 \end{pmatrix} + \begin{pmatrix} 1 \\ 0 \\ 1 \end{pmatrix}.$$

The vectors $\begin{pmatrix} 1 \\ 1 \\ 0 \end{pmatrix}, \begin{pmatrix} 0 \\ 1 \\ 0 \end{pmatrix} \in U_1$ and $\begin{pmatrix} 0 \\ 0 \\ 1 \end{pmatrix}, \begin{pmatrix} 1 \\ 0 \\ 1 \end{pmatrix} \in U_2$. Hence, the decomposition of $\mathbf{v}$ as the sum of two vectors from U_1 and U_2 is not unique.

Definition 4.4.3. Let $U_1, U_2, \ldots, U_n$ be linear subspaces of V. The vector space V is the **direct sum** of U_i, $i = 1, \ldots, U_n$ and we write

$$V = U_1 \oplus U_2 \oplus \ldots \oplus U_n,$$

if every element $\mathbf{v} \in V$ has a unique decomposition as

$$\mathbf{v} = \mathbf{u}_1 + \mathbf{u}_2 + \cdots + \mathbf{u}_n \text{ for every } \mathbf{u}_i \in U_i, i = 1, \ldots, n.$$

Proposition 4.4.4. *Let U_1 and U_2 be linear subspaces of V. Then $V = U_1 \oplus U_2$ if and only if the following holds true:*

(1) $V = U_1 + U_2$
(2) $U_1 \cap U_2 = \{0\}$.

Proof. ($\Rightarrow$) Suppose $V = U_1 \oplus U_2$, then $V = U_1 + U_2$. Let $\mathbf{v} \in U_1 \cap U_2$. Then

$$\mathbf{v} = \underbrace{\mathbf{v}}_{\in U_1} + \underbrace{0}_{\in U_2} = \underbrace{0}_{\in U_1} + \underbrace{\mathbf{v}}_{\in U_2}.$$

Hence $\mathbf{v}$ has two different representations as a sum of vectors from U_1 and U_2. It follows $\mathbf{v} = 0$ and $U_1 \cap U_2 = \{0\}$.

($\Leftarrow$) Let us suppose that (1) and (2) are true. Then, for every $\mathbf{v} \in V$, this vector has a decomposition as

$$\mathbf{v} = \mathbf{u}_1 + \mathbf{u}_2, \, \mathbf{u}_1 \in U_1, \, \mathbf{u}_2 \in U_2.$$

For the uniqueness, suppose

$$\mathbf{v} = \mathbf{u}'_1 + \mathbf{u}'_2, \text{ with } \mathbf{u}'_1 \in U_1, \, \mathbf{u}'_2 \in U_2.$$

Substracting these equalities, we have $0 = (\mathbf{u}_1 - \mathbf{u}'_1) + (\mathbf{u}_2 - \mathbf{u}'_2)$. It follows $\mathbf{u}_1 - \mathbf{u}'_1 = \mathbf{u}'_2 - \mathbf{u}_2 \in U_1 \cap U_2$. Since $U_1 \cap U_2 = \{0\}$, we have that $\mathbf{u}_1 = \mathbf{u}'_1$ and $\mathbf{u}_2 = \mathbf{u}'_2$, the uniqueness of the decomposition is proven.

Definition 4.4.4. Let V be a K-vector space. A subset $A \subseteq V$ is called **affine subspace** of V if there is a linear subspace $U \leq V$ and a vector $\mathbf{v} \in V$ such that

$$A = \mathbf{v} + U := \{\mathbf{v} + \mathbf{u} \,|\, \mathbf{u} \in U\}.$$

Example 4.4.3.
(1) The affine subspaces of $\mathbb{R}^2$ are the points, lines and the entire space $\mathbb{R}^2$.
(2) The affine subspaces of $\mathbb{R}^3$ are the points, lines, planes and the entire space $\mathbb{R}^3$.
(3) Affine subspaces are obtained via translation of a linear subspace with a given vector.

4.5 Linearly Independent Vectors. Generator Systems. Basis

Definition 4.5.1. Let V be a K-vector space and $\mathbf{v}_1, \mathbf{v}_2, \ldots, \mathbf{v}_n \in V$. Every element of the type

$$\mathbf{v} = \lambda_1\mathbf{v}_1 + \lambda_2\mathbf{v}_2 + \cdots + \lambda_n\mathbf{v}_n = \sum_{i=1}^{n}\lambda_i\mathbf{v}_i,$$

with $\lambda_i \in K$, $i = 1, \ldots, n$ is called **linear combination** of the vectors $\mathbf{v}_1, \mathbf{v}_2, \ldots, \mathbf{v}_n$. We denote the set of all linear combinations of the vectors $\mathbf{v}_1, \mathbf{v}_2, \ldots, \mathbf{v}_n$ with $\langle \mathbf{v}_1, \mathbf{v}_2, \ldots, \mathbf{v}_n \rangle$.

Proposition 4.5.1. *Let V be a K-vector space and* $\mathbf{v}_1, \mathbf{v}_2, \ldots, \mathbf{v}_n$ *be vectors from V. The set* $\langle \mathbf{v}_1, \mathbf{v}_2, \ldots, \mathbf{v}_n \rangle$ *of all linear combinations of* $\mathbf{v}_1, \mathbf{v}_2, \ldots, \mathbf{v}_n$ *is a linear subspace of V.*

Proof. Let us verify the linear subspace axioms:
  (1) $0 = 0\mathbf{v}_1 + 0\mathbf{v}_2 + \cdots + 0\mathbf{v}_n \in \langle \mathbf{v}_1, \mathbf{v}_2, \ldots, \mathbf{v}_n \rangle$.
  (2) Let $\mathbf{v}, \mathbf{w} \in \langle \mathbf{v}_1, \mathbf{v}_2, \ldots, \mathbf{v}_n \rangle$. Then $\mathbf{v}$ and $\mathbf{w}$ are linear combinations

$$\mathbf{v} = \lambda_1\mathbf{v}_1 + \lambda_2\mathbf{v}_2 + \cdots + \lambda_n\mathbf{v}_n$$

$$\mathbf{w} = \mu_1\mathbf{v}_1 + \mu_2\mathbf{v}_2 + \cdots + \mu_n\mathbf{v}_n.$$

By summing up, we have $\mathbf{v} + \mathbf{w} = (\lambda_1 + \mu_1)\mathbf{v}_1 + (\lambda_2 + \mu_2)\mathbf{v}_2 + \cdots + (\lambda_n + \mu_n)\mathbf{v}_n$, hence $\mathbf{v} + \mathbf{w} \in \langle \mathbf{v}_1, \mathbf{v}_2, \ldots, \mathbf{v}_n \rangle$.
  (3) Let $\mathbf{v} \in \langle \mathbf{v}_1, \mathbf{v}_2, \ldots, \mathbf{v}_n \rangle$, then $\mathbf{v} = \lambda_1\mathbf{v}_1 + \lambda_2\mathbf{v}_2 + \cdots + \lambda_n\mathbf{v}_n$ and let $\lambda \in K$ be arbitrary chosen. Then $\lambda_n = \lambda\lambda_1\mathbf{v}_1 + \lambda\lambda_2\mathbf{v}_2 - \lambda\lambda_n\mathbf{v}_n$ wherefrom follows $\lambda\mathbf{v} \in \langle \mathbf{v}_1, \mathbf{v}_2, \ldots, \mathbf{v}_n \rangle$.

Remark 44 (1) $\mathbf{v}_1 = 1\mathbf{v}_1 + 0\mathbf{v}_2 + \cdots + 0\mathbf{v}_n \in \langle \mathbf{v}_1, \mathbf{v}_2, \ldots, \mathbf{v}_n \rangle$. *Similarly,* $\mathbf{v}_2, \ldots, \mathbf{v}_n \in \langle \mathbf{v}_1, \mathbf{v}_2, \ldots, \mathbf{v}_n \rangle$.
  (2) $\langle \mathbf{v}_1, \mathbf{v}_2, \ldots, \mathbf{v}_n \rangle$ *is the smallest linear subspace containing* $\mathbf{v}_1$, $\mathbf{v}_2, \ldots, \mathbf{v}_n$. *If* $U \leq V$ *is another linear subspace containing* $\mathbf{v}_1$, $\mathbf{v}_2, \ldots, \mathbf{v}_n$, *we deduce from the subspace properties that it will also contain all linear combinations of these vectors, hence* $\langle \mathbf{v}_1$, $\mathbf{v}_2, \ldots, \mathbf{v}_n \rangle \subseteq U$.
  (3) *The subspace* $\langle \mathbf{v}_1, \mathbf{v}_2, \ldots, \mathbf{v}_n \rangle$ *is called the linear subspace spanned by the vectors* $\mathbf{v}_1, \mathbf{v}_2, \ldots, \mathbf{v}_n$. *These vectors are called the* **generator system** *of the spanned subspace.*
  (4) *If* $\mathbf{w}_1, \mathbf{w}_2, \ldots, \mathbf{w}_k \in \langle \mathbf{v}_1, \mathbf{v}_2, \ldots, \mathbf{v}_n \rangle$, *then* $\langle \mathbf{w}_1, \mathbf{w}_2, \ldots, \mathbf{w}_k \rangle \subseteq \langle \mathbf{v}_1$, $\mathbf{v}_2, \ldots, \mathbf{v}_n \rangle$.

Example 4.5.1

(1) $\mathbb{R}^n$ is spanned by

$$\mathbf{e}_1 = \begin{pmatrix} 1 \\ 0 \\ 0 \\ \vdots \\ 0 \end{pmatrix}, \mathbf{e}_2 = \begin{pmatrix} 0 \\ 1 \\ 0 \\ \vdots \\ 0 \end{pmatrix}, \dots, \mathbf{e}_n = \begin{pmatrix} 0 \\ \vdots \\ 0 \\ 0 \\ 1 \end{pmatrix}.$$

(2) $\mathbb{R}^2$ is spanned by $\mathbf{e}_1$, $\mathbf{e}_2$ and also by $\mathbf{v}_1 = \begin{pmatrix} 1 \\ 1 \end{pmatrix}$ and $\mathbf{v}_2 \begin{pmatrix} 1 \\ -1 \end{pmatrix}$.

This proves that a subspace can be spanned by different generator systems.

(3) The vector space $P_n(\mathbb{R})$ of polynomial functions of degree at most n is generated by

$$p_j(x) := x^j \text{ for every } x \in \mathbb{R}, j = 1, \dots, n.$$

Proposition 4.5.2. *Let $G \subseteq V$ and $\mathbf{v} \in V$. We define $G' := G \cup \{\mathbf{v}\}$. Then $\langle G \rangle = \langle G \rangle'$ if and only if $\mathbf{v} \in \langle G \rangle$.*

Proof. ($\Rightarrow$) $\langle G \rangle'$ is the set of all linear combinations of vectors from G and $\mathbf{v}$. It follows that $\mathbf{v}$ is a linear combination of vectors from G, hence v is an element *of $\langle G \rangle$.*

($\Leftarrow$) Suppose $\mathbf{v} \in \langle G \rangle$. Then $\mathbf{v}$ is a linear combination of vectors from G. It follows that $G' \subseteq \langle G \rangle$, hence $\langle G \rangle' \subseteq \langle G \rangle$. The inverse inclusion holds true, so $\langle G \rangle = \langle G \rangle'$.

Definition 4.5.2. A vector space V is called **finitely generated**, if it has a finite generator system.

Remark 45 (1) *Let V be a K-vector space and $\mathbf{v} \in V$. Then $\langle \mathbf{v} \rangle = K\mathbf{v} = \{\lambda \mathbf{v} \mid \lambda \in K\}$.*

(2) *If $\mathbf{v}_1, \mathbf{v}_2, \dots, \mathbf{v}_n \in V$, then $\langle \mathbf{v}_1, \mathbf{v}_2, \dots, \mathbf{v}_n \rangle = K\mathbf{v}_1 + K\mathbf{v}_2 + \dots + K\mathbf{v}_n$.*
(3) *If V is finitely generated and $K = \mathbb{R}$, then it is the sum of the lines $K\mathbf{v}_i$, $i = 1, \dots, n$, defined by the generator vectors $\langle \mathbf{v}_1, \dots, \mathbf{v}_n \rangle$.*
(4) $\langle \varnothing \rangle = \{0\}$.

Definition 4.5.3. The vectors $v_1, v_2, \ldots, v_n \in V$ are called **linearly independent** if none of the vectors v_j can be written as a linear combination of the other vectors v_i, $i \neq j$. In the other case, these vectors are called **linearly dependent**.

Proposition 4.5.3. *The following are equivalent:*

(1) $v_1, v_2, \ldots, v_n$ *are linearly independent.*

(2) $\lambda_1 v_1 + \lambda_2 v_2 + \cdots + \lambda_n v_n = 0$ *implies* $\lambda_1 = \lambda_2 = \cdots = \lambda_n = 0$.

(3) $\lambda_1 v_1 + \lambda_2 v_2 + \cdots + \lambda_n v_n = \lambda'_1 v_1 + \lambda'_2 v_2 + \cdots + \lambda'_n v_n$ *implies* $\lambda_i = \lambda'_i$ *for every* $i = 1, \ldots, n$.

Proof. $(1 \Rightarrow 2)$ Suppose there exist $i \in \{1, \ldots, n\}$ such that $\lambda_i \neq 0$ and $\lambda_1 v_1 + \lambda_2 v_2 + \cdots + \lambda_n v_n = 0$. Then

$$v_i = -\lambda_1 \lambda_i^{-1} v_1 - \lambda_2 \lambda_i^{-1} v_2 - \cdots - \lambda_{i-1} \lambda_i^{-1} v_{i-1} - \lambda_{i+1} \lambda_i^{-1} v_{i+1} - \cdots - \lambda_n \lambda_i^{-1} v_n,$$

i.e., v_i is a linear combination of v_j, $j \neq i$.

$(2 \Rightarrow 3)$ If $\lambda_1 v_1 + \lambda_2 v_2 + \cdots + \lambda_n v_n = \lambda'_1 v_1 + \lambda'_2 v_2 + \cdots + \lambda'_n v_n$, then $(\lambda_1 - \lambda'_1) v_1 + (\lambda_2 - \lambda'_2) v_2 + \cdots + (\lambda_n - \lambda'_n) v_n = 0$. It follows $\lambda_i = \lambda'_i$, for every $i = 1, \ldots, n$.

$(3 \Rightarrow 1)$ Suppose $v_1, v_2, \ldots, v_n$ are linearly dependent. Then there exist $i \in \{1, \ldots, n\}$ such that v_i is a linear combination of v_j, $j \neq i$. Then

$$v_i = \lambda_1 v_1 + \lambda_2 v_2 + \cdots + \lambda_{i-1} v_{i-1} + \lambda_{i+1} v_{i+1} + \cdots + \lambda_n v_n$$

where from follows that

$$\lambda_1 v_1 + \lambda_2 v_2 + \cdots + \lambda_{i-1} v_{i-1} - v_i + \lambda_{i+1} v_{i+1} + \cdots + \lambda_n v_n = 0 = 0 v_1 + \cdots + 0 v_n.$$

From the hypothesis follows $\lambda_j = 0$, $j \neq i$ and $-1 = 0$, which is a contradiction.

Remark 46 (1) *Any permutation of the elements of a linearly independent set is linearly independent.*

(2) *For every* $v \in V$, *the set* $\{v\}$ *is linearly independent.*

(3) *The vectors* v_1, v_2 *are linearly dependent if and only if* $v_1 = v_2$ 0 *or* v_2 *is a scalar multiple of* v_1.

(4) *If* $0 \in \{v_1, v_2, \ldots, v_n\}$, *the set* $\{v_1, v_2, \ldots, v_n\}$ *is linearly dependent.*

(5) *If* $v_1, v_2, \ldots, v_n$ *are linearly independent, then they are pairwise distinct.*

Example 4.5.2.

(1) Any two non-parallel vector in the plane are linearly independent. The vectors $\mathbf{e}_1 = \begin{pmatrix} 1 \\ 0 \end{pmatrix}$ and $\mathbf{e}_2 = \begin{pmatrix} 0 \\ 1 \end{pmatrix}$ are linearly independent, as well as $\mathbf{v}_1 = \begin{pmatrix} 2 \\ 1 \end{pmatrix}$ and $\mathbf{v}_2 = \begin{pmatrix} 1 \\ 2 \end{pmatrix}$.

(2) Any three vectors in $\mathbb{R}^2$ are linearly dependent.

(3) Any three vectors in $\mathbb{R}^3$ which do not lie in the same plane are linearly independent. In particular, the vectors $\mathbf{v}_1 = \begin{pmatrix} 1 \\ 1 \\ 0 \end{pmatrix}$, $\mathbf{v}_2 = \begin{pmatrix} 1 \\ 0 \\ 1 \end{pmatrix}$, $\mathbf{v}_3 = \begin{pmatrix} 0 \\ 1 \\ 1 \end{pmatrix}$ are linearly independent.

(4) Any collection of 4 vectors in $\mathbb{R}^3$ are linearly dependent.

(5) The vectors $\mathbf{e}_1, \mathbf{e}_2, \ldots, \mathbf{e}_n$ are linearly independent in K_n.

(6) The polynomial functions $p_0, p_1, \ldots, p_n$ are linearly independent in $P_n(\mathbb{R})$.

(7) The non zero rows of a triangle shaped matrix are linearly independent.

Remark 47 *The notion of linear independence can be defined for infinite sets of vectors too.*

Definition 4.5.4. An arbitrary family of vectors $(\mathbf{v}_i)_{i \in I}$ in V is **linearly independent** if none of its vectors can be written as a linear combination of a finite subset of the family $\mathbf{v}_i$, $i \in I$.

Definition 4.5.5. A **basis** of a vector space V is a family of vectors $(\mathbf{v}_i)_{i \in I}$ in V satisfying:

(1) $\langle \{\mathbf{v}_i \mid i \in I\} \rangle = V$,

(2) The family $(\mathbf{v}_i)_{i \in I}$ is linearly independent.

Remark 48 *The vectors $\mathbf{v}_1, \mathbf{v}_2, \ldots, \mathbf{v}_n$ are a basis of V if and only if they are linearly independent and every vector in V is a linear combination of $\mathbf{v}_1, \mathbf{v}_2, \ldots, \mathbf{v}_n$.*

Example 4.5.3

(1) $\mathbf{e}_1 = \begin{pmatrix} 1 \\ 0 \end{pmatrix}$ and $\mathbf{e}_2 = \begin{pmatrix} 0 \\ 1 \end{pmatrix}$ are a basis of $\mathbb{R}^2$, as well as the vectors

$\mathbf{v}_1 = \begin{pmatrix} 1 \\ 1 \end{pmatrix}$ and $\mathbf{v}_2 = \begin{pmatrix} 1 \\ -1 \end{pmatrix}$.

(2) $\mathbf{e}_1, \mathbf{e}_2, \ldots, \mathbf{e}_n$ are a basis of K^n, called *canonical basis*.

(3) $p_0, p_1, \ldots, p_n$ are a basis for the real vector space $P_n(\mathbb{R})$.

4.5.1 Every vector space has a basis

Let V be a K-vector space. If V has a basis $B: =(\mathbf{v}_i)_{i \in I}$, then every vector $\mathbf{w} \in V$ has a unique decomposition as a linear combination of the vectors of B. We ask whether every vector space has a basis. The following result uses the choice axiom.

Theorem 4.5.4. *Let V be a K-vector space and $X \subseteq V$. If $V = \langle X \rangle$ and the subset $X_1 \subseteq X$ is linearly independent, then there exists a basis B of V such that $X_1 \subseteq B \subseteq X$.*

Proof. In order to find this basis, we need a maximal generator set whose elements are linearly independent. For this, consider the set

$$C := \{X' \mid X_1 \subseteq X' \subseteq X, X' \text{ is linearly independent}\}.$$

The set C is not empty, since $X_1 \in C$. We want to prove that this set has a maximal element. Consider an arbitrary non empty chain $\mathfrak{L} \subseteq C$ and denote

$$X_0 := \cup \{X' \mid X' \in C\}.$$

We prove that X_0 is linearly independent. For this, choose a finite linearly independent subset $\mathbf{v}_1, \mathbf{v}_2, \ldots, \mathbf{v}_n \in X_0$. From the definition of X_0 we obtain that for every $i \in \{1, \ldots, n\}$ there exist subsets $X'_1, X'_2, \ldots, X'_n$ in $\mathfrak{L}$ such that $\mathbf{v}_i \in X'$, $i = 1, \ldots, n$. Since C is a chain, it follows the existence of $i_0 \in \{1, \ldots, n\}$ with $X'_i \subseteq X'_{i_0}$, $i = 1, \ldots, n$ implying $\mathbf{v}_i \in X'_{i_0}$, $i = 1, \ldots, n$. The subset $X_{i_0} \in C$ is linearly independent, hence any finite subset is linearly independent too, particularly, $\{\mathbf{v}_1, \mathbf{v}_2, \ldots, \mathbf{v}_n\}$. The vectors $\mathbf{v}_1, \mathbf{v}_2, \ldots, \mathbf{v}_n$ being arbitrarily chosen, we conclude the linearly independence of

X_0 and $X_1 \subseteq X_0 \subseteq X$, i.e., $X_0 \in C$ and X_0 is an upper bound for $\mathcal{L}$. From Zorn's lemma follows the existence of a maximal element B in $\mathcal{L}$. Since $B \in C$, it follows that B is linearly independent.

Now we are going to prove that B is a generator system for V. Suppose $V \neq \langle B \rangle$. Then $X \not\subset \langle B \rangle$. Let $\mathbf{v} \in X_n \backslash \langle B \rangle$. Then, we can prove that $B \cup \{\mathbf{v}\}$ is linearly independent, being in contradiction with the maximality of B. For this, consider a linear combination

$$\lambda_1 \mathbf{v}_1 + \lambda_2 \mathbf{v}_2 + \cdots + \lambda_n \mathbf{v}_n + \lambda \mathbf{v} = 0, \; \lambda_i, \lambda \in K, \; \mathbf{v}_i \in B, \; i = 1, \ldots, n,$$

Then $\mathbf{v}$ can be written as a linear combination of vectors $\mathbf{v}_1, \mathbf{v}_2, \ldots, \mathbf{v}_n$ from B

$$\mathbf{v} = -\lambda_1 \lambda^{-1} \mathbf{v}_1 - \lambda_2 \lambda^{-1} \mathbf{v}_2 - \cdots - \lambda_n \lambda^{-1} \mathbf{v}_n$$

hence $\mathbf{v} \in \langle B \rangle$, which is a contradiction.

Corollary 4.5.5.

(1) *Every vector space has a basis.*
(2) *Any linearly independent subset of a vector space can be completed to a basis.*
(3) *The basis of a vector space V are exactly the maximal linearly independent subsets of V.*
(4) *From any generator system of a vector space, we can extract a basis of V.*
(5) *The basis of a vector space V are exactly the minimal generator systems of V.*
(6) *If X is a generator system of V and B is a linearly independent subset of V, then B can be completed with vectors from X to a basis B of V.*

Proposition 4.5.6. *Let V be a finite generated vector space and $\mathbf{v}_1$, $\mathbf{v}_2, \ldots, \mathbf{v}_m$ a generator system. Every subset containing $m+1$ vectors from V is linearly dependent.*

Proof. We are going to prove this assertion by induction. Consider the assertion $P(m)$: *If the vector space V is spanned by m vectors, $V = \langle \mathbf{v}_1, \mathbf{v}_2, \ldots, \mathbf{v}_m \rangle$, then any subset of $m+1$ vectors $\{\mathbf{w}_1, \mathbf{w}_2, \ldots, \mathbf{w}_m, \mathbf{w}_{m+1}\}$ from V is linearly dependent.*

Verification step for $P(1)$: Let $\{\mathbf{w}_1, \mathbf{w}_2\} \subseteq \langle \{\mathbf{v}_1\} \rangle$. There are some scalars $\lambda_1, \lambda_2 \in K$, such that $\mathbf{w}_1 = \lambda_1 \mathbf{v}_1, \mathbf{w}_2 = \lambda_2 \mathbf{v}_1$. It follows that $\mathbf{v}_1 = \lambda_1^{-1} \mathbf{w}_1$ and $\mathbf{w}_2 = \lambda_2 \lambda_1^{-1} \mathbf{w}_1$, hence the vectors $\mathbf{w}_1$ and $\mathbf{w}_2$ are linearly dependent.

Induction step $P(m-1) \Rightarrow P(m)$ Suppose that some m vectors from a $m-1$ vectors generated space are linearly dependent and prove this assertion for m vectors. Let $\{\mathbf{w}_1, \mathbf{w}_2, \ldots, \mathbf{w}_m, \mathbf{w}_{m+1}\} \subseteq \langle\{\mathbf{v}_1, \mathbf{v}_2, \ldots, \mathbf{v}_m\}\rangle$.

Then

$$\mathbf{w}_1 = \lambda_{11}\mathbf{v}_1 + \lambda_{12}\mathbf{w}_2 + \cdots + \lambda_{1m}\mathbf{v}_m$$
$$\mathbf{w}_2 = \lambda_{21}\mathbf{v}_1 + \lambda_{22}\mathbf{w}_2 + \cdots + \lambda_{2m}\mathbf{v}_m$$
$$\vdots$$
$$\mathbf{w}_{m+1} = \lambda_{m+1,1}\mathbf{v}_1 + \lambda_{m+1,2}\mathbf{w}_2 + \cdots + \lambda_{m+1,m}\mathbf{v}_m$$

We distinguish the following cases

Case 1. $\lambda_{11} = \lambda_{12} = \cdots = \lambda_{m+1,m} = 0$. The vectors $\mathbf{w}_2, \ldots, \mathbf{w}_{m+1} \in \langle\{\mathbf{v}_2, \ldots, \mathbf{v}_m\}\rangle$ and from the induction hypothesis follow their linearly dependence. The vectors $\{\mathbf{w}_1, \mathbf{w}_2, \ldots, \mathbf{w}_{m+1}\}$ are linearly dependent too.

Case 2. At least one of the coefficients $\lambda_{i1} \neq 0$, $i = 1, \ldots, m+1$. Suppose $\lambda_{11} \neq 0$. Then

$$\mathbf{w}'_2 = (\lambda_{22} - \lambda_{21}\lambda_{11}^{-1}\lambda_{12})\mathbf{v}_2 + \cdots + (\lambda_{2m} - \lambda_{21}\lambda_{11}^{-1}\lambda_{1m})\mathbf{v}_m$$
$$\vdots$$
$$\mathbf{w}'_{m+1} = (\lambda_{m+1,2} - \lambda_{m+1,1}\lambda_{11}^{-1}\lambda_{12})\mathbf{v}_2 + \cdots + (\lambda_{m+1,m} - \lambda_{m+1,1}\lambda_{11}^{-1}\lambda_{1n})$$

Then $\mathbf{w}'_2, \ldots, \mathbf{w}'_{m+1} \in \langle\{\mathbf{v}_2, \ldots, \mathbf{v}_{m+1}\}\rangle$. From the induction hypothesis follows the linearly dependence of $\mathbf{w}'_2, \ldots, \mathbf{w}'_{m+1}$ There exist scalars $\mu_2, \ldots, \mu_{m+1} \in K$, not all of them being zero, such that

$$\mu_2\mathbf{w}'_2 + \cdots \mu_{m+1}\mathbf{w}'_{m+1} = 0.$$

Since $\mathbf{w}'_i = \mathbf{w}_i - \lambda_{i1}\lambda_{11}^{-1}\mathbf{w}_1$, we have

$$\mu_2(\mathbf{w}_2 - \lambda_{21}\lambda_{11}^{-1}\mathbf{w}_1) + \cdots \mu_{m+1}(\mathbf{w}_{m+1} - \lambda_{m+1,1}\lambda_{11}^{-1}\mathbf{w}_1) = 0$$

and

$$\left(\sum_{i=2}^{m+1}\mu_i\lambda_{i1}\lambda_{11}^{-1}\right)\mathbf{w}_1 + \mu_2\mathbf{w}_2 + \cdots + \mu_{m+1}\mathbf{w}_{m+1} = 0.$$

Since not all scalars $\mu_2, \ldots, \mu_{m+1}$ are zero, the vectors $\mathbf{w}_1, \mathbf{w}_2, \ldots, \mathbf{w}_{m+1}$ are linearly independent.

Corollary 4.5.7. *If a vector space has a generator system* $\mathbf{v}_1, \ldots, \mathbf{v}_n$, *any set of linearly independent vectors has at most n elements.*

Theorem 4.5.8. *Let* $\mathbf{v}_1, \mathbf{v}_2, \ldots, \mathbf{v}_m \in V$ *being linearly independent. If* $\mathbf{w}_1, \mathbf{w}_2, \ldots, \mathbf{w}_r \in V$ *are chosen such that*

$$\langle \{\mathbf{v}_1, \mathbf{v}_2, \ldots, \mathbf{v}_m, \mathbf{w}_1, \mathbf{w}_2, \ldots, \mathbf{w}_r\} \rangle,$$

then we can find indices $1 \le i_1 < i_2 < \cdots < i_s \le r$ *such that*

$$\mathbf{v}_1, \mathbf{v}_2, \ldots, \mathbf{v}_m, \mathbf{w}_{i1}, \mathbf{w}_{i2}, \ldots, \mathbf{w}_{is}$$

is a basis of V.

Proof. Consider the set

$$C := \{\{\mathbf{v}_1, \mathbf{v}_2, \ldots, \mathbf{v}_m, \mathbf{w}_{i1}, \mathbf{w}_{i2}, \ldots, \mathbf{w}_{is}\} \mid$$

$$V = \langle \{\mathbf{v}_1, \mathbf{v}_2, \ldots, \mathbf{v}_m, \mathbf{w}_{i1}, \mathbf{w}_{i2}, \ldots, \mathbf{w}_{is}\} \rangle, 1 \le i_1 < i_2 < \cdots < i_s \le r\}.$$

The vectors $\mathbf{v}_1, \mathbf{v}_2, \ldots, \mathbf{v}_m, \mathbf{w}_1, \mathbf{w}_2, \ldots, \mathbf{w}_r$ are satisfying this condition, hence $C \ne \varnothing$. We order the set C by inclusion. Then, there exists in C a minimal generator set. It can be proved that this minimal generator set is a basis of V. It is sufficient to check the linearly independence of the minimal system

$$\mathbf{v}_1, \mathbf{v}_2, \ldots, \mathbf{v}_m, \mathbf{w}_{i1}, \mathbf{w}_{i2}, \ldots, \mathbf{w}_{is}.$$

Suppose one can find scalars $\lambda_1, \ldots, \lambda_m, \mu_1, \ldots, \mu_s \in K$, not all of them being zero with

$$\lambda_1 \mathbf{v}_1 + \cdots + \lambda_m \mathbf{v}_m + \mu_1 \mathbf{w}_{i1} + \cdots + \mu_s \mathbf{w}_{is} = 0.$$

Then there exist an index $j \in \{1, \ldots, s\}$ with $\mu_j \ne 0$. If for all j, $\mu_j = 0$, then $\lambda_1 \mathbf{v}_1 + \cdots + \lambda_m \mathbf{v}_m = 0$ and from the linearly independence of $\mathbf{v}_1, \ldots, \mathbf{v}_m$ one obtains $\lambda_i = 0$, $i \in \{1, \ldots, m\}$, which is a contradiction.

If $\mu_j \ne 0$, then $\mathbf{w}_{ij} = \lambda_1 \mu_j^{-1} \mathbf{v}_1 - \cdots - \lambda_m \mu_j^{-1} \mathbf{v}_m - \mu_1 \mu_j^{-1} \mathbf{w}_{i1} - \cdots - \mu_s \mu_j^{-1} \mathbf{w}_{is}$ is a linear combination of

$$\mathbf{v}_1, \ldots, \mathbf{v}_m, \mathbf{w}_{i1}, \mathbf{w}_{1j-1}, \mathbf{w}_{1j+1}, \ldots, \mathbf{w}_{is}.$$

We have obtained a generator system for

$$V = \langle \{\mathbf{v}_1, \ldots, \mathbf{v}_m, \mathbf{w}_{i1}, \mathbf{w}_{1j-1}, \mathbf{w}_{1j+1}, \ldots, \mathbf{w}_{is}\} \rangle$$

with less elements than the minimal generator system we have previously considered. Contradiction!

The following results are consequences of this theorem:

Theorem 4.5.9. (Basis choice theorem). *If the vector space V is generated by the vectors $\mathbf{w}_1, \ldots, \mathbf{w}_r$, then one can find indices $1 \leq i_1 < i_2 < \cdots < i_n \leq r$ such that the vectors $\mathbf{w}_{i1}, \ldots, \mathbf{w}_{in}$ are a basis of V.*

Proof. Apply the above theorem for $m = 0$.

Theorem 4.5.10. [Exchange theorem (Steinitz)]. *Let V be a vector space, $\mathbf{v}_1, \ldots, \mathbf{v}_m$ linearly independent vectors from V and $\mathbf{w}_1, \mathbf{w}_2, \ldots, \mathbf{w}_n$ a generator system of V. Then $m \leq n$ and*

$$V = \langle \mathbf{v}_1, \ldots, \mathbf{v}_m, \mathbf{w}_{m+1}, \ldots, \mathbf{w}_n \rangle.$$

Proof. We proceed by induction after m. If $m = 0$, then $0 \leq n$ and the conclusion is obviously true.

Suppose the assertion is true for $m - 1$ linearly independent vectors and let us prove it for m vectors. Let $\mathbf{v}_1, \mathbf{v}_2, \ldots, \mathbf{v}_m \in V$ be m linearly independent vectors. Then $m - 1$ of them will be linearly independent too, hence $m - 1 \leq n$ and

$$V = \langle \mathbf{v}_1, \mathbf{v}_2, \ldots, \mathbf{v}_{m-1}, \mathbf{w}_1, \mathbf{w}_2, \ldots, \mathbf{w}_n \rangle.$$

If $m - 1 = n$, then $V = \langle \mathbf{v}_1, \ldots, \mathbf{v}_{m-1} \rangle$ and $\mathbf{v}_m \in \langle \mathbf{v}_1, \ldots, \mathbf{v}_{m-1} \rangle$, contradicting the linearly independence of the vectors $\mathbf{v}_1, \ldots, \mathbf{v}_m$, wherefrom follows that $m - 1 < n$ and $m \leq n$.

On the other hand, from $V = \langle \mathbf{v}_1, \ldots, \mathbf{v}_{m-1}, \mathbf{w}_m, \ldots, \mathbf{w}_n \rangle$, follows that $\mathbf{v}_m$ can be expressed as a linear combination of these generators

$$\mathbf{v}_m = \lambda_1 \mathbf{v}_1 + \cdots + \lambda_{m-1} \mathbf{v}_{m-1} + \lambda_m \mathbf{w}_m + \cdots + \lambda_n \mathbf{w}_n.$$

Since the vectors $\mathbf{v}_1, \ldots, \mathbf{v}_m$ are linearly independent, it follows that not all scalars $\lambda_m, \ldots, \lambda_n$ are zero. Without restricting generality, one can suppose that $\lambda_m \neq 0$. Hence

$$\mathbf{w}_m = -\lambda_1 \lambda_m^{-1} \mathbf{v}_1 - \cdots - \lambda_{m-1} \lambda_m^{-1} \mathbf{v}_{m-1} + \lambda_m^{-1} \mathbf{v}_m - \lambda_{m+1} \lambda_m^{-1} \mathbf{w}_{m+1} - \cdots - \lambda_n \lambda_m^{-1} \mathbf{w}_n.$$

The vector $\mathbf{w}_m$ can be written as a linear combination of the other vectors, hence

$$V = \langle \mathbf{v}_1, \ldots, \mathbf{v}_m, \mathbf{v}_{m+1}, \ldots, \mathbf{w}_n \rangle.$$

Corollary 4.5.11. *All base of a finitely generated vector space are finite and have the same cardinality.*

Theorem 4.5.12. *Let V be K-vector space and X an infinite basis. Then, for every generator system Y of V, we have*

$$|X| \le |Y|.$$

Proof. Let $\mathbf{w} \in Y$ and $X_\mathbf{w}$ the set of those elements from X which occur in the decomposition of $\mathbf{w}$ as a linear combination of vectors from X. Then

$$X = \bigcup_{\mathbf{w} \in Y} X_\mathbf{w}.$$

Obviously, $\bigcup_{\mathbf{w} \in Y} X_\mathbf{w} \subseteq X$. Suppose the inclusion is strict, i.e., there exists $\mathbf{v} \in X \setminus \bigcup_{\mathbf{w} \in Y} X_\mathbf{w}$. Since V is generated by Y, $\mathbf{v}$ has a decomposition as

$$\mathbf{v} = \lambda_1 \mathbf{w}_1 + \cdots + \lambda_n \mathbf{w}_n, \lambda_i \in K, \mathbf{w}_i \in Y, i \in \{1, \ldots, n\}.$$

Every $\mathbf{w}_i$ is a linear combination of elements of $X_{\mathbf{w}i}$, $i \in \{1, \ldots, n\}$, hence $\mathbf{v}$ is a linear combination of elements of $\bigcup_{\mathbf{w} \in Y} X_\mathbf{w}$. Contradiction!

The sets $X_\mathbf{w}$ are finite for all vectors $\mathbf{w} \in Y$. The set X being infinite, it follows that Y is infinite too and

$$|X| \le \sum_{\mathbf{w} \in Y} |X\mathbf{w}| = |Y|$$

Corollary 4.5.13.

(1) *If a vector space V has an infinite basis, this basis is infinite, having the same cardinality as X.*
(2) *If the vector space V has an infinite basis, then V is not finitely generated.*

Definition 4.5.6. Let V be a K-vector space. The **dimension** of V is the cardinality of the basis of V. We denote the dimension of V by dim V. The space V is called **finite dimensional** if it has a finite basis. The space V is called **infinite dimensional** if it has an infinite basis.

Example 4.5.4.

(1) The dimension of $\mathbb{R}^2$ is 2. The dimension of $\mathbb{R}^3$ is 3.
(2) If K is a field, dim $K^n = n$, since

$$\mathbf{e}_1 = \begin{pmatrix} 1 \\ 0 \\ \vdots \\ 0 \end{pmatrix}, \ \mathbf{e}_2 = \begin{pmatrix} 0 \\ 1 \\ \vdots \\ 0 \end{pmatrix}, \ \dots, \ \mathbf{e}_n = \begin{pmatrix} 0 \\ 0 \\ \vdots \\ 1 \end{pmatrix}$$

is a basis, called *canonical basis* of K^n.

(3) $\dim P_n(\mathbb{R}) = n+1$.

(4) The vector spaces $P(\mathbb{R})$ and $F(X, \mathbb{R})$ are infinite dimensional.

As a direct consequence of these the previous results we get:

Remark 49 (1) *Let V be an n dimensional vector space. Then any linearly independent system has at most n vectors.*

(2) *Let V be a finitely generated vector space. Then V is finite dimensional and its dimension is smaller than the cardinality of any of its generator systems.*

(3) *If V is a finite dimensional vector space then any minimal generator system is a basis.*

(4) *In a finite dimensional vector space, any maximal linearly independent set of vectors is a basis.*

(5) $\dim V = n$ *if and only if there exist in V exactly n linearly independent vectors and any $n + 1$ vectors from V are linearly dependent.*

(6) *If $U \leq V$ is a linear subspace of V, then $\dim U \leq \dim V$. If V is finite dimensional, then $U = V$ if and only if $\dim U = \dim V$.*

Proposition 4.5.14. *Let V be a finite dimensional vector space and $\dim V = n$. Let $\mathbf{v}_1, \dots, \mathbf{v}_n \in V$ be a family of vectors. The following are equivalent:*

(1) $\mathbf{v}_1, \dots, \mathbf{v}_n \in V$ *are linearly independent.*

(2) $\mathbf{v}_1, \dots, \mathbf{v}_n \in V$ *is a generator system of V.*

(3) $\mathbf{v}_1, \dots, \mathbf{v}_n \in V$ *is a basis of V.*

Theorem 4.5.15. (Dimension theorem.) *Let U_1 and U_2 be linear subspaces of a vector space V. Then*

$$\dim(U_1 + U_2) + \dim(U_1 \cap U_2) = \dim U_1 + \dim U_2.$$

Proof. If $\dim U_1 = \infty$ or $\dim U_2 = \infty$, then $\dim(U_1 + U_2) = \infty$. Suppose U_1 and U_2 are finite dimensional and $\dim U_1 = m_1$, $\dim U_2 = m_2$.

The intersection of U_1 and U_2 is a subspace and $U_1 \cap U_2 \leq U_1$, $U_1 \cap U_2 \leq U_2$. We denote $\dim U_1 \cap U_2 = m$ and let $\mathbf{v}_1, \ldots, \mathbf{v}_m$ be a basis of $U_1 \cap U_2$. From the Steinitz theorem, we deduce the existence of the vectors $\mathbf{u}_{m+1}, \ldots, \mathbf{u}_{m_1}$ and $\mathbf{w}_{m+1}, \ldots, \mathbf{w}_{m_2}$, such that $\mathbf{v}_1, \ldots, \mathbf{v}_m$, $\mathbf{u}_{m+1}, \ldots, \mathbf{u}_{m_1}$ is a basis of U_1 and $\mathbf{v}_1, \ldots, \mathbf{v}_m$, $\mathbf{w}_{m+1}, \ldots, \mathbf{w}_{m_2}$ is a basis of U_2. Then $\mathbf{v}_1, \ldots, \mathbf{v}_m$, $\mathbf{u}_{m+1}, \ldots, \mathbf{u}_{m_1}$, $\mathbf{w}_{m+1}, \ldots, \mathbf{w}_{m_2}$ is a basis of $U_1 + U_2$. Hence

$$\dim (U_1 + U_2) = m + (m_1 - m) + (m_2 - m)$$
$$= m_1 + m_2 - m$$
$$= \dim U_1 + \dim U_2 - \dim (U_1 \cap U_2).$$

4.5.2 Algorithm for computing the basis of a generated sub-space

Let V be a K-vector space and $\mathbf{v}_1, \ldots, \mathbf{v}_m \in V$. Let U denote the subspace generated by these vectors, $U = \langle \mathbf{v}_1, \ldots, \mathbf{v}_n \rangle$. We search for a basis for U and an efficient algorithm to perform this task.

Proposition 4.5.16. *Let* $\mathbf{v}_1, \ldots, \mathbf{v}_n$ *and* $\mathbf{w}_1, \ldots, \mathbf{w}_k$ *be two families of vectors of a K-vector space V. If* $\mathbf{w}_1, \ldots, \mathbf{w}_k \in \langle \mathbf{v}_1, \ldots, \mathbf{v}_n \rangle$, *then*

$$\langle \mathbf{w}_1, \ldots, \mathbf{w}_k \rangle \subseteq \langle \mathbf{v}_1, \ldots, \mathbf{v}_n \rangle.$$

Corollary 4.5.17. *If*

$$\mathbf{w}_1, \ldots, \mathbf{w}_k \in \langle \mathbf{v}_1, \ldots, \mathbf{v}_n \rangle$$

and

$$\mathbf{v}_1, \ldots, \mathbf{v}_n \in \langle \mathbf{w}_1, \ldots, \mathbf{w}_k \rangle,$$

then

$$\langle \mathbf{v}_1, \ldots, \mathbf{v}_n \rangle = \langle \mathbf{w}_1, \ldots, \mathbf{w}_k \rangle.$$

Remark 50 *If the vectors* $\mathbf{w}_p, \ldots, \mathbf{w}_k$ *are linear combinations of* $\mathbf{v}_p$, $\ldots, \mathbf{v}_n$ *and* $\mathbf{v}_p, \ldots, \mathbf{v}_n$ *can be written as linear combinations of* $\mathbf{w}_p$, $\ldots, \mathbf{w}_k$, *the two families are generating the same subspace.*

We deduce that the following operations on a vector system $\mathbf{v}_1$, $\ldots, \mathbf{v}_n$ do not modify the generated subspace:

(1) Addition of a scalar multiple of a vector $\mathbf{v}_j$ to any other vector $\mathbf{v}_k$, $k \neq j$, of this family.

$$\langle \mathbf{v}_1, \ldots, \mathbf{v}_k, \ldots, \mathbf{v}_j, \ldots, \mathbf{v}_n \rangle = \langle \mathbf{v}_1, \ldots, \mathbf{v}_k + \lambda \mathbf{v}_j, \ldots, \mathbf{v}_j, \ldots, \mathbf{v}_n \rangle.$$

(2) Permuting the order of the vectors $\mathbf{v}_j$ and $\mathbf{v}_k$:

$$\langle \mathbf{v}_1, \ldots, \mathbf{v}_k, \ldots, \mathbf{v}_j, \ldots, \mathbf{v}_n \rangle = \langle \mathbf{v}_1, \ldots, \mathbf{v}_j, \ldots, \mathbf{v}_k, \ldots, \mathbf{v}_n \rangle.$$

(3) Multiplying a vector $\mathbf{v}_j$ with a scalar $\lambda \neq 0$:

$$\langle \mathbf{v}_1, \ldots, \mathbf{v}_j, \ldots, \mathbf{v}_n \rangle = \langle \mathbf{v}_1, \ldots, \lambda_{\mathbf{v}_j}, \ldots, \mathbf{v}_n \rangle$$

Let now $K = \mathbb{R}$ or $K = \mathbb{C}$, $n \in \mathbb{N}$, and consider the K-vector space K^n.

The following algorithm allows the computation of a basis for a subspace generated by $\mathbf{v}_1, \ldots, \mathbf{v}_m \in K^n$, subspace denoted by $U = \langle \mathbf{v}_1, \ldots, \mathbf{v}_m \rangle$. This algorithm is grounded on a successive filtration of redundant elements from the list $\mathbf{v}_1, \ldots, \mathbf{v}_m$ until we obtain an equivalent generator system, i.e., a system which generates the same subspace, consisting only of linear independent vectors.

ALGORITHM:

Step 1: Build the matrix $A \in M_{m \times n}(K)$ whose rows consist of the column vectors $\mathbf{v}_i$, $i \in \{1, \ldots, n\}$.

Step 2: Apply the Gauss-Jordan algorithm to get A in triangle shape by elementary transformations. Denote by B the obtained triangle matrix.

Step 3: Ignore all zero rows in B and denote by $\mathbf{w}_1, \ldots, \mathbf{w}_k$ the non zero rows of B. These vectors are linear independent and generate the same subspace as the initial vectors.

Step 4: The basis consists of the vectors $\mathbf{w}_1, \ldots, \mathbf{w}_k$.

Example 4.5.5. In $\mathbb{R}^4$ consider

$$\mathbf{v}_1 = \begin{pmatrix} 1 \\ 2 \\ 0 \\ 1 \end{pmatrix}, \mathbf{v}_2 = \begin{pmatrix} 2 \\ 3 \\ 1 \\ -1 \end{pmatrix}, \mathbf{v}_3 = \begin{pmatrix} -1 \\ 1 \\ 1 \\ -1 \end{pmatrix}, \mathbf{v}_4 = \begin{pmatrix} 2 \\ 6 \\ 2 \\ -1 \end{pmatrix}, \mathbf{v}_5 = \begin{pmatrix} 0 \\ 2 \\ 2 \\ -3 \end{pmatrix}.$$

We apply the algorithm to compute a basis for the subspace $\langle \mathbf{v}_1, \mathbf{v}_2, \mathbf{v}_3, \mathbf{v}_4, \mathbf{v}_5 \rangle$.

1	2	0	1	$\mathbf{v}_1$					1	0	0	0	0
2	3	1	−1		$\mathbf{v}_2$				0	1	0	0	0
−1	1	1	−1			$\mathbf{v}_3$			0	0	1	0	0
2	6	2	−1				$\mathbf{v}_4$		0	0	0	1	0
0	2	2	−3					$\mathbf{v}_5$	0	0	0	0	1

1	2	0	1	$\mathbf{v}_1$					1	0	0	0	0
0	−1	1	−3	$-2\mathbf{v}_1$	$+\mathbf{v}_2$				−2	1	0	0	0
0	3	1	0	$\mathbf{v}_1$		$+\mathbf{v}_3$			1	0	1	0	0
0	2	2	−3	$-2\mathbf{v}_1$			$+\mathbf{v}_4$		−2	0	0	1	0
0	2	2	−3					$\mathbf{v}_5$	0	0	0	0	1

1	2	0	1	$\mathbf{v}_1$					1	0	0	0	0
0	−1	1	−3	$-2\mathbf{v}_1$	$+\mathbf{v}_2$				−2	1	0	0	0
0	0	4	−9	$-5\mathbf{v}_1$	$+3\mathbf{v}_2$	$+\mathbf{v}_3$			−5	3	1	0	0
0	0	4	−9	$-6\mathbf{v}_1$	$+2\mathbf{v}_2$		$+\mathbf{v}_4$		−6	2	0	1	0
0	0	4	−9	$-4\mathbf{v}_1$	$+2\mathbf{v}_2$			$+\mathbf{v}_5$	−4	2	0	0	1

1	2	0	1	$\mathbf{v}_1$					$\mathbf{w}_1$	1	0	0	0	0
0	−1	1	−3	$-2\mathbf{v}_1$	$+\mathbf{v}_2$				$\mathbf{w}_2$	−2	1	0	0	0
0	0	4	−9	$-5\mathbf{v}_1$	$+3\mathbf{v}_2$	$+\mathbf{v}_3$			$\mathbf{w}_3$	−5	3	1	0	0
0	0	0	0	$-\mathbf{v}_1$	$-\mathbf{v}_2$	$-\mathbf{v}_3$	$+\mathbf{v}_4$			−1	−1	−1	1	0
0	0	0	0	$\mathbf{v}_1$	$-\mathbf{v}_2$	$-\mathbf{v}_3$		$+\mathbf{v}_5$		1	−1	−1	0	1

We obtain the basis

$$\mathbf{w}_1 = \begin{pmatrix} 1 \\ 2 \\ 0 \\ 1 \end{pmatrix}, \mathbf{w}_2 = \begin{pmatrix} 0 \\ -1 \\ 1 \\ -3 \end{pmatrix}, \mathbf{w}_3 = \begin{pmatrix} 0 \\ 0 \\ 4 \\ -9 \end{pmatrix}.$$

Moreover, one can express these basis vectors as ε linear combination of the original generator vectors:

$$\mathbf{w}_1 = \mathbf{v}_1, \ \mathbf{w}_2 = -2\mathbf{v}_1 + \mathbf{v}_2, \ \mathbf{w}_3 = -5\mathbf{v}_1 + 3\mathbf{v}_2 + \mathbf{v}_3.$$

The last two rows indicate the decomposition of the vectors $\mathbf{v}_4$ and $\mathbf{v}_5$ as linear combinations of the vectors $\mathbf{v}_1$, $\mathbf{v}_2$, $\mathbf{v}_3$:

$$\mathbf{v}_4 = \mathbf{v}_1 + \mathbf{v}_2 + \mathbf{v}_3,$$

$$\mathbf{v}_5 = -\mathbf{v}_1 + \mathbf{v}_2 + \mathbf{v}_3.$$

We have obtained two basis for the subspace spanned by $\mathbf{v}_{\,\cdot}$, $\mathbf{v}_2$, $\mathbf{v}_3$, $\mathbf{v}_4$, $\mathbf{v}_5$, namely $B_1 : \mathbf{w}_1$, $\mathbf{w}_2$, $\mathbf{w}_3$ and $B_2 : \mathbf{v}_1$, $\mathbf{v}_2$, $\mathbf{v}_3$.

CHAPTER 5

Conceptual Knowledge Processing

5.1 Introduction

Investigating knowledge structures has a long tradition. First, it appeared in philosophy, then it has been expanded to Artificial Intelligence, where several tools have been developed in the last few years. Even if procedures are different, as well as the grounding formalism, all these methods converge to the same scope: investigating a huge amount of data in a way that is machine tractable and understandable by humans. Conceptual Knowledge Processing proposes an approach based on an elaborate understanding of knowledge and its structures. It is something which promotes both critical discourse and full decision support for a human expert browsing through knowledge structures in his field of expertise.

Understanding what knowledge really is, is quite different from performing computations on data sets enhanced with different types of logic or semantic structures. Discovering knowledge is an ambitious target, "transforming knowledge itself into *ambitious knowledge* and is situated in opposition to cognitive-instrumental understanding of knowledge" (Wille 2006).

In view of this problematic situation [of rational argumentation] it is better not to give up reasoning entirely, but rather to break with the concept of reasoning which is orientated by the pattern of logic-mathematical proofs. In accordance with a new foundation of critical rationalism, Kant's question of transcendental reasoning has to be taken up again as the question about the normative conditions of the possibility of discursive communication and understanding (and discursive criticism as well). Reasoning then appears primarily not as deduction of propositions within an objectivizable system of propositions in which one has already abstracted from the actual pragmatic dimension of argumentation, but as answering of why-questions of all sorts within the scope of argumentative discourse (Apel 1989)

R. Wille defined Conceptual Knowledge Processing to be an applied discipline dealing with ambitious knowledge which is constituted by conscious reflexion, discursive argumentation and human communication on the basis of cultural background, social conventions and personal experiences. Its main aim is to develop and maintain methods and instruments for processing information and knowledge which support rational thought, judgment and action of human beings and therewith promote critical discourse (see also (Wille 1994, Wille 1997, Wille 1999, Wille 2000)).

On the basis of this, knowledge is closely related to the person. The relation between data, information and knowledge is expressed by K. Devlin in (Devlin 1999) as follows:

Data = symbols + syntax
Information = data + semantics
Knowledge = internalized information + ability to use them

As stated by its founder, R. Wille (Wille 2006), Conceptual Knowledge Processing is a particular approach to knowledge processing, underlying the constitutive role of thinking, arguing and communicating among human beings in dealing with knowledge and its processing. The term processing also underlines the fact that gaining or approximating knowledge is a process which should always be conceptual in the above sense. The methods of Conceptual Knowledge Processing have been introduced and discussed by R. Wille (Wille 2006), based on the pragmatic philosophy of Ch. S. Peirce, continued by K.-O. Apel and J. Habermas.

The mathematical theory underlying the methods of Conceptual Knowledge Processing is Formal Concept Analysis, providing a powerful mathematical tool to understand and investigate knowledge, based on a set theoretical semantics, developing methods for representation, acquiring, retrieval of knowledge, but even for further theory building in several other domains of science.

Formal Concept Analysis (for short FCA) is a mathematical theory that has been introduced at the end of the 1980s as an attempt to restructure lattice theory in a form that is suitable for data analysis. Having a profound philosophical background, Formal Concept Analysis is the mathematization of the traditional understanding of the concept of being a unit of thought which is homogeneous and coherent.

FCA has a big potential to be applied to a variety of linguistic domains as a method of knowledge representation and FCA applications provide a very suitable alternative to statistical methods. It is somewhat surprising that FCA is not used more frequently in linguistics (Priss 2005). The reason could be that the notion of "concept" in FCA does not exactly correspond to the notion of "concept" as developed in Computational Linguistics. However, the notion of "synset" in WordNet (Fellbaum 1998) could be assimilated with the former, if the semantical relations in WordNet are considered as subconcept-superconcept relations in FCA. It is stated that: "it is entirely safe to use FCA for linguistic applications as long as it is emphasized that the mathematical vocabulary should be separated from the linguistic vocabulary (Priss 2005)."

5.2 Context and Concept

There are two basic notions in FCA, namely formal context and formal concept. The term formal emphasizes that we have moved into a more abstract setting, using a precise mathematical formalism in order to introduce the language of FCA. Nevertheless, the adjective *formal* will often be left out.

The basic data set in FCA is the *formal context*. Almost any real life data set can be represented by a formal context, while more general data sets can be represented by the more general notion of a many-valued context (see Definition 5.3.1).

Definition 5.2.1. A **formal context** $K := (G, M, I)$ consists of two sets G and M and a binary relation I between G and M. The elements of G are called **objects** (in German *Gegenstände*) and the elements of M are called **attributes** (in German *Merkmale*). The relation I is called the incidence relation of the formal context, and we sometimes write gIm instead of $(g,m) \in I$. If gIm holds, it is said that *the object g has the attribute m*.

A small context is usually represented by a cross table, i.e., a rectangular table of crosses and blanks, where the rows are labeled by the objects and the columns are labeled by the attributes. A cross in entry (g,m) indicates gIm, i.e., object g has attribute m.

Example 5.2.1. The following example comes from Ganter et al. (2005). The formal context partly represents the lexical field *bodies of water*

	natural	artificial	stagnant	running	inland	maritime	constant	temporary
tarn	×		×		×		×	
trickle	×			×	×		×	
rill	×			×	×		×	
beck	×			×	×		×	
rivulet	×			×	×		×	
runnel	×			×	×		×	
brook	×			×	×		×	
burn	×			×	×		×	
stream	×			×	×		×	
torrent	×			×	×		×	
river	×			×	×		×	
channel				×	×		×	
canal		×		×	×		×	
lagoon	×		×			×	×	
lake	×		×		×		×	
mere		×	×		×		×	
plash	×		×		×			×
pond		×	×		×		×	
pool	×		×		×		×	
puddle	×		×		×			×
reservoir		×	×		×		×	
sea	×		×			×	×	

Collecting information from a context means reading off the context, row by row or column by column. For instance, a *tarn* is a *natural, stagnant, inland* and *constant* body of water. If two or more bodies of water are considered, we have to look at their common defining properties. For example, all common properties characteristic for *rivulet* and *runnel* are *natural, running, inland, constant*.

This fact is formally modeled by the introduction of two operators between $\mathfrak{P}(G)$ and $\mathfrak{P}(M)$. These operators are called the *derivation operators* and are used to gather information in order to extract knowledge from the data set represented as a context.

Definition 5.2.2. Let K := (G, M, I) be a formal context. For a set $A \subseteq G$ of objects we define

$$A' := \{m \in M \mid gIm \text{ for all } g \in A\}$$

the set of all attributes common to the objects in A. Similarly, for a set $B \subseteq M$ of attributes we define

$$B' := \{g \in G \mid gIm \text{ for all } m \in B\}$$

the set of all objects which have all attributes in B.

Remark 51 (1) *If $g \in G$ is an object, then we write g' for $\{g\}'$. Similarly, if $m \in M$ is an attribute, then $m' := \{m\}'$.*

(2) *g' is just the set of all attributes of g, i.e., we run the row of g in the context and collect all attribute names of the columns marked by an X.*

(3) *Similarly, m' is the set of all objects sharing m as a common attribute.*

(4) *For $A \subseteq G$, A' is just the set of all attributes shared by all $g \in A$.*

(5) *Similarly, for $B \subseteq M$, B' is the set of all objects sharing all attributes in B.*

Example 5.2.2. For the above context, we obtain

(1) *$\{lagoon\}' = \{natural, stagnant, maritime, constant\}$,*

(2) *$\{natural, running, inland, constant\}' = \{trickle, rill, beck, rivulet, runnel, brook, burn, stream, torrent, river\}$,*

(3) *$G' = \varnothing$, $M' = \varnothing$. The set G' is the set of all commonly shared attributes. In the above context, this set is empty. Similarly, M' is the set of all objects, sharing all attributes in M. This set*

is also empty in this context. Moreover, $G' \neq \varnothing$ if and only if there exists a column full of X marks. $M' \neq \varnothing$ if and only if there exists a row full of X marks.

Remark 52 *Choose in the definition of the derivation operator the subset A to be empty. Then*

$$\varnothing' = M.$$

Similarly, if we consider the empty subset of M in the same definition, we obtain

$$\varnothing' = G.$$

Using Remark 51, one can easily see that

Proposition 5.2.1. *Let* $\mathrm{K} := (G, M, I)$ *be a formal context. Let* $A \subseteq G$ *be a set of objects,* $B \subseteq M$ *be a set of attributes. Then*

$$A' = \bigcap_{g \in A} g',$$

and

$$B' = \bigcap_{m \in B} m'.$$

R. Wille (Wille 1995) asserts that *concepts* can be philosophically understood as the basic units of thought formed in dynamic processes within social and cultural environments. A concept is constituted by its *extent*, comprising all objects which belong to the concept, and its *intent*, including all attributes of objects of the extent. A concept can be thought of as a basic knowledge unit. In the following, we are going to formalize what a (formal) concept is, and describe the process of knowledge extraction and acquisition from a given data set, formalized as a (formal) context.

Example 5.2.3. As has been seen before,

{*rivulet, runnel*}' = {*natural, running, inland, constant*}.

Are there any other common properties of these bodies of water (in the context under consideration)? The answer is NO. This can easily be seen from the cross table describing the bodies of water context (5.1). There are no X marks commonly shared by the rivulet and the runnel, i.e., there is no other column in the context providing X marks in the rows of rivulet and runnel, besides the above listed column names, collected by derivation.

Is this information complete? Do we have the entire knowledge about rivulet and runnel which can be extracted from this data set? Knowledge often means the ability to answer questions. Are there any other bodies of water sharing exactly the same attributes as the rivulet and the runnel? The answer is YES. All objects from trickle to the river have these attributes in common. Hence, there is a maximal knowledge structure which can be extracted from the context. The object side contains all objects from trickle to river, while the attribute side contains the properties natural, running, inland and constant. This has been done by applying the derivation operator twice. In the first step, {*rivulet, runnel*} has been derived. The result of this derivation was {*natural, running, inland, constant*}. Answering the second question is equivalent to a second derivation, i.e., those of {*natural, running, inland, constant*}.

This construction is now maximal and contains only X marks, i.e., it is a maximal rectangle of crosses in the context. The information contained is complete, there is no other object sharing exactly these attributes since there is no row containing exactly the same X marks as these objects do.

Hence, we have extracted a first unit of knowledge from the bodies of water context. This unit of knowledge will be called a concept.

The derivation operators defined before form a Galois connection between the power-set lattices $\mathfrak{P}(G)$ and $\mathfrak{P}(M)$. They can be considered as concept forming operators, as the following definition states:

Definition 5.2.3. A **formal concept** of the context K := (G, M, I) is a pair (A, B) where $A \subseteq G$, $B \subseteq M$, $A' = B$, and $B' = A$. The set A is the **extent** and B the **intent** of the concept (A, B). The set of all concepts of the context (G, M, I) is denoted by $\mathfrak{B}(G, M, I)$.

For a given object $g \in G$, we define the object concept to be $\gamma g := (g'', g')$ and the attribute concept to be $\mu m := (m', m'')$. The object concept γg is the smallest concept whose extent contains g, while the attribute concept is the largest concept whose intent contains m.

Example 5.2.4. The formal context of the lexical field *bodies of water* has 24 concepts. Here are some examples:

(1) ({*tarn, lake, pool, sea, lagoon*}, {*natural, stagnant, constant*}),
(2) ({*canal*}, {*running, artificial, inland, constant*}).

The derivation operators are satisfying the following conditions:

Proposition 5.2.2. *If (G, M, I) is a formal context, $A, A_1, A_2 \subseteq G$ are sets of objects, then*

1) *If $A_1 \subseteq A_2$ then $A'_2 \subseteq A'_1$;*
2) *$A \subseteq A''$, i.e., A is included in the extent generated by A;*
3) *$A' = A'''$.*

Dually, the same rules hold for subsets $B, B_1, B_2 \subseteq M$ of attributes.

Moreover,
4) *$A \subseteq B \Leftrightarrow B \subseteq A' \Leftrightarrow A \times B \subseteq I$.*

Proof.
1) Let $A_1 \subseteq A_2$ be two subsets of G. Let $m \in A'_2 = \{g \in G \mid \forall m \in A_2$. $gIm\}$. In particular, gIm for all $m \in A_1$, hence $m \in A'_1$.
2) Let $g \in A$. Then, for all $m \in A'$, gIm, i.e., $g \in A''$.
3) Since $A \subseteq A''$, by 1) and 2), we have $A' \subseteq A'''$ and $A' \subseteq A'''$, i.e., the desired equality.
4) Using the definition of the derivation, it follows immediately that $A \subseteq B \Leftrightarrow B \subseteq A' \Leftrightarrow A \times B \subseteq I$.

Furthermore, the intersection of any number of extents (intents) is always an extent (intent) as the following proposition shows:

Proposition 5.2.3. *Let $(A_k)_{k \in T}$ be a family of subsets of G. Then*

$$\left(\bigcup_{k \in T} A_k \right)' = \bigcap_{k \in T} A'_k.$$

The same holds for sets of attributes.

Proof. Let $m \in (\bigcup_{k \in T} A_k)'$. Then, for every $g \in \bigcup_{k \in T} A_k$, we have gIm, i.e., gIm for all $g \in A_k$, $k \in T$. By the definition of the derivation operator, $m \in A'_k$ for every index $k \in T$, hence $m \in \bigcap_{k \in T} A'_k$.

Let now $K := (G, M, I)$ be a formal context. The set $\mathfrak{B}(K)$ of all formal concepts of K will be ordered by the so-called **subconcept-superconcept relation,** defined by

$$(A_1, B_1) \le (A_2, B_2) :\Leftrightarrow A_1 \subseteq A_2 \; (\Leftrightarrow B_1 \supseteq B_2).$$

In this case we call (A_1, B_1) a **subconcept** of (A_2, B_2) and (A_2, B_2) is a **superconcept** of (A_1, B_1). The set of all concepts of K ordered in this way is denoted by $\mathfrak{B}(K)$ and is called the **concept lattice** of the context K.

The two derivation operators are closure operators of the form

$$X \to X'',$$

acting on the power sets $\mathfrak{P}(G)$ and $\mathfrak{P}(M)$, respectively. For a given subset $A \subseteq G$, the set A'' is called the **extent closure** of A. Similarly, for $B \subseteq M$, the set B'' is called **intent closure** of B.

Remark 53 (1) *Whenever all objects from a set $A \subseteq G$ have a common attribute m, all objects from A'' have that attribute.*

(2)   *Whenever an object $g \in G$ has all attributes from $B \subseteq M$, this object also has all attributes from B''.*

Proposition 5.2.4. *For subsets $A, A_1, A_2 \subseteq G$, we have*

(1) $A_1 \subseteq A_2$ *implies* $A''_1 \subseteq A''_2$,
(2) $(A'')'' = A''$.

Dually, for subsets $B, B_1, B_2 \subseteq M$, we have

(1) $B_1 \subseteq B_2$ *implies* $B''_1 \subseteq B''_2$,
(2) $(B'')'' = B''$.

Proposition 5.2.5. *For a context (G, M, I) and $A \subseteq G$ a set of objects, A'' is a concept extent. Conversely, if A is an extent of (G, M, I), then $A = A''$. Similarly, if $B \subseteq M$ is a set of attributes, then B'' is a concept intent, and every intent B satisfies $B'' = B$.*

Proof. For each subset $A \subseteq G$, the pair (A'', A') is a formal concept. Similarly, for every $B \subseteq M$, the pair (B', B'') is a concept. The conclusion follows immediately from this.

Remark 54 *The closed sets of the closure operator $A \to A''$, $A \subseteq G$ are precisely the extents of (G, M, I), and the closed sets of the operator $B \to B''$, $B \subseteq M$, are precisely the intents of (G, M, I).*

Remark 55 *All concepts of (G, M, I) are of the form (A'', A') for $A \subseteq G$. Dually, all concepts of (G, M, I) can be described by (B', B'') for $B \subseteq M$.*

The following Proposition gives the correspondence between the incidence relation and the subconcept-superconcept relation.

Proposition 5.2.6. *Let* $g \in G$ *and* $m \in M$. *Then*

$$gIm \Leftrightarrow \gamma g \le \mu m.$$

Proof. Suppose gIm. Then $g \in m'$, i.e., $\{g\} \subseteq m'$. By the properties of derivation, $m'' \subseteq g'$ wherefrom follows $g'' \subseteq m'$ which means $\gamma g \le \mu m$. The converse implication follows immediately.

The set of all *formal concepts* of a *formal context* together with the subconcept-superconcept order relation $\le$ forms o complete lattice called the *concept lattice* or *conceptual hierarchy*, and is denoted by $\mathfrak{B}(K)$.

The Basic Theorem on Concept Lattices *Let* $K := (G, M, I)$ *be a formal context. The concept lattice* $\mathfrak{B}(K)$ *is a complete lattice in which infimum and supremum are given by:*

$$\bigwedge_{t \in T}(A_t, B_t) = (\bigcap_{t \in T} A_t, (\bigcup_{t \in T} B_t)''),$$

$$\bigvee_{t \in T}(A_t, B_t) = ((\bigcup_{t \in T} B_t)'', \bigcap_{t \in T} B_t).$$

A complete lattice V *is isomorphic to* $\mathfrak{B}(K)$ *if and only if there are mappings* $\tilde{\gamma}: G \to V$ *and* $\tilde{\mu}: M \to V$ *such that* $\tilde{\gamma}(G)$ *is supremun-dense in* V, $\tilde{\mu}(M)$ *is infimum-dense in* V *and* gIm *is equivalent to* $\tilde{\gamma}g \le \tilde{\mu}m$ *for all* $g \in G$ *and all* $m \in M$. *In particular,* $V \simeq \mathfrak{B}(V, V, \le)$.

Remark 56 (1) *If* V *is a complete lattice, then a subset* $A \subseteq V$ *is supremum dense in* V *if and only if for each* $v \in V$ *there exist a subset* $X \subseteq A$ *such that* $v = \bigvee X$, *i.e., every element* v *of* V *is supremum of some elements of* A. *The dual holds for infimum dense subsets, i.e., a subset* $A \subseteq V$ *is infimum-dense in* V *if and only if every element* v *of* V *is infimum of some elements of* A.

(2) *If a complete lattice has supremum or infimum dense subsets, because of the precedent remark, they can be considered as kind of building blocks of the lattice* V.

(3) *Consider the concept lattice* $\mathfrak{B}(K)$. *Since this is a complete lattice, choose in the second part of the basic theorem for* V *the concept lattice. Then there exist two mappings*

$$\gamma\colon G \to \underline{\mathfrak{B}}(\mathrm{K}) \text{ and } \mu\colon M \to \underline{\mathfrak{B}}(\mathrm{K})$$

satisfying the properties stated in the basic theorem. The map γ is the object concept map, i.e., for a given object $g \in G$, $\gamma(g) = (g'', g')$, the object concept of g, while for a given attribute $m \in m$, the map μ is the attribute concept map, i.e., $\mu(m) = (m', m'')$, the attribute concept of m. Hence, every concept $(A, B) \in \underline{\mathfrak{B}}(\mathrm{K})$ is supremum of some object concepts and infimum of some attribute concepts. This is true since

$$(A,B) = \bigvee_{g \in A} (g'', g'') \text{ and } (A,B) = \bigwedge_{m \in B} (m', m'').$$

(4) *The second part of the basic theorem describes how to label a concept lattice so that no information is lost. This labeling has two rules. Every object concept $\gamma(g)$ is labeled by g, and every attribute concept $\mu(m)$ is labeled by m.*

Every other extent and intent in a labeled order diagram is obtained as follows. Consider a formal concept (A, B) corresponding to node c of a labeled diagram of the concept lattice $\underline{\mathfrak{B}}(\mathrm{K})$. An object g is in the extent A if and only if a node with label g lies on a path going from c downwards. An attribute m is in the intent B if and only if a node with label m lies on a path going from c upwards.

The entire information contained in a formal context can easily be recovered from the concept lattice. The object set G appears as an extent of the largest or most general concept $(G, G') = (\varnothing', \varnothing'')$. The attribute set M appears as an intent of the smallest or most specific concept $(M', M) = (\varnothing', \varnothing'')$.

The incidence relation is recovered from

$$I = \cup \; \{A \times B \mid (A, B) \in \underline{\mathfrak{B}}(G, M, I)\}.$$

In a context, there is lots of redundant information with respect to concept building. The following operations on contexts do not alter the structure of the concept system.

Definition 5.2.4. A context (G, M, I) is called **clarified**, if for every two objects $g, h \in G$, from $g' = h'$ follows $g = h$ and dually for attributes.

Remark 57 *Clarifying a context means merging together exactly those objects (attributes) which share exactly the same attributes (objects).*

Example 5.2.5. The clarified bodies of water context:

	natural	artificial	stagnant	running	inland	maritime	constant	temporary
tarn	×		×		×		×	
trickle, ..., river	×			×	×		×	
channel				×	×		×	
canal		×		×	×		×	
lagoon	×		×			×	×	
lake	×		×		×		×	
mere		×	×		×		×	
plash	×		×		×			×
pond		×	×		×		×	
pool	×		×		×		×	
puddle	×		×		×			×
reservoir		×	×		×		×	
sea	×		×			×	×	

Attributes and objects might also be combinations of other attributes and objects, respectively. Suppose $m \in M$ is an attribute and $A \subseteq M$ is a set of attributes which does not contain m but has the same extent as m: $m \notin A$, $m' = A'$. Then the attribute concept μm can be written as the infimum of μa, $a \in A$, hence the set $\mu(M \setminus \{m\})$ is again infimum-dense in $\underline{\mathfrak{B}}((G, M, I))$. According to the Basic Theorem, we can remove m from M and the concept lattices are still the same.

The removing of **reducible attributes** (attributes with $\wedge$-reducible attribute concepts) and **reducible objects** (objects with $\vee$-reducible object concepts) is called **reducing** the context.

Definition 5.2.5. Let (G, M, I) be a clarified context. It is called *row reduced* if every concept object is $\vee$-irreducible and **column reduced** if every attribute concept is $\wedge$-reducible. A **reduced** context is a context which is row and column reduced.

5.3 Many-valued Contexts

The most general situation for data sets is when attributes have values like *intensity, grade, colour* or just some numerical values. We call them many-valued attributes.

Definition 5.3.1. A tuple (G, M, W, I) is called **many-valued context** if G, M and W are sets and $I \subseteq G \times M \times W$ is a ternary relation with

$$(g, m, w) \in I \text{ and } (g, m, v) \in I \Rightarrow w = v.$$

The set G is called **set of objects**, M is called **set of attributes** and W is the **set of attribute values**. $(g, m, w) \in I$ is read *object g has attribute m with values w*. Sometimes we write $m(g) = w$ for $(g, m, w) \in I$.

Many-valued contexts can also be represented by tables, the rows being labeled by object names from G, the columns labeled by attribute names from M while the entry in row g and column m is the value w of m, provided $(g, m, w) \in I$.

Concepts can be assigned to many-valued contexts by a process called **conceptual scaling**. The process of scaling is not unique, i.e., the concept system depends on the scales we choose. In this process of scaling, each attribute is interpreted by a context called **scale**.

Definition 5.3.2. Let (G, M, W, I) be a many-valued context and $m \in M$ an attribute. A **scale** for m is a formal context $S_m := (G_m, M_m, I_m)$ with $m(G) \subseteq G_m$. The objects of a scale are called **scale values,** its attributes are called **scale attributes.**

Remark 58 *From this definition follows that every context can act as a scale. The term* **conceptual** *points out that we choose special scales, bearing a certain meaning in order to transform the many-valued context into a one-valued preserving the semantics of the attributes.*

The simplest case is that of *plain scaling*. Every attribute is scaled, the result of scaling of all attributes is put together without interconnection.

Definition 5.3.3. Let (G, M, W, I) be a many-valued context and S_m, $m \in M$ are scales. The **derived scaled context** with respect to plain scaling is the context (G, N, J) with

$$N := \bigcup_{m \in M} M_m,$$

and

$$g J(m, n) :\Leftrightarrow m(g) = w \text{ and } w I_m n.$$

For more information regarding scaling we refer to (Ganter and Wille 1999).

5.4 Finding all Concepts

Since formal concepts represents the knowledge which is encoded in the formal context, the problem of finding all concepts is a major one. Geometrically speaking, formal concepts can be defined as being maximal rectangle of crosses in the representation of a formal context as a cross-table. Formally, this means:

Definition 5.4.1. A rectangle of crosses in (G, M, I) is a pair (A, B) such that $A \times B \subseteq I$. For two rectangles (A, B) and (C, D), we define $(A, B) \subseteq (C, D)$ if and only if $A \subseteq C$ and $B \subseteq D$.

Example 5.4.1. Consider the following context (Ganter and Wille 1999):

	a	b	c	d	e	f	g	h	i
leech (1)	×	×					×		
bream (2)	×	×					×	×	
frog (3)	×	×	×				×	×	
dog (4)	×		×				×	×	×
spike-weed (5)	×	×		×		×			
reed (6)	×	×	×	×		×			
bean (7)	×		×	×	×				
maize (8)	×		×	×		×			

In this context, a maximal rectangle of crosses is {1, 2, 3}, {a, b, g}, while {1, 2, 3, 5}, {a, b} is not maximal with respect to $\subseteq$, since the crosses of row 6 (reed) can be added to those of {1, 2, 3, 5}.

The idea of a maximal rectangle of crosses, suggests a first (naive) way to compute all concepts. Every formal concept is a maximal rectangle of crosses, while every maximal rectangle of crosses is a formal concept.

Mining all maximal rectangles of crosses provides the following formal concepts that are discovered in this context:

$C_0 := (\{1,2,3,4,5,6,7,8\},\{a\}), C_1 := (\{1,2,3,4\},\{a,g\}), C_2 := (\{2,3,4\},\{a,g,h\}),$

$C_3 := (\{5,6,7,8\},\{a,d\}), C_4 := (\{5,6,8\},\{a,d,f\}), C_5 := (\{3,4,6,7,8\},\{a,c\}),$

$C_6 := (\{3,4\},\{a,c,g,h\}), C_7 := (\{4\},\{a,c,g,h,i\}), C_8 := (\{6,7,8\},\{a,c,d\}),$

$C_9 := (\{6,8\},\{a,c,d,f\}), C_{10} := (\{7\},\{a,c,d,e\}), C_{11} := (\{1,2,3,5,6\},\{a,b\}),$

$C_{12} := (\{1,2,3\},\{a,b,g\}), C_{13} := (\{2,3\},\{a,g,g,h\}), C_{14} := (\{5,6\},\{a,b,d,f\}),$

$C_{15} := (\{3,6\},\{a,b,c\}), C_{16} := (\{3\},\{a,b,c,g,h\}), C_{17} := (\{6\},\{a,b,c,d,f\}),$

$C_{18} := (\{\},\{a,b,c,d,e,f,g,h,i\}).$

The corresponding concept lattice $\mathfrak{B}(G, M, I)$ is given in the figure 5.4.1.

It is somehow amazing that a relative simple contextual structure of data provides a list of 18 knowledge units. Moreover, these are indeed all maximal combinations objects-attributes which can be formed through the incidence relation.

Remark 59 *The derivation operators provide another method of constructing all formal concepts. For $A \subseteq G$ and $B \subseteq M$, we can build the formal concepts (A'', A') and (B', B''). This approach is more methodical than the previous one, but since we have to check and store every subset A of G, for instance, it becomes quickly highly exponential, since we have to test $2^{|G|}$ subsets.*

A major improvement would be to generate the intents (or dually the extents) of the concept system in a specific order, avoiding redundancy and unnecessary storage. This idea is based on the fact that the derivation operators are closure operators. B. Ganter introduced in (Ganter and Wille 1999) the NextClosure algorithm which describes the process of finding all closures of a closure operator. More algorithms can be found in (Carpineto and Romano

2007). Starting with a context $K := (G, M, I)$, the NextClosure algorithm will compute the list $\text{Int}((G, M, I))$ of all intents or dually, the list $\text{Ext}((G, M, I))$ of all extents.

If we have computed the system of all intents, $\text{Int}((G, M, I))$, then

$$\mathfrak{B}(G, M, I) = \{(B', B) \mid B \in \text{Int}((G, M, I))\}.$$

Hence, the entire conceptual hierarchy $\mathfrak{B}(K)$ can be reconstructed from the set of intents (dually, extents).

We start by giving a linear ordering on the attributes set, i.e., $M := \{1, 2, \ldots, n\}$. This is not restricting at all, since even writing them in the cross-table means to order them.

We define the lectic order on the set of attributes as follows:

Definition 5.4.2. For $A, B \subseteq M$ and $i \in \{1, 2, \ldots, n\}$, we define $A <_i B \Leftrightarrow i \in B \setminus A$ and $A \cap \{1, \ldots, i-1\} = B \cap \{1, \ldots, i-1\}$,

$$A < B \Leftrightarrow A <_i B \text{ for some } i \in \{1, 2, \ldots, n\}.$$

For $i = 1$, we put $\{1, \ldots, i-1\} = \varnothing$.

Example 5.4.2. Let $M := \{a, b, c, d, e, f, g, h\}$. We consider the sets $\{g\}, \{e\}, \{e, h\}, \{e, g\}, \{d, e, g, h\}, \{c\}$. We have

(1) $\{g\} <_5 \{e\}$ because $g \in \{g\} \setminus \{e\}$ and $A \cap \{a, b, c, d\} = B \cap \{a, b, c, d\}$.

(2) $\{e, h\} <_5 \{e, g\}$ because $e \in \{e, h\} \setminus \{e, g\} = \{e\}$ and $A \cap \{a, b, c, d\} = B \cap \{a, b, c, d\}$.

(3) All sets are ordered lectically:

$$\{g\} <_5 \{e\} <_5 \{e, h\} <_5 \{e, g\} <_4 \{d, e, g, h\} <_3 \{c\}.$$

By this example, the idea of the algorithm becomes more clear. We start with the lectically smallest closure, i.e., $\varnothing''$ and then we generate the lectically next closure. For this we need the following Definition.

Definition 5.4.3. For $A \subseteq M$, $i \in \{1, 2, \ldots, n\}$, put

$$A \oplus i := ((A \cap \{1, \ldots, i-1\}) \cup \{i\})''.$$

Example 5.4.3. We illustrate how $A \oplus i$ works on the following context

	1	2	3	4	5
a	×	×		×	×
b	×	×	×	×	
c		×			×
d	×				×

- $A := \{1, 2, 5\}$, $i = 2$.

×	×	·	·	×

First, we eliminate all elements which are greater or equal 2: ($\{1, 2, 5\} \cap \{1\}$)

×	·	·	·	·

Then we add i: ($\{1, 2, 5\} \cap \{1\}$) $\cap \{2\} = \{1\} \cap \{2\}$

×	×	·	·	·

and now we build the closure: $A \oplus i = ((\{1, 2, 5\} \cap \{1\}) \cup \{2\})'' = (\{1\} \cup \{2\})'' = \{1, 2\}'' = \{1, 2, 4\}$.

- $A = \{5\}$, $i = 1$.

$$A \oplus i = ((\{5\} \cap \varnothing) \cup \{1\})'' = \{1\}'' = \{1\}$$

The following Lemma (Belohlavek 2008) gives information about the behavior of the $<_i$ relation.

Lemma 5.4.1. *Let (G, M, I) be a context and $X, Y, Z_1, Z_2 \subseteq M$ subsets of attributes. Then*

(1) *If $X <_i Z_1$, $Z <_j Z_2$, and $i < j$ then $Z_2 <_i Z_1$;*
(2) *If $i \notin X$ then $X < X \oplus i$;*
(3) *If $X <_i Y$ and Y is an intent then $X \oplus i \subseteq Y$;*
(4) *If $X <_i Y$ and Y is an intent then $X <_i Y \oplus i$.*

The following results are explaining how to find the lectically next closed set and hereby all closed sets.

Proposition 5.4.2. *Let $X \to X''$ be a closure operator on a set M, and $A, B \subseteq M$. Then*

(1) *If $i \notin A$ then $A < A \oplus i$.*
(2) *If B is closed an $A <_i B$ then $A \oplus i \subseteq B$, in particular $A \oplus i \leq B$.*
(3) *If B is closed and $A <_i B$ then $A <_i A \oplus i$.*

Using this, it is easy to characterize the "next" closed set:

Theorem 5.4.3. *For $A \subseteq G$, $A \neq G$, the smallest closed set that is larger (with respect to <) than A, is $A^+ := A \oplus i$, where i is the largest element of G satisfying $A <_i A \oplus i$.*

An algorithm to compute all closures is grounded on this theorem. This algorithm calles two functions, FirstClosure and NextClosure in order to compute all closures (Ganter and Wille 1999).

Algorithm All Closures

Input: A closure operator $X \to X''$ on a finite set M.
Output: All closed sets in lectic order.
Begin
    First Closure;
    repeat
        Output A;
        Next Closure;
    until not *success*;
end.

Algorithm First Closure
Input: A closure operator $X \to X''$ on a finite set M.
Output: The closure A of the empty set.
Begin
$A := \varnothing ''$;
end.

Algorithm Next Closure
Input: A closure operator $X \to X''$ on a finite set M, and a subset $A \subseteq M$.
Output: A is replaced by the lectically next closed set.
Begin
    for all $i \in$ M starting downwards with the largest element of M do;
        if $i \in A$ then
            $A := A \backslash \{i\}$
else
    $B := (A \cup \{i\})''$
    if $B \backslash A$ does not contain any element $< i$ then return B
return $\perp$
end.

Remark 60 *For any formal context (G, M, I), the set of extents and the set of intents are closure systems on G and M, respectively. The Next Closure algorithm can be applied to compute all extents or all intents, hence all formal concepts of a formal context. If the algorithm is used to compute all extents, it is called Next Extent, if it is used to compute all intents, it is called Next Intent.*

This algorithm creates every concept only once, no extra space and time are needed to store and retrieve the generated concepts. On the other hand, the algorithm does not support the construction of the conceptual hierarchy. There is no relationship between the lectic order and the conceptual order given by the subconcept-superconcept relationship.

Example 5.4.4. (NextClosure algorithm—simulation). We are going to simulate the NextClosure algorithm on the example clarifyied bodies of water context, where the object names and attribute names have been substituted by numbers from 1–13 and letters from a to h, respectively:

	a	b	c	d	e	f	g	h
1	×		×		×		×	
2	×			×	×		×	
3				×	×		×	
4		×		×	×		×	
5	×		×			×	×	
6	×		×		×		×	
7		×	×		×		×	
8	×		×		×			×
9		×	×		×		×	
10	×		×		×		×	
11	×		×		×			×
12		×	×		×		×	
13	×		×			×	×	

The simulation starts with the first intent, $\emptyset''$, computing all intents in the lectic order:

Last generated intent	A	i	$B = (A \cup \{i\})''$	smallest new element	success?
	$\emptyset$		$\emptyset'' =$ first intent		yes
$\emptyset$	$\emptyset$	h	$\{a,c,e,h\}$	a	no
	$\emptyset$	g	g	g	yes
$\{g\}$	$\{g\}$	h	M	a	no
	$\emptyset$	f	$\{a,c,f,g\}$	a	no
	$\emptyset$	e	$\{e\}$	e	yes
$\{e\}$	$\{e\}$	h	$\{a,c,e,h\}$	a	no
	$\{e\}$	g	$\{e,g\}$	g	yes
$\{e,g\}$	$\{e,g\}$	h	M	a	no
	$\{e\}$	f	M	a	no
	$\emptyset$	d	$\{d,e,g\}$	d	yes
$\{d,e,g\}$	$\{d,e,g\}$	h	M	a	no
	$\{d,e\}$	f	M	a	no
	$\emptyset$	c	$\{c,g\}$	c	yes
$\{c\}$	$\{c\}$	h	$\{a,c,e,h\}$	a	no
	$\{c\}$	g	$\{c,g\}$	g	yes
$\{c,g\}$	$\{c,g\}$	h	M	a	no
	$\{c\}$	f	$\{a,c,f,g\}$	a	no
	$\{c\}$	e	$\{c,e\}$	e	yes
$\{c,e\}$	$\{c,e\}$	h	$\{a,c,e,h\}$	a	no
	$\{c,e\}$	g	$\{c,e,g\}$	g	yes
$\{c,e,g\}$	$\{c,e,g\}$	h	M	a	no
	$\{c,e\}$	f	M	a	no
	$\{c\}$	d	M	a	no
	$\emptyset$	b	$\{b,e,g\}$	b	yes

Last generated intent	A	i	$B = (A \cup \{i\})''$	smallest new element	success?
$\{b, e, g\}$	$\{b, e, g\}$	h	M	a	no
	$\{b, e\}$	f	M	a	no
	$\{b\}$	d	$\{b, d, e, g\}$	d	yes
$\{b, d, e, g\}$	$\{b, d, e, g\}$	h	M	a	no
	$\{b, d, e\}$	f	M	a	no
	$\{b\}$	c	$\{b, c, e, g\}$	c	yes
$\{b, c, e, g\}$	$\{b, c, e, g\}$	h	M	a	no
	$\{b, c, e\}$	f	M	a	no
	$\{b, c\}$	d	M	a	no
	$\emptyset$	a	$\{a\}$	a	yes
$\{a\}$	$\{a\}$	h	$\{a, c, e, h\}$	c	no
	$\{a\}$	g	$\{a, g\}$	g	yes
$\{a, g\}$	$\{a, g\}$	h	M	b	no
	$\{a\}$	f	$\{a, c, f, g\}$	c	no
	$\{a\}$	e	$\{a, e\}$	e	yes
$\{a, e\}$	$\{a, e\}$	h	$\{a, c, e, h\}$	c	no
	$\{a, e\}$	g	$\{a, e, g\}$	g	yes
$\{a, e, g\}$	$\{a, e, g\}$	h	M	b	no
	$\{a, e\}$	f	M	b	no
	$\{a\}$	d	$\{a, d, e, g\}$	d	yes
$\{a, d, e, g\}$	$\{a, d, e, g\}$	h	M	b	no
	$\{a, d, e\}$	f	M	b	no
	$\{a\}$	c	$\{a, c\}$	c	yes
$\{a, c\}$	$\{a, c\}$	h	$\{a, c, e, h\}$	e	no
	$\{a, c\}$	g	$\{a, c, g\}$	g	yes

Last generated intent	A	i	$B = (A \cup \{i\})''$	smallest new element	success?
$\{a,c,g\}$	$\{a,c,g\}$	h	M	b	no
	$\{a,c\}$	f	$\{a,c,f,g\}$	f	yes
$\{a,c,f,g\}$	$\{a,c,f,g\}$	h	M	b	no
	$\{a,c\}$	e	$\{a,c,e\}$	e	yes
$\{a,c,e\}$	$\{a,c,e\}$	h	$\{a,c,e,h\}$	h	yes
$\{a,c,e,h\}$	$\{a,c,e\}$	g	$\{a,c,e,g\}$	g	yes
$\{a,c,e,g\}$	$\{a,c,e,g\}$	h	M	b	no
	$\{a,c,e\}$	f	M	b	no
	$\{a,c\}$	d	M	b	no
	$\{a,\}$	b	M	b	yes
M					

Remark 61 *The* NextClosure *algorithm is called* NextExtent *if it is used for extents, and* NextIntent *if it is used for concept intents.*

The following table shows the intents lectically ordered:

	a	b	c	d	e	f	g	h
1								
2							X	
3					X			
4					X		X	
5				X	X		X	
6			X					
7			X				X	
8			X		X			
9			X		X		X	

	a	b	c	d	e	f	g	h
10		X			X		X	
11		X		X	X		X	
12		X	X		X		X	
13	X							
14	X						X	
15	X				X			
16	X				X		X	
17	X			X	X		X	
18	X		X					
19	X		X				X	
20	X		X			X	X	
21	X		X		X			
22	X		X		X			X
23	X		X		X		X	
24	X	X	X	X	X	X	X	X

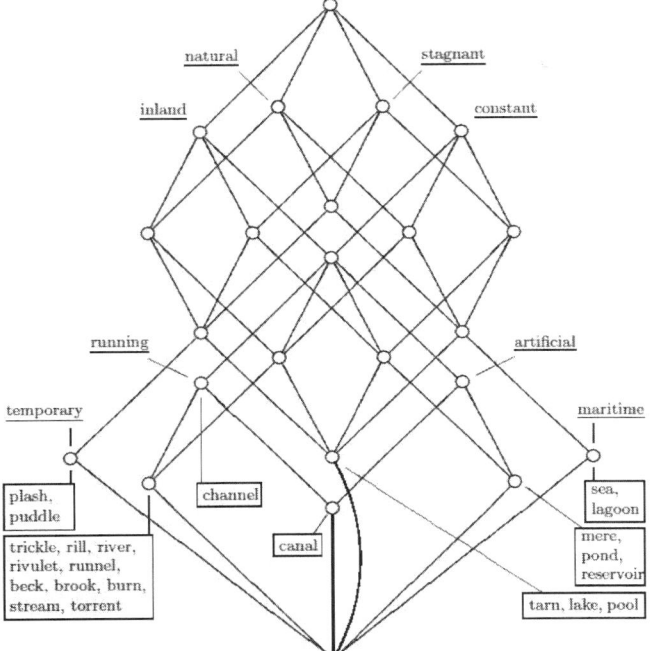

Figure 5.1: Concept lattice for bodies of water context

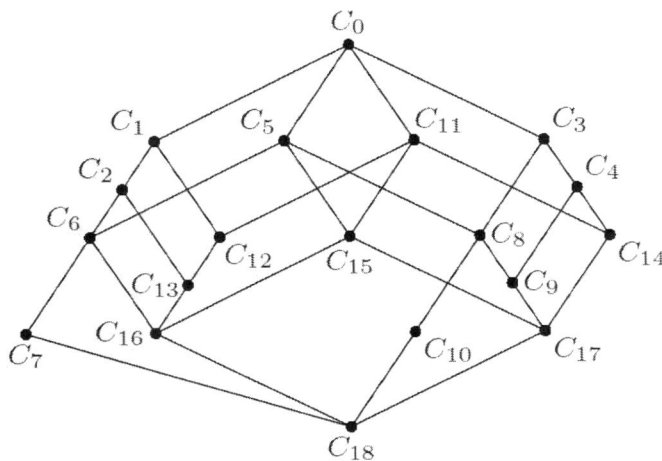

Figure 5.2:   Concept lattice for context 5.4.1

PART III

Knowledge Representation for NLP

CHAPTER 6

Measuring Word Meaning Similarity

6.1 Introduction

The question of how to measure similarity and distance as well as represent the idea of relevance among words and concepts, has led to the definition of space models and the way they are structured, together with inherent metrics and logic. Answers to this question have long been sought, in order to provide appropriate solutions to problems in Information Retrieval (IR), where set–theoretical approaches failed to cope with "relatedness" and "aboutness" (van Rijsbergen 2004), as well as problems in computational linguistics, knowledge discovery and text mining, where word meaning disambiguation and clustering pose a challenge.

In the following, a study of related work about space models as well as the inherent metrics is undertaken. This is done for the purpose of showing the impact a chosen structure of space model might have on the type of metrics, operators and perhaps, logic. The study focuses on the related work on measuring semantic word and concept similarity as provided by the classification of the implementation methods. Another focal point is related work concerning measurement of word and concept similarity with agnostic, statistical methods. This contrasts with the work on measuring semantic relatedness of words and concepts, in that no semantic network of words or concepts exists. This technique applies mostly to text mining, words and concepts clustering methods.

6.2 Baseline Methods and Algorithms

6.2.1 Intertwining space models and metrics

High dimensional mathematical spaces such as graphs and hierarchies have always been introduced in order to measure distances and similarities. The most standard distance measures in mathematics are called metrics, which must satisfy certain conditions or axioms such as being symmetric and transitive. For example, if it holds that A is the same as B and B is the same as C, then it follows that A is the same as C, according to the transitivity axiom. Similarly, if it holds that A is close to B, then, according to the symmetry axiom, B is close to A too. A very important similarity measure for points in the plane and higher dimensions is the cosine measure. The cosine measure assigns a high similarity to points that are in the same direction from the origin, zero similarity to points that are perpendicular to one another and negative similarity to those that are pointing in opposing directions.

It has been a matter of concern, however, that these mathematical techniques are actually inappropriate when natural language and measure of distance and similarity among words or concepts is being dealt with. Some of the properties of metrics are not always ideal for describing distances between words and concepts. For example, the axioms of symmetry and transitivity do not hold. It is more obvious to say that my car is close to Regent Park instead of saying that Regent Park is close to my car, if the axiom of symmetry would have been valid. Moreover, saying that the handle is part of the door and the door part of the apartment does not necessarily follow that the handle is part of the apartment.

Even when words and concepts are put on graphs and taxonomies, one should be able to measure similarity or distance by finding the lengths of paths among concepts or words. However, there are at least three problems with this. First of all, finding the shortest path is often computationally expensive and might take a long time. Secondly, one may not have a reliable taxonomy. The fact that there is a short path between two words in a graph does not necessarily mean that they are very similar, because the links in this short path may have arisen from very different contexts. Thirdly, the meanings of words encountered in documents and corpora might be very different from those given by a general taxonomy such as WordNet. For example,

WordNet 2.0 only gives two meanings of the word apple, i.e., fruit and tree, which is a decided contrast with the top 10 pages currently returned by the Google search engine when doing an internet search with the query apple, which are all about Apple Computers.

Going beyond individual words or concepts, another limitation of current methods and metrics is that they have focused purely on individual concepts, mainly single words. Ideally, one should be able to find the similarity between two collections of words. Therefore, some process is needed for semantic composition, i.e., working out how to represent the meanings of a sentence or document based on the meaning of the words it contains. It might also happen that words belong to more than one collection, since they are ambiguous. These words can sometimes behave like semantic wormholes, accidentally transporting one from one area of meaning to another. But this effect can also be used positively, because finding those words which have this strange wormhole effect when measuring distances helps one to recognize which words are ambiguous in the first place.

The technique of using vectors to represent points in space is rather an old one. Those interested in the mathematical development of vectors could start with the pioneering work of Descartes, 1637. This addresses the idea of describing points by measuring their coordinates on two chosen lines. A good historical account can be found in (Boyer and Merzbach 1991). Hermann Grassmann (1809–1877), however, with his work on extension theory (Ausdehnungslehre) (Grassmann 1862) lays the foundation of the modern notion of vectors in any number of dimensions. It is not surprising that vector spaces have been introduced as a space model for words. A typical contemporary representative of such a vector space is the Infomap vector space model (Widdows 2004), or WORDSPACE, as pioneered by Heinrich Schütze (Schütze 1998). The model works by mapping words to points in a high dimensional space, by recording the frequency of co occurrence between words in the text, for example, the number of times two words appear in the same document. Each document might be conceived as a dimension. The distribution of co-occurrences between a word and some set of content bearing terms serves as a profile of the word's meaning and can associate the word with other words of similar meanings.

A word can thus be described by a spectrum of related words. In the context of search engines, comparing the profiles of query words to those generated for each document, articles which are supposed to be conceptually related to the query words, are returned, even if the words themselves do not appear in the text. To this extent, a logic with vectors as well as vector negation has also been introduced in WORDSPACE as an operator in order to cope with word disambiguation by manipulating vectors in WORDSPACE so that they are perpendicular or orthogonal to one another. Moving the vectors so that they are at right angles to one another effectively, makes the meanings of two word vectors irrelevant to one another. Logical operators can all be defined and built in a vector space such as WORDSPACE and these are the natural logical operations in this space, just as the Boolean logic gates are the natural logical operations for set theory.

Classical set theory is binary and discrete, whereas vector spaces are smooth and continuous. You find this difference between continuous and discrete interpretations of logic in very simple sentences. If somebody says "you should take bus 60 or bus 70," and you later discover that they meant you to take bus 67, you would think you had been very badly advised. If, however, you were told that the journey would take "60 or 70 minutes," and it took 67 minutes, you would agree that you had been advised perfectly. Therefore, it has been attempted to show that these properties are reflected in the logic that arise in these spaces. As an example, the vector negation in WORDSPACE is inspired by quantum logic, which was introduced by Garrett Birkhoff (Birkhoff 1967), the founder of modern lattice theory, and John von Neumann [BvN36], the founder of modern computer architecture.

It is acknowledged by this logic that a highly ambiguous word within a word space resembles the status of a particle in Physics, where a particle could be in more than one state simultaneously according to Quantum Mechanics. A continuous interpretation of logic is also subject to two other models, the probabilistic and the fuzzy set models, which pose an alternative to vector space models. These three together constitute the principal mathematical models used for information retrieval. The fuzzy set model is an adaptation of the traditional Boolean model to cope with continuous values. A traditional thesaural relationship in information retrieval is that of a "Broader Term". Fuzzy sets for information retrieval are discussed at length by Miyamoto (Miyamoto 1990). A probabilistic model, on the other side, is based upon networks using probability theory to find, for instance, important paths in directed graphs and the path, which leads between a query and relevant documents. One of the most interesting probabilistic models is the inference network used in the INQUERY system [Callan et al 1992, Turtle and Croft 1989].

Despite the practical differences among the models, there is a unifying theme behind normalised vectors, fuzzy sets and probability distributions: they all replace binary values (0 and 1), which signify 'not belonging' and 'belonging', with continuous values, which signify 'partially belonging' or 'probably belonging'. To this extent, they all seem to be different implementations of one underlying space model. An attempt has been made in 20th century mathematics to weave together the threads of hierarchies, vector spaces and Boolean logic as separate mathematical systems. They are conceived as different variants of one underlying structure called Lattice (Birkhoff 1967). The distinctive property of a lattice is that two elements can be disjoined (allowed to spread out and diversify) and conjoined (woven together). Consequently, concept lattices (Ganter and Wille 1999) are used to describe a way in which words are represented, e.g., different kinds of a horse. Ordered sets form the most basic building block of a lattice, where an ordering relationship, e.g., for natural numbers, is assumed. There are also two key operators, necessary to turn an ordered set into a lattice: the *join* and *meet* operators as a complementary pair of operators, which correspond to disjunction and conjunction, within a lattice. In other words, the paths upwards or downwards from any two nodes always meet.

It turns out, however, that concept lattices are quite complicated as a data structure in terms of the number of concepts generated. In order to simplify a generated lattice, association rules are used for clustering concepts on the lattice. To this extent, concept hierarchies and conceptual scaling have been introduced. Moreover, in order to cope with uncertain information for conceptual clustering and conceptual hierarchy generation, fuzzy logic has recently been incorporated into Formal Concept Analysis (FCA)

(Thanh et al 2006), since traditional FCA-based conceptual clustering approaches are hardly able to represent vague information. The use of coordinates and dimensions has also become a cornerstone in representing mental processes in cognitive science. In particular, the formalization of conceptual spaces in order to address different stimuli such as taste, colour, etc., as pointers in a conceptual space has been subject to (Gärdenfors 2000).

6.2.2 Measuring similarity

In general, it is considered that similarity measure is the converse of a distance function. Similarity functions take a pair of points and return a large similarity value for nearby points, a small similarity value for distant points. One way to transform a distance function towards a similarity measure is to take the reciprocal. This is the standard method for transforming resistance and conductance in physics and electronics. There is an extensive literature in measuring semantic similarity, in general and word similarity, in particular. An attempt to classify the related work of measuring semantic similarity of words and concepts relies on the implementation methods. These are known as edge counting methods, information content methods, feature based methods and hybrid methods.

A well known paper and typical representative of feature based methods is the conducted work by Tversky (Tversky 1977). Feature based methods measure the similarity among terms as a function of their properties or are based on their relationships to other similar terms in a taxonomy. In particular, Tversky uses set theory and defines similarity measure in terms of a matching process. This is close to the information theory based definition of similarity. The matching model is not forced to satisfy metric properties such as minimality, symmetry and triangle inequality. For example, the similarity between office building and buildings can be greater than the similarity between buildings and office building.

Edge counting methods rely on counting of words or edges in word networks such as taxonomies in order to determine semantic similarity. Typical representatives of these methods are (Rada et al 1989, Leacock and Chodorow 1998, Li et al 2003). In the recent years, information content methods, which rely on information theoretic concepts and techniques for measuring semantic similarity of words, have emerged. Typical advocates of these methods are (Jiang and Conrath 1998, Resnik 1999, Lord et al 2003). These methods rely on the premise of how informative a particular word or concept is supposed to be according to information theory and the quantitative measure of the information content of a concept.

However, edge counting and information content based methods have a common characteristic: they both work with information on the structure and position of the terms in the structure. Another common property is that they assign higher similarity to terms, which are close together and lower in the hierarchy (more specific terms) than those which are equally close together but higher in the hierarchy (more general terms). Recently, a hybrid method has also been introduced (Rodriguez and Egenhofer 2003), which aims at combining all three previously mentioned methods. In particular, the introduced method uses the combination of three matching processes: word-matching, feature matching and semantic neighbourhood matching, which constitute specification components of the similarity by inserting respective weights. Because the structure and information content among different ontologies are not directly comparable, edge counting and information content based semantic similarity methods are suited mostly for comparing terms from the same ontology. Cross ontology similarity measures rely on feature based methods and hybrid method, the latter because of the embedded feature based and word matching processes in the absence of a common structure. Therefore, another classification scheme single versus cross ontology could be considered for all methods of semantic similarity measurement.

In contrast to the previous work on measuring semantic relatedness of words and concepts, as well as in the absence of semantic network of words or concepts, a distributional hypothesis that similar words appear in similar contexts is being followed. Therefore, this approach usually addresses corpora to examine the context in which each word appears and calculate the similarity between context distributions. Distributional similarity, however, is at best an approximation to semantic similarity. Attempts to improve effectiveness of distributional similarity and come closer to semantic similarity have been made. For example, it is claimed that a context-weighted similarity metric, as introduced in (Curran 2003), outperforms every distributional similarity found in the literature.

In conjunction with an underlying space model for words and concepts, various formal and semi-formal techniques have been developed for extraction of an ontology, or clusters of words or concepts, from various data types. For example, knowledge discovery techniques address data types such as textual data (Maedche and Staab 2001), dictionary (Hearst 1992), knowledge based semi-structured schemata (Deitel et al 2001), as well as relational schemata (Johannesson 1994). Compared to other types of data, ontology generation from textual data has attracted the most attention.

The extraction of a domain-specific ontology from texts with data mining techniques is discussed in (Kietz et al 2000, Maedche and Staab 2000a, Maedche and Staab 2000b), whereby (Kietz et al 2000, Maedche and Staab 2000b) concentrate on the core mechanism, which relies on the frequency of concepts in the texts, while the emphasis in (Maedche and Staab 2000a) is on the discovery of semantic relations by using association rules. The semantic richness and diversity of corpora does not lend itself to full automation, so that the involvement of a domain expert becomes necessary (Maedche and Staab 2000c).

Unsupervised concept clustering methods help one to learn semi-automatically sub-categorization frames of verbs and ontologies. However, in some cases they have not used text unit clustering and did not use cluster quality criteria. In "RELFIN" (Schaal et al 2005), however, the focus has been on the extraction of specific views based on a context given by a web document collection. An appropriate corpus from the web is being extracted and clustering is being performed over it in order to establish a set of clusters over which a semi-automated extraction of parts of an ontology can be performed.

6.3 Summary and Main Conclusions

In a nutshell, regardless of the intended application domain, be it a search engine or clustering in knowledge discovery and data mining, the accomplished tasks of determining distance or similarity among words and concepts are almost entirely mathematical rather than linguistic and are based on numbers rather than meanings of numbers and measurements. Using its distribution in a corpus, the meaning of each word is represented by a characteristic list of numbers, and the numbers representing a whole document are then given simply by averaging the numbers for the words in the document.

On the other hand, measuring word similarity by taking into consideration meaning in terms of information content of words or concepts, presupposes the existence of a taxonomic or existing WordNet relationships. Expanding or augmenting, however, a WordNet under construction from a text archive requires the definition of a space model and metrics, where both worlds, information content and frequency based measurements are smoothly accommodated.

CHAPTER 7

Semantics and Query Languages

7.1 Introduction

Providing information out of data has always been the major concern in information science. Addressing and disseminating the acquired information, however, has been the major focus of research activities in query languages and question answering systems. Despite the variety of approaches taken so far, two central questions can be raised when we ask for information: *how* to ask for information and *what* kind of queries the system can answer. Although the second question has been exhaustively examined by theories and/or practical solutions referring to model- or proof- theoretic semantics, closed versus open world assumptions, fuzziness or uncertainty of query results, etc., the way *how* to ask for available information still remains a major challenge, especially when end-users are involved in the querying process. The roots of the challenge can be found at the assumption we make that an end-user:

1. learns the query language syntax
2. understands the meaning of the underlying database schema in terms of adequate interpretations of data constructs such as relations, classes and collections,
3. understands the meaning of attributes and/or values,
4. is aware of the context which relates to the data as expressed in terms of measurement units, explanation or definitions, etc.

In this sense, the key answer to the question of *how to ask* for the available information goes through the role of semantics the lack of which becomes more and more apparent in existing information systems, i.e., a *semantic gap*. Furthermore, the end-user does not really know what he is looking for or how to ask for available information, unless she/he fully understands it, i.e. seeing, interpreting it. Even if the user knows what he wants, he is confronted with complex syntax formalisms, since typical systems rely only on low-level features as expressed by the data model on which the data repository relies.

The problem becomes more acute when the semantic complexity of a particular application domain increases. Acquisition of scientific information, for example, requires great familiarity with the domain science which is agnostic to end-users other than domain experts. Even within a particular domain science, using an information system as communication platform for knowledge/ information exchange among domain experts becomes very difficult, or even impossible, without an appropriate interpretation of data and/or its context. Additionally, querying information systems which rely on large data repositories in terms of large or complex database schemas, becomes cumbersome, since most of the querying interfaces they provide are oriented towards application programmers and not end-users.

On the other side, the growth of the Internet as an easy-to-access information repository providing various types of information ranging from unstructured data to the traditional record-oriented one poses an additional querying challenge. Proposals coming from different areas, namely databases, artificial intelligence and human-computer interaction, share the final goal of making the web a huge, easy-to-access, information repository. Therefore, the problem definition resembles the definition of a *bridge of the semantic gap* in terms of moving from low-level features to high-level semantics.

In order to meet this requirement and overcome the end-user/ queried system communication problem, various advanced database or information system query interfaces have been elaborated which aimed mainly at meeting the challenges (1) and partly (2).

However, the challenges (3) and (4) could not be adequately met by these query techniques, since they still lack the exploitation of *meaning* of vocabulary terms during formulation of a query expression. Furthermore, these solutions shifted the problem from understanding the data to understanding the user interface, and in most cases, they are still bound to the world of *symbols* chosen for the underlying data/knowledge representation model. Thus interpretation of an underlying database schema or knowledge model in terms of linguistic elements of a particular natural language (partly expressed by challenge (2)) such as English or German has not been the case.

In addition, it has been the holy grail of many approaches to enable construction of reasonable or meaningful queries in more than one natural language by meeting all the challenges described above. It is expected that the system actually helps in formulating a reasonable query by taking into consideration the meaning and context of data. This presupposes that a considerable amount of knowledge concerning the nature of terms to be used within a query must be provided by the system. An exemplary approach, in terms of the meaning driven querying methodology, is being discussed in this chapter as a guideline and framework for similar approaches.

7.2 Baseline Methods and Algorithms

7.2.1 The methodology

The **meaning driven querying methodology** relies upon the interactive guidance of end-users in constructing meaningful or reasonable multi-lingual queries through system suggestions of terms to accomplish a query statement. This is achieved through the exploration of the *knowledge space* of vocabulary terms as given by a particular application domain.

Having the system guide the end-user to the construction of a reasonable query (meeting challenges (1) and (2)) presupposes that only reasonable suggestions of query terms are made for further consideration in the query construction process (meeting challenges (3) and (4)). This has been conceived as a transition from a particular query state (context)—set of terms currently being members of the query state—to a new query state. Thereby, the new terms which can extend the current query context are determined by the meaning of terms and/or the query context. In order to meet the challenge (3), the meaning of terms is defined by the natural language based interpretation of symbols used by the data repository system, the data context such as measurement units, explanations or definitions, images, etc., where, in order to meet the challenge (4), the meaning of the query context is defined by the semantic dependencies holding among terms, i.e., which terms can be put together in a reasonable query statement.

The meaning of terms determines the *view of terms*, and consequently, the degree of relevance for further consideration by an end-user selection out of the set of system suggested terms. On the other hand, the semantic dependencies determine which terms *should appear* (suggested) to the end-user when a particular query context is given. Therefore, the suggested terms are inferred by checking the semantic consistency between the current query context and the terms to extend it such that

- only those attributes and/or values are suggested which are related to those already participating in the current query context,
- only those attributes and/or values are suggested for which no semantic conflicts with the already participating attribute/value terms are detected,
- only those operations are suggested which are applicable to the terms according to the data contents which they represent.

Therefore, the query construction process through an end-user/ system interaction can be formally defined as a *state automaton*, where a state is equivalent to a current query context—set of terms already participating in the query state—and a state transition function where the meaning of terms and/or query context determines the next query statement. It turns out that the specification of a meaning driven state automaton leads to the specification of a *Meaning Driven Data Querying Language* MDDQL where the automaton acts as an acceptor of the query statements. Since the final statement is determined on the basis of already checked semantic consistencies, the query is regarded as an already accepted one, which means that syntactic and semantic parsing of the query become part of the query construction process. This is particularly useful when more than one natural language is considered for representing the interpretation of symbols in the data repository system.

A system based on the meaning driven querying methodology can be implemented as follows. Its main components are a) the knowledge base, represented by an undirected graph expressing a semantic space for the representation of meaning of vocabulary (query) terms and b) an inference engine operating upon the knowledge space and providing the state transition (inference) algorithms for suggestions of semantically consistent terms. The conditioned connectionism model within the formed semantic space enables a (recursive) structural definition of the conceptualization of the domain of discourse. This enables the assignment of properties/attributes to concepts as well as the assignment of well-defined value domains to properties/attributes in a multi-lingual mode. In order to exclude values and/or attributes from system suggestions which are not consistent with the current set of attribute/value pairs, however, preconditions have also been expressed and assigned to nodes which must be satisfied.

The specified and implemented algorithms infer the set of terms to be suggested to the end-user given a particular query context and the representation model of meaning. They implement all possible state transition functions applying for the construction of specific families of query statements according to the wished complexity, therefore, having an impact on the expressiveness of the query language.

In addition, the system could be augmented with add-on components such as a) various visual query interfaces with which the end-user/system interaction takes place, b) an application server where transformations of the constructed queries with MDDQL towards database specific query languages such as SQL take place. Given that more than one data or knowledge repository are given, the system can be extended by semantic mediators which enable the construction of an answer from more than one repository.

7.2.2 The theory on semantics

The term semantics stands for meaning or the study of meaning. The meaning of a natural language word or sentence is the entity or action it denotes. In computer understanding of language, semantics is the process of determining the meaning of the input. Traditionally, this presupposes that one has, first, a computational representation for meaning, and, second, a method for deriving the representation for a given input.

The nature of meaning and the means for its representation have been topics of interest in linguistics and the philosophy of language. Scholars in the field of artificial intelligence developed semantic theories which are more or less influenced by scholars of philosophy of language. One particular concern in all semantic theories is compositionality. The principle of compositionality is that the meaning of the whole sentence is some systematic function of the meaning of its components. This is intuitively reasonable, but there are two alternatives: a) the meaning of the whole is a systematic function of not only the meaning of the parts but also of the context and situation in which it is formulated—ideally a semantic theory should be able to account for this as well, b) the meaning of the whole is a non-systematic function of the meaning or form of the parts.

Another major debating subject has been the separation of syntactic and semantic parsing processes. Usually, systems dealing with simple input do not consider separation between syntactic and semantic parsing processes. It is clear, however, that syntactic knowledge interacts with semantic knowledge-semantic analysis requires information about input structure but in order to provide this, the parser requires information about the meaning of the constituting parts. This is because many sentences in natural language are structurally ambiguous. A complete separation of the two processes has been experienced in the system LUNAR (Woods 1973), where the disadvantage was that the semantic interpreter was forced to make a decision on semantic well formedness without knowing what alternatives there are (possibly the second best).

However, in all semantic theories, the major focus has been the semantic interpretation during or after syntactic parsing of a sentence or question and not how to exploit represented knowledge (meaning) concerning a particular application domain in order to construct a meaningful query. Moreover, since different natural languages are supposed to be used for expressing the same query, providing syntactic or semantic parsing for each natural language is rather impractical or even impossible.

The semantic theories developed in the past focussed primarily on the means of representation of meaning in order to understand a *syntactic structure*. For example, *decompositional semantics* (Marcus 1984) attempted to represent the meaning of each word by decomposing it into a set of semantic primitives. It turned out that such a lexical decomposition is problematic, since it is extremely difficult, even impossible, to find a suitable, linguistically universal collection of semantically primitive elements in which all words (of all languages) can be decomposed into their necessary properties. Furthermore, *decompositional semantics* is also problematic in its notion of how a sentence is represented; it is necessary to decide how such lexical representations must be combined into the representation of the whole sentence. Consequently, there must be corresponding methods to infer the context of the resultant structure.

Attempts to make *decompositional semantics* usable have been made by (Schank 1973, Sowa 1984). In particular, Schank injected his primitives in the form of *conceptual-dependencies* into a structure that represented a sentence upon which some sort of reasoning has been performed. However, the set of conceptual dependency primitives is incomplete and unable to capture particular nuances of meaning. Sowa proposed a system of *conceptual graphs* where a decompositional approach is combined with that of *frames* in order to gain advantage of both.

Montague semantics (Thomason 1974) refers to *truth-conditional* and *model-theoretic semantics*, where the first is related to the meaning of a sentence as a set of necessary and sufficient conditions for the sentence to be true, i.e., to correspond to a *state of affairs* in the world. Model-theoretic means that the theory uses a formal mathematical model of the world in order to set up relationships between linguistic elements and their meanings. Montague employs not just one model of the world but rather a set of possible worlds. The truth of a sentence is then relative to a chosen possible world at a particular time, where a world-time pair constitutes an index. However, Montague's system contains a set of syntactic rules and a set of semantic rules in one-to-one correspondence. Semantic interpretation is done in terms of *semantic objects* such as individuals in (the model) of the world, individual concepts, properties of individual concepts and higher-order functions. The meaning of a sentence is a truth condition relative to a world-time index.

Situation semantics (Barwise and Perry 1983) attempts to formalize the idea of a situation in the real world as a suitable semantic object. They see linguistic semantics as just a special case of meaning in the world. The viewpoint taken is that *truth-conditional semantics* is inadequate because the content of the sentence is lost, since only truth values in a possible world are of relevance. In *procedural semantics* (Joshi et al. 1981), production rules translate the parsed input into procedure calls that operate upon the database. The rules have been very specific and very powerful. They are applied by looking at specific words and therefore, the whole system is not readily adaptable. However, the meaning of a sentence is the procedure into which the sentence is compiled, either in the computer or in the mind. The procedure itself can be seen as the intension of the sentence, the concept or idea behind it and the result of the execution as the extension, the particular entity denoted. The manipulated items are still *uninterpreted symbols*.

Knowledge-based semantics (Brachman and Schmolze 1985) has been an approach in order to avoid many of the problems of decompositional and procedural semantics. *Frames* based representation of meaning has been used in order to resolve ambiguities of structured input when a sentence is being parsed. Therefore, syntactic and semantic analysis work very closely together. However, knowledge based semantics is more powerful for full services of a retrieval or inference system. It is thus more powerful but also less self-contained; it is more suited in a large, general AI system and less in a natural-language oriented interface to a database.

In 1957 an event occurred that not only revolutionized the world of linguistics but left a lasting impression on philosophy, psychology and other areas. It was the publication of a short monograph by Noam Chomsky titled *Syntactic Structures* (Chomsky 1957). This monograph explored the implications of automata theory for natural languages. Noam Chomsky argued that the sentences of a natural language cannot be meaningfully generated by a finite-state machine or by any context-free grammar, or at least that "...any grammar that can be constructed, will be extremely complex, ad hoc, and unrevealing". He, then, proposed a theory of Transformational Grammar TG (Chomsky 1957, 1965).

At the most abstract level, the theory of TG involves specifying a set of "kernel" sentences of a language; an assortment of "transformations" such as tense of verb and passive voice as well as the order in which transformations should be carried out. Despite the fact that TG has been proposed due to its efficacy of a transformational component, Chomsky also recognized that TG would have to be "formulated properly in terms that must be developed in a full scale theory of transformations". Although the TG had an impact on Computational Linguistics (CL), it centered mostly around matters of syntax. In the long run, the hypothesis of TG most significant for work in CL is that an understanding of the syntax or structure of natural language sentences can be arrived at solely on a grammatical basis without considering the real world properties, e.g. meanings, of the terms being discussed. This notion, sometimes known as the "autonomy of syntax" continues to provide a useful division in categorizing current work in CL as the debate continues as to what interactions are desirable or necessary, between the structural (syntactic) and interpretive (semantic, pragmatic) components of a theory or implementation.

7.2.3 Automata theory and (query) languages

Since query languages underlie a formal syntax specification, one can embark, in the following, on the theoretical considerations of language syntax formalisms. This is meant to show that languages and their formal specification are mostly observed only from a syntactical point of view. *Formal languages* and *computability* are part of the foundations of theoretical computer science. In particular, one defines a *language* over a finite set Σ called *alphabet* to be a subset of Σ^*, where Σ^* is the set of all *words* which can be constructed out of the elements of Σ. For example, if $\Sigma = \{a, b\}$, then $\{a^n b^n \mid n \geq 0\}$ is a language over Σ. A *word* over alphabet Σ is a finite sequence $a_1, \ldots, a_n$, where $a_i \in \Sigma$, $1 \leq i \leq n$, $n \geq 0$.

An important type of computation over words is *acceptance*. The objective is to accept precisely the words that belong to some language of interest. Specifications of languages rely on the definitions of *different kinds of acceptors*. The simplest form of an *acceptor* is a *finite-state automaton (fsa)*. An *fsa* processes words by scanning the word and remembering only a restricted of information about what has already been scanned. This can be formalized by computation allowing a finite set of states and transitions among the states driven by the input. In other words, an *fsa* M over alphabet Σ is a 5-tuple $\{S, \Sigma, \delta, s_0, F\}$, where

- S is a finite set of states;
- Σ is an alphabet called the *input alphabet*;
- δ, the transition function, is a mapping $S \times \Sigma \rightarrow S$;
- s_0 is a particular state of S called the *start state*;
- $F \subseteq S$ is the set of *accepting or final states*.

Given an input word $\omega = a_1, \ldots, a_n$, an *fsa* reads one symbol at a time from left to right. This can be visualized as a tape on which the input word is written and the *fsa* with a head that reads symbols from the tape one at a time. The *fsa* starts in state s_0. A move from state si to a state s_{i+1}, i.e. $\delta(si , a)$, is done by reading the current symbol a and moving the head to the next symbol on the right. If the *fsa* is in an accepting state after the last symbol in ω has been read, ω is accepted. We also say that the *fsa* has a *finite control*. The language accepted by the *fsa* M is denoted L(M).

A language accepted by some *fsa* is called a *regular language*. Not all languages are regular, since for some languages there might be no *fsa* accepting them. For instance, there is no *fsa* which can be specified for the language $\{a^n\, b^n \mid n \geq 0\}$, since no fsa can remember the number of scanned as in order to compare it to the number of bs due the boundness of the memory. An alternative to specify regular languages is by using the so-called *regular expressions*. Such an expression over Σ is written using the symbols in Σ and the operations *concatenation, union* and *repeat*. This alternative is more convenient, if the alphabet is small. However, one of the most useful features of *regular languages* is that they have a dual characterization, one using *fsa* and another using regular expressions. Indeed, Kleene's theorem says that a language L is regular iff it can be specified by a *regular expression*.

There are two important variations of *fsa* that do not change their accepting power. The first allows scanning the input back and forth any number of times, yielding two-way automata. The second is *non-determinism*. A non-deterministic *fsa* allows several possible next states in a given move, i.e., the transition function δ takes the form $S \times \Sigma \rightarrow P\,(S)$, where $P\,(S)$ is the *power set* of the set of states S. Thus several computations are possible on a given input and a *word* is accepted, if there is at least one computation that ends in an accepting state. Non-deterministic *fsa* (nfsa) accept the same set of languages as *fsa*. However, the number of states in the equivalent deterministic *fsa* may be exponential in the number of states in a non-deterministic one. Therefore, non-determinism can be viewed as a convenience allowing much more succinct specification of some regular languages. The equivalence between *fsa* and *nfsa* has been proved by Rabin and Scott.

A *Turing Machine* (TM) can be defined as an *acceptor* of a language L(T M), if we change the functionality of the head of a *fsa* in that not only are symbols being read but also written—overwriting of symbols with elements of the given alphabet, when the head moves in both directions and the length of tape is infinite. Formally, a *Turing machine* is denoted $M = (S, \Sigma, \Gamma, \delta, s_0, B, F)$, where

- S is a finite set of states,
- Γ is the finite set of allowable *tape symbols,*
- $B \in \Gamma$ is the *blank* symbol,
- $\Sigma \subset \Gamma$ not including B is the set of *input symbols,*
- δ, is the *next move function,* a mapping $S \times \Gamma \rightarrow S \times \Gamma \times \{L, R\}$, with δ maybe undefined for some arguments and L is the move of the head to the *left,* R is the move of the head to the *right,*
- s_0 is a particular state of S called the *initial state,*
- $F \subseteq S$ is the set of *accepting or final states.*

T M s can also be viewed as *generators* of words rather than simple *acceptors.* Typically, this is a non-terminating computation generating an infinite language. The set of words generated by a T M is denoted G(T M). A language that can be accepted by a *Turing machine* is said to be *recursively enumerable (r.e.).* In other words, if L(M) is such a language, then any Turing machine recognizing L(M) must fail to halt on some input not in L(M). As long as M is running on some input ω, we can never tell whether M will eventually accept ω if we let it run long enough or whether M runs for ever.

Finally, T M s are viewed as computing a function from input to output and, therefore, they provide the classical formalization of computation. They are also used to develop classical complexity theory. Thus a function f from Σ^* to Σ^* is computable if some TM computing it exists. Church's thesis states that any function computable by some reasonable computing device is also computable by a *turing machine.* Therefore, the definition of computability by TM s is robust. Variations such as non-deterministic T M and allowing multiple tapes make no changes to the accepting power.

A further specification mode of languages is provided by a different approach to *fsa* and T M, where the *generation* of words is emphasized rather than *acceptance,* despite the fact that this can be turned into an accepting mechanism by *parsing.* The major representative of this approach is *context-free grammars* (CFG) (Chomsky 1956). It is defined as a 4-tuple {N, Σ, S, P}, where

- N is a finite set of *non-terminal symbols;*
- Σ is a finite alphabet of *terminal symbols,* disjoint from N;

- S is a distinguished symbol of N , called the *start symbol*;
- P is a finite set of *productions* of the form $\xi \rightarrow \omega$, where $\xi \in N$ and $\omega \in (N \cup \Sigma)^*$

Thus a CFG $G = \{N, \Sigma, S, P\}$ defines a language L(G) consisting of all words in Σ^* that can be derived from S by repeated applications of the productions P. An application of the production $\xi \rightarrow \omega$ to a word v containing ξ consists of replacing one occurrence of ξ by ω. The specification power of CFGs lies between that of *fsa's* and of T M's. All *regular* languages are *context-free* and all *context-free* languages are *recursive*. The opposite does not always hold. For example, $\{a^n b^n \mid n \geq 0\}$ is *context-free* but not *regular*, whereas $\{a^n b^n c^n \mid n \geq 0\}$ is *recursive* but not context free.

Similar to regular expressions which have an equivalent automaton, i.e., the finite automaton, *context-free grammars* have their machine counterpart, the *push-down automaton* (PDA). The equivalence, in this case, is less satisfactory, since the PDA is a non-deterministic device and the deterministic version accepts only a subset of all context-free languages (CFL's). The PDA is actually a finite state automaton with control of both an input tape, from which the input symbols are read, and a *stack* or "first-in-last-out" list, where symbols, from some alphabet possibly disjoint from the input alphabet, may be entered or removed from the top of the list. Depending on the *input symbol*, the *top symbol on the stack* and the *state of the finite control*, a number of choices are possible. Formally speaking, the PDA M can be defined as a system $\{S, \Sigma, \Gamma, \delta, s_0, Z_0, F\}$, where

- S is a finite set of states;
- Σ is an alphabet called the *input alphabet*;
- Γ is an alphabet called the *stack alphabet*;
- δ, the transition function, is a mapping $S \times (\Sigma \cup \{\epsilon\}) \times \Gamma \rightarrow S \times \Gamma^*$;
- s_0 is a particular state of S called the *initial state*;
- $Z_0 \in \Gamma$ is a particular stack symbol called the *start symbol*;
- $F \subseteq S$ is the set of *accepting or final states*.

Placing restrictions on productions $\xi \rightarrow \omega$ of a phrase structure grammar, we receive a grammar, which is called *context-sensitive* and its language *context-sensitive language* (CSL). For instance, a production rule of the form $a_1 A a_2 \rightarrow a_1 \beta a_2$ with $\beta \neq \epsilon$ (ϵ is the *empty* set) looks almost like a context-free production rule, but it permits replacement of A only in the *context* a_1Aa_2. The counterpart automaton characterizing CSL is a *linear bounded automaton* (LBA).

Finally, the four classes of *recursive enumerable (r.e.)* languages, CSL, CFL and *regular* are characterized as languages of type 0,1,2,3, respectively. It has been shown that except for the empty string, the type-i languages properly include the type-$(i + 1)$ languages, for i = 0, 1, 2. This means that a CFL is equivalent to a CSL, but not every CSL is equivalent to a CFL. The same relation holds between CFLs and regular languages, as discussed previously, and between regular expressions and CSL. This is due to the *hierarchy theorem* and Chomsky's hierarchy which defines these classes of languages as the only potential models of natural languages.

The definition of a meaning driven query language automaton: Any meaning based constructed query in MDDQL is always a syntactically and semantically correct query. This is due to the fact that queries are synthesized from system suggested terms whose semantic consistency is a consequence of reaching the current query state. The MDDQA accepting MDDQL queries is defined as an abstract machine M = {Sq , Σq , Γq , δq , q_0, Pq , Fq }, where:

- Sq is a finite set of query states q_i, $i \geq 0$,
- Σq is the input alphabet
- Γq is a subset of Σq, called the *current query state*, realized by a *stack*,
- Pq is the set of propositional formulas pq each one expressing a *term-including precondition* for which a *truth assignment* holds
- δq , the transition function
- q_0 is a particular state of Sq called the *initial query state*,
- $F_q \subseteq S_q$ is the set of *accepting or final query states*.

A state transition diagram is depicted in Figure 1 in order to illustrate a state automaton in terms of query states. Since each query state is conceived as a subset of the set of alphabet terms T $\equiv I_A$, all query states {q_0, q_1 , q_2, q_3 , q_4, q_5} include alphabet terms presented with linguistic elements of some natural language. q_0 is the initial query state and q_f is the final one. Consequently, each subsequent state includes additional terms (application domain or operational) and is considered to be a more restrictive description of the previous instance(s).

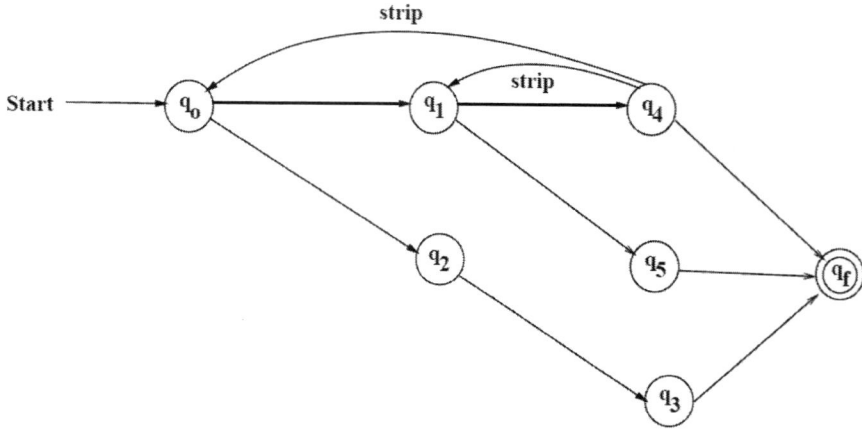

Figure 1: An exemplar state transition diagram for MDDQL.

However, the cardinality of the set of all potential query states is bounded by the countable finite set of all *proper directed paths* as expressed by the holding links of the graph representing the space of information objects I, as depicted by Figure 2.

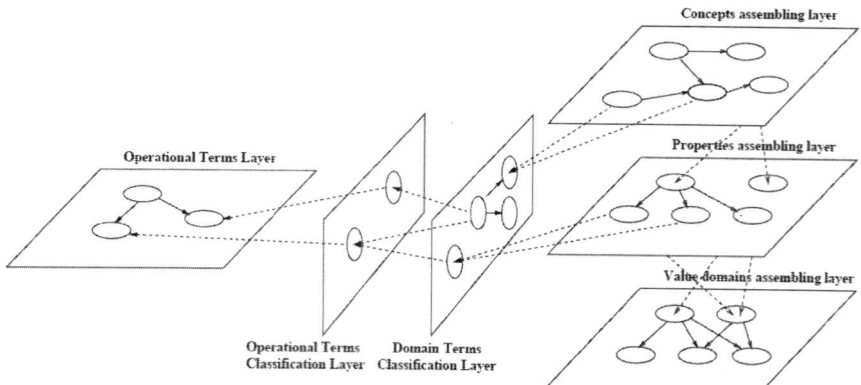

Figure 2: Connectivity of query language vocabulary terms.

The specified automaton can be considered an *abstract machine* which consists of an *input tape*, a *finite control* and a *head* which is moved by the finite control mechanism to the left or right on the input tape and sets the automaton to a particular (query) state. *Stacks* are also given, since the automaton needs to remember the set of current terms as well as those of the set of previously visited query states. This is similar to devices specified for *push-down automata*. Strip move is also allowed where the finite control mechanism sets the automaton to a previous query state and moves the head to the left of the input tape. This is similar to *deletion* operations over the current set of query terms.

Since query state transitions are based on semantically consistent input terms as suggested by the system as well as on end-user choices based on term interpretation and meaning, the MDDQA is conceived as an *interactive acceptor* of meaningful queries formulated in MDDQL. This allows *syntax/semantic* parsing of a query to be replaced with end-user/system interaction, whereby the semantics of query alphabet terms are taken into account during query construction. In the following table(s), an example of the specification of an automaton is provided, where the transition states are determined by the semantic roles of nodes and edges rather than any other values. This example, within the context of our MDDQL case study, illustrates how an automaton can act as an acceptor and language generator as well. In our case, the language being generated is a subset of SQL. Symbols S, F and W stand for the keywords SELECT, FROM and WHERE, respectively.

Type of INPUT EDGE (Parent, Child)	q_k = state to move on $k \geq 0, k \in N$	Auxiliary Stack (AS)
ETN (root)	$q_k = \{S_{k-1},$ $F_{k-1} \cup (t_{ETN}, token_{ETN}),$ $W_{k-1} \cup (C^\lambda_{tETN})^o = \{\}\}$ $\lambda = \lambda_{ETN}, o = o_{ETN}$	**push** ETN(root) to *AS*
(ETN_i, ETN_j) $i \neq j, j > i$ $i, j \geq 0, i, j \in N$	$q_k = \{S_{k-1},$ $F_{k-1} \cup (t_{ETNj}, token_{ETNj}),$ $W_{k-1} \cup (C^\lambda_{tETNj})^o = \{\}\}$ $\lambda = \lambda_{ETNi}, o = o_{ETNj}$	**push** ETN_j to *AS*
(ETN, PTN)	$q_k = \{S_{k-1}, F_{k-1}, W_{k-1}\}$	**push** PTN to *AS*
(PTN_i, PTN_j) $i \neq j, j > i$ $i, j \geq 0, i, j \in N$	$q_k = \{S_{k-1}, F_{k-1} W_{k-1}\}$	**push** PTN_j to *AS*
(PTN, nil)	$q_k =$ $\{S_{k-1} \cup (t_{peakETN}, PTN')^\omega$ $F_{k-1}, W_{k-1}\},$ **if** $\exists(\omega'), \omega' \neq \varnothing$ $\omega' \in popP^\omega$ **then** $\omega = \omega'$	$\forall PTN \in AS, pop(AS)$ $popP = \{PT_i,..., PT_j\}$ $PTN' = conc(token_{PTi},..., token_{PTj})$ $popP^\omega = \{PT_i^\omega,..., PT_j^\omega\}$ $peakETN = peak(AS)$

contd....

contd.

Type of INPUT EDGE (Parent, Child)	q_k = state to move on $k \geq 0$, $k \in N$	Auxiliary Stack (AS)
(PTN, VTN)	$q_k = \{S_{k-1}, F_{k-1}, W_{k-1}\}$	**Push** VTN to AS
(VTN_i, VTN_j) $i \neq j$, $j > i$ $i, j \geq 0$, $i, j \in N$	$q_k = \{S_{k-1}, F_{k-1}, W_{k-1}\}$	**Push** VTN_j to AS
(VTN, nil)	$q_k = \{S_{k-1}, F_{k-1},$ $W_k = \{C^\lambda_{tpeakETN} \cup$ $\cup((t_{peakETN}, PTN')^\lambda,$ $VTN_c{}^\omega)^\sigma\}$ **if** $\exists(\omega')$, $\omega' \in popV^\omega$ **then** $\omega = \omega'$ **if** $\exists(\sigma)$, $\sigma' = 0$, $\sigma' \in \{popP^\sigma \cup popV^\sigma\}$ **then** $\sigma = \sigma'$ **if** $\exists(\lambda')$, $\lambda' = < OR >$, $\lambda' \in popP^\lambda$, **then** $\lambda = \lambda'$	$\forall VTN \in AS$, $pop(AS)$ $popV = \{VT_i,..., VT_j\}$ $VTN_c = conc(token_{VTi},...,$ $token_{VTj})$ $popV^\omega = \{VT_i^\omega,..., VT_j^\omega\}$ $popV^\sigma = \{VT_i^\sigma,..., VT_j^\sigma\}$ $\forall PTN \in AS$, $pop(AS)$ $popP = \{PT_i,..., PT_j\}$ $PTN' = conc(token_{PTi},...,$ $token_{PTj})$ $popP^\sigma = \{PT_i^\sigma,..., PT_j^\sigma\}$ $popP^\lambda = \{PT_i^\lambda,..., PT_j^\lambda\}$ $peakETN = peak(AS)$
(ETN(*root*), RTN)	$q_k = \{S_{k-1},$ $F_{k-1} \cup (t_{RTN}, token_{RTN}),$ $W_{k-1} \cup (C^\lambda_{tRTN})^\sigma = \{\}\}$ $\lambda = \lambda_{RTN}$, $\sigma = \sigma_{RTN}$	**Push** RTN to AS
(RTN_i, RTN_j) $i \neq j$, $j > i$ $i, j \geq 0$, $i, j \in N$	$q_k = \{S_{k-1},$ $F_{k-1} \cup (t_{RTNj}, token_{ETNj}),$ $W_{k-1} \cup (C^\lambda_{tRTNj})^\sigma = \{\}\}$ $\lambda = \lambda_{RTNi}$, $\sigma = \sigma_{RTNj}$	**Push** RTN_j to AS
(RTN, PTN)	$q_k = \{S_{k-1}, F_{k-1}, W_{k-1}\}$	**Push** PTN to AS

ETN: Entity Term Node
RTN: Relation Term Node
PTN: Property Term Node
VTN: Value Term Node
PT: Property Term
VT: Value Term

7.2.4 Exemplary algorithms and data structures

A query construction is a move from a current query state to a more descriptive or refined one which is guided by decisions (state transitions) that depend on the current query context as well as the model used to represent meaning of terms. The end-user, however, is part of the finite control mechanism, and therefore partly determines which next state to move to through choices of terms based on the intentional meaning of terms. Therefore, the end-user can choose only those query states, which are semantically consistent with the current one.

The MDDQL query tree is defined as a set of interconnected *query term nodes* which are represented as objects and carry on all semantic information necessary for the inferences for further terms as well as the interpretation of the query tree (a submitted query) toward the generation of a database specific query language such as SQL. Operators or operations are also represented as objects. However, they are assigned to the query term nodes rather than being tree nodes themselves. The query tree consists of nodes which have been classified either as *Entity Set*, or *Relationship*, or *Property*, or *Value* node. The structure of a query tree, however, underlies some constraints as far as this classification schema is concerned.

In order to infer which terms make sense to be included in the set of proposed terms, two major steps need to be taken: 1) the location of the term within the vocabulary structure (e.g., cyclic graph), upon which a suggestion request has been activated on the blackboard, 2) determination of the context of the located term. The latter is determined in terms of a) the neighbourhood of the term under consideration as defined by a distance measure, b) the constraints which might hold for the inferred neighbouring terms. Since location of the related terms as well as the neighboured terms and their constraints are given in terms of unique term identifiers only, terms expressed with same words but having different meanings can be taken into consideration. The validity of constraints, however, can be checked either at the time of the inference of meaningful terms or at the time of submission of the query. This holds, for example, when mutually exclusive terms need to be detected when the query has been completed. For instance:

- Entity sets which contradict with entity sets to which they are connected through the logical operator *AND* within the query. For instance, *Discharged* and *Passed away* as sub-categories of *Patients* which are not semantically consistent within the same query, since a patient either survived the hospitalization or passed away. The same holds for values such as *done* AND *unknown* within the same set of suggested value terms.
- Property terms to be used in conditional clauses which are only valid under certain circumstances. For instance, *Reasoning for denial of thrombolysis* is valid only if *Thrombolysis* has not been used as restriction indicating that we are interested in patients who received thrombolysis as a kind of immediate therapy. Validity of properties is also checked upon the targeting location (hospital name) of the query.
- Operator terms which do not comply with the nature of the term under consideration.

In the following, state transition algorithms are described which realize the δ function among semantically consistent query states. In particular, we distinguish among three major cases of interaction strategy for the query construction:

- starting query construction (formulation) with *entity terms*
- starting with any arbitrary term from the set of domain alphabet
- terms $t \in I \, D$, attribute/value pairs for premises of classification rules.

General characteristics of the state transition algorithms are that

- all algorithms work with entry points represented by nodes belonging to one of the *concepts- properties-value domains* layers (Figure 2), i.e., only terms which belong to the domain alphabet are considered first.

The token < SELEC T > refers to the choices made by the end-user. Note that < SELEC T > tokens indicate the possible choice of the end-user for a term to extend the current context (state) of the query out of a set of suggested (inferred) terms. Since the alphabet terms are represented as *information objects*, it is also possible to have a look at additional properties of the terms such as *definition, measurement unit, icons*, etc. which provide a deeper insight into the *intentional meaning* of a particular term (concept, property, value), before selection.

Starting with entities. The first algorithm refers to the function of the automaton when the starting terms (entry points of the graph) are concept terms classified as *entities*, i.e., possible entry points are those nodes of I A which belong to the *concepts (assembling) layer*. In other words, a top- down navigation strategy of the *concepts-properties-value domains* assembling layers of the graph (Figure 2) is realized. The algorithm is illustrated with an example from the application domain of 'Information Management System for Mines Actions' some terms of which will accompany the algorithm description.

BEGIN Algorithm 1.0

Start query construction.
Suggest set of natural languages
e.g., {English, German}
< *SELECT* > a natural language in which terms should appear.
e.g., {English}

Suggest intitial entity terms to start with.
e.g., {Explosive devices, Country, Incidents}
< *SELECT* > an initial *entity term*
e.g., {Explosive devices}
Move to the initial query state q_0.
e.g., $q_0 = {Explosive devices}$
Set initial query state q_o to be the current state q_c

Repeat
< *SELECT* > a term in current query state $t_s \in q_c$
while there are outgoing edges from $t \in q_c$ to some nodes (terms)
with $t \in I_D$ **do**
get connected node
if connected node is not already in q_c **then**
    if there is a set of preconditions assigned to node **then**
        if set of preconditions is satisfied
            add node to list of inferred nodes (terms) T_I
        else ignore node
    else add node to list of suggested nodes
else ignore node

end of while

Suggest inferred terms T_I to end-user for selection as information objects
e.g., {*caused by, manufactured in, category, nomenclature, material, shape, purpose*} *when* $t_s \in q_c$ *is* (*Explosive devices*}
< *SELECT* > term(s) from T_I
e.g., {*manufactured in, purpose*}
Add selected term(s) to the current query state q_c by assigning them to t_s
e.g., {*Explosive devices, manufactured in purpose*}
Move to q_j which is a subsequent query state of q_c
Make q_j the current query state $q_c \equiv q_j$

until final query state is reached
e.g., {*Explosive devices, purpose, manufactured in, country, name, United Kingdom, Switzerland*}

END Algorithm 1.0

Following this example, the term {*country*} belongs to the set of suggested (inferred) terms T_I when t_s = {*manufactured in*}, $t_s \in q_c$ and $q_c \equiv$ {*Explosive devices, purpose, manufactured in*}, whereas the term {*name*} belongs to the set of suggested (inferred) terms T_I when t_s = {country}, $t_s \in q_c$ and $q_c \equiv$ {*Explosive devices, purpose, manufactured in, country*}. Similarly, the terms {*United Kingdom, Switzerland*} belong to the set of suggested (inferred) terms TI when t_s = {name}, $t_s \in q_c$ and $q_c \equiv$ {Explosive devices, purpose, manufactured in, country, name}.

Note that the final query statement of the given example, conceived as a set of query terms, is equivalent to the expression of the question *What is the purpose of explosive devices manufactured in countries named as United Kingdom and Switzerland?* formulated in English. However, starting with terms classified as *entities* reflects the philosophy of establishing a central theme or subject of discussion such as *talking about "explosive devices"* before proceeding with further terms which might refine the intended question or query.

The general case. The second state transition algorithm refers to the function of the automaton when any domain alphabet term can be considered as entry point, i.e. the end-user can start with an arbitrary query term which belongs to the set of domain alphabet terms. In other words, the entry node of the graph (Figure 6.3) could be any node which belongs to one of the assembling layers of the graph, i.e., *concepts, properties, value domains*. The token < T Y P E > also refers to a term typing action of the end-user.

This algorithm reflects the philosophy of constructing a query when the meaning of the entry term has already been clarified completely. For instance, starting with the term *purpose*, i.e., let's talk about "purpose", the system suggests *Mines clearing action* and *Explosive devices* as potential clarifications of "purpose", i.e., to give an answer to the question *are we talking about the purpose of Mines clearing action or that of Explosive devices?* before proceeding with the further refinement of the query/question.

BEGIN Algorithm 2.0

Start query construction.
Suggest set of natural languages
e.g., {English, German}
< *SELECT* > a natural language in which terms should appear.
e.g., {English}

< *TYPE* > the name of the term to be clarified (entry term E_t)
e.g., {purpose}
while there is an entry term E_t **do**
    while there are term nodes in I_D the name of which matches
E_t **do**
e.g., {TUI=500, TUI=600}
get matched term node
get all connected nodes with *incoming edges* - navigating
backwards I_D
put into set of clarifying term nodes C_T
end of while
suggest set of clarifying term nodes C_T to the end-user
e.g., {Mines clearing action, Explosive Devices}
< *SELECT* > a clarifying term C_T out of set C_T
e.g., {Explosive Devices}
    Add pair of connected term nodes to list of *visited path*
    Set C_T to term to be clarified E_t
end of while
Make out of list of term nodes from *visited path* the current query
state q_c *e.g., {Explosive Devices, purpose}*

and continuing similar to the previous case (algorithm 1.0):

Repeat
$< SELECT >$ a term in current query state $t_s \in q_c$
e.g., {Explosive Devices}
while there are outgoing edges from $t \in q_c$ to some nodes (terms) with $t \in I_D$ **do**

> **get** connected node
> **if** connected node is not already in q_c **then**
> > **if** there is a set of preconditions assigned to node **then**
> > > **if** set of preconditions is satisfied
> > > > **add** node to list of inferred nodes (terms) T_I
> > > **else** ignore node
> > **else add** node to list of suggested nodes
> **else** ignore node

end of while

Suggest inferred terms T_I to end-user for selection as information objects
e.g., {caused by, manufactured in, category, nomenclature, material, shape}
$< SELECT >$ term(s) from T_I
e.g., {manufactured in}
Add selected term(s) to the current query state q_c by assigning them to t_s
e.g., {Explosive devices, manufactured in, purpose}
Move to q_j which is a subsequent query state of q_c
Make q_j the current query state $q_c \equiv q_j$

until final query state is reached
e.g., {Explosive devices, purpose, manufactured in, country, name, United Kingdom, Switzerland}

END Algorithm 2.0

   Satisfaction of preconditions. In the following, a description of the algorithm to support the truth value semantics is given, as related to preconditions assigned to the domain of alphabet terms. Assuming that a node with assigned precondition(s) has been reached during navigation through the space of information objects I_D as described by algorithm **1.0**, consideration of this node within the set of suggested (inferred) nodes (terms) T_I depends upon the determined truth value semantics of preconditions as follows.

An example will also be given on the basis of a scenario where it is assumed that

- the terms {*Patient, Coronary angiography, atypical angina, gender, female*} are within the current query state q_c,
- the precondition {*(Patient, atypical angina)* $\wedge \neg$ *(gender, female)*} is assigned to node (term) {*risk factors*},
- the terms {*gender, stress test, risk factors*} have incoming edges from T_s = {*atypical angina*}.

BEGIN Algorithm 3.0

get term circumstances (set of assigned preconditions) P_T
e.g., $P_T \equiv \{p_t = $ *(Patient, atypical angina)* $\wedge \neg$ *(gender, female)*}
get normal form of P_T (conjunctive or disjunctive)
e.g., $P_T \rightarrow C\,N\,F$
assign default truth value *true* to P_T, i.e. $P_T \rightarrow true$
while there are preconditions in P_T (circumstances) assigned to term T and *extLooping = true*
do

      get precondition p_t
      e.g., $p_t = $ *(Patient, atypical angina)* $\wedge \neg$ *(gender, female)*
      get normal form of precondition (conjunctive or disjunctive)
      e.g., $p_t \rightarrow C\,N\,F$
      assign default truth value *true* to p_i, i.e. $p_i \rightarrow true$
      while there are literals $p_i \in p_t$, $i \geq 0$ and *intLooping = true*
         get literal p_i

Following the example given above, despite the fact that there is an incoming edge to the term {*risk factors*} from the term {*atypical angina*}, the term {*risk factors*} will not be considered as an element of the set of suggested (inferred) terms. This is due to the fact that T_s = {*atypical angina*} is the activated term by the end-user *within the current query state* q_c = {*Patient, Coronary angiography, atypical angina, gender, female*} and the assigned precondition to the term {*risk factors*} is not satisfied within this query state. This reflects the situation of exceptional handling of assignments as found in the real world. In other words, the assigned precondition as presented previously would express the fact that *risk factors are considered as property of patients with atypical angina only if their gender is male* or the equivalent expression *risk factors is considered as property of patients with atypical angina only if their gender is NOT female.*

> *e.g., p_0 = (Patient, atypical angina) (first pass)*
> *e.g., p_1 = ¬ (gender, female) (second pass)*
> **if** it holds that $t_n \in p_i$ and $t_n \in q_c$ **then**
> > **if** p_i is *negated* **then**
> > > **assign** truth value *false* to p_i
> > > *e.g., $p_1 \rightarrow$ false (second pass)*
> >
> > **else**
> > > **assign** truth value *true* to p_i
> > > *e.g., $p_0 \rightarrow$ true (first pass)*
> >
> > **if** $p_i \rightarrow$ *false* and normal form of precondition p_t is conjunctive **then**
> > > **set** *intLooping = false*
> > > **assign** $p_t \rightarrow$ *false*
> > > *e.g., $p_t \rightarrow$ false (second pass)*
> >
> > **else if** $p_i \rightarrow$ *true* and normal form of precondition p_t is disjunctive **then**
> > > **set** *intLooping = false*
> >
> > **else**
> > > **assign** to p_t the truth value of p_i
> **end of while**

> **if** $p_t \rightarrow$ *false* and normal form of P_T is conjunctive **then**
> > **set** *extLooping = false*
> > **assign** $P_T \rightarrow$ *false*
> > *e.g., $p_t \rightarrow$ false*
>
> **else if** $p_t \rightarrow$ *true* and normal form of P_T is disjunctive **then**
> > **set** *extLooping = false*
>
> **else**
> > **assign** to P_T the truth value of p_t

end of while
END Algorithm 3.0

Terms dependency graphs. From the nature of preconditions as a set of nodes (terms) within the space of information objects I_D, it becomes obvious that preconditions render edges or nodes of the multi-layered graph underlying the connectionism model of the representation of context of *non-deterministic* terms. Their validity depends upon the satisfaction of these preconditions given a particular query state q_c. However, the time of examination of satisfaction of preconditions is a matter of efficiency and terms presentation (suggestion) policy. Since preconditions are examined in order to decide if terms appearing in the same query statement are semantically consistent, there are, in general, two alternatives:

- A progressive strategy where all nodes (terms) are inferred and suggested to the end-user by taking into consideration only the connecting edges, whereby all preconditions related to the query state are examined retrospectively,
- A conservative strategy, where preconditions are examined during inference of nodes (terms) to be suggested, i.e., only those nodes (terms) are inferred (suggested) which have satisfied assigned preconditions.

In both cases, the final query state includes only those terms which are semantically consistent with each other. However, the *conservative strategy* is more complicated as far as *decidability* of satisfaction of preconditions is concerned. Since preconditions are expressed in terms of other nodes within the same graph, the time needed to examine the satisfaction of preconditions depends on the size of sub-graphs which are involved as a basis of the current query context.

Formally speaking, all "terms dependency graphs" are defined as set of *proper paths* of nodes (terms) $\{t_1, \ldots, t_n\}$ which belong to some final query state $q_f \in F_q$, with $t_i = t_j$ for all i, j $\leq$ n, i.e., no cycles are allowed in order to guarantee decidability of satisfaction of preconditions. However, the decision which nodes (terms) constitute the "terms dependency graphs" is taken on the basis of expression of preconditions. The TDGs provide the basis for the timely decision of satisfaction of preconditions. The following algorithm provides an insight into the construction of "term dependency graphs" (TDGs).

BEGIN Algorithm 4.0
For each node or term in a given final query state
to which a precondition has been assigned

> **put** this node t_p into *TDG*
> **For each** literal in the precondition
> > **get** first node (term) t_l
> > **put** t_l into *TDG*
> > **connect** t_l to t_p within *TDG*

END Algorithm 4.0

In case a *progressive* strategy has been chosen, all we need is to traverse the TDGs backwards, i.e., starting from leaf nodes, as related to the final query state. For each leaf or inner node within a particular TDG, we examine the preconditions assigned to them by applying algorithm **3.0**. In order to illustrate the effects of examining preconditions retrospectively, consider the scenario of having in a query (final) state the terms *{Patient, Coronary angiography, atypical angina, gender, female, risk factors, less than 2}*. A retrospective examination of the precondition *{(Patient, atypical angina) ∧ ¬ (gender, female)}* will cause the elimination of *{risk factors, less than 2}* from the final query state, since *{risk factors}* is not semantically consistent within this query state.

In case a *conservative* strategy has been chosen, we need to consider examination of preconditions in advance. Therefore, in order to avoid undecidability, a "dependency order" of examination of preconditions must be taken into account. The decision as to which preconditions, and therefore, which nodes (terms) are decidable is taken on the basis of the partial graph as resulted by the outgoing edges of *concept terms* of a particular query state within the information space ID and the elaborated TDGs. We will call the resultant partial graph the *scope of relevant preconditions* when a particular query state is given.

In general, the following algorithm determines "dependency orders" within sets of inferred (suggested) nodes (terms).

BEGIN Algorithm 5.0
Construct partial graph of I_D as *scope of relevant preconditions* S_p
e.g., the partial graph resulted from {Patient, Coronary angiography, Atypical angina}
get TDGs
while there is a TDG
get TDG
if all nodes (terms) in *TDG* are members of S_p **then**
e.g., {Gender → Risk factors → Stress test}
   exclude all inner or leaf nodes in TDG from set of suggested nodes (terms)
   e.g., exclude {Risk factors, Stress test}

turn all incoming edges to inner or leaf nodes of *TDG* within S_p to "**red**"
e.g., turn edges {Atypical angina → Risk factors} and {Atypical angina → Stress test} to "red"

while there are outgoing edges from root nodes of *TDG*
    set intermediately connected nodes to *potential nodes*
    e.g., potential node(s) are {Risk factors} (first pass)
    e.g., potential node(s) are {Stress test} (second pass)
    set *potential nodes* to root nodes of *TDG*
    e.g., root node is {Risk factors} (first pass)
    e.g., root node is {Stress test} (second pass)
    turn all "red" incoming edges for *potential nodes* within S_p to "green"
    e.g., edge {Atypical angina → Risk factors} turned to green (first pass)
    e.g., edge {Atypical angina → Stress test} turned to green (second pass)
    i.e., enforcement of node consideration (precondition) checking by
    next activation of t_s = {atypical angina}
  end of while

end of while

END Algorithm 5.0

Meaningful operations: So far, the behavior of the inference engine and/or meaning driven querying automaton has been described in terms of the query alphabet (application) domain, where inferences concerning operational terms have been excluded. Allowing operational terms to be members of a particular query state, the inference engine needs to be extended by algorithms operating over a given query state of domain alphabet terms and also infer (suggest) meaningful operations to the end-user.

Inferences of sets of meaningful operations or operators are drawn on the basis of connecting edges holding among *classes or collections* of term instances. Therefore, the inference of operational term instances is a matter of consideration of connecting edges leading to the classification layers of domain and operational terms, respectively, as well as the preconditions assigned to operational term instances as information objects.

In the following, we refer to the *class or collection* of a term instance as the *role of term*. Since more than one term or role might be relevant for an operation/operator, we refer to the *context of roles* as the set of all roles assigned to domain terms. Algorithm **6.0** provides the inference mode of sets of operational terms to be suggested for assignment to domain terms. Illustrating examples are given based on the various application domains.

BEGIN Algorithm 6.0

If q_c is the current query state
e.g., {Patient, Coronary angiography, atypical angina, age, 48, gender, male, stress test, positive, risk factors, less than 2}
and t_s is the activated term by the end-user
e.g., t_s = {48}
get the *context of roles* r_{ts} of t_s
e.g., r_{ts} = {Domain Value, Atomic Value, Quantitative Value}
get the *roles* $R_{tf} = \cup r_{tf}$ of operational terms associated with r_{ts}
e.g., R_{tf} = {Comparison Operator}
set the operational term instances $t_f R_{tf}$ to list of candidate operational terms
e.g., {OP3, OP4, OP5, OP6, OP7, OP9, OP10, OP11}
while there are operational term instances t_f in the list of candidate operational terms
        if there is an assigned precondition p'_t to t_f **then**
        e.g., p_{op10} = (Atomic Value ∧ Quantitative Value)
        e.g., p_{op11} = ¬ (Quantitative Value)
            if p'_t is satisfied due to definitions 2.5, 2.6
            then
                add to list of inferred (suggested) operational terms
        else
            ignore t_f
            e.g., ignore {OP10, OP11}
        else
            add to list of inferred (suggested) operational terms
end of while
suggest list of inferred (suggested) operational terms
e.g., {OP3, OP4, OP5, OP6, OP7, OP9}
< SELECT > operational term
e.g., {OP4}

END Algorithm 6.0

In this case, the final query state would be *something* equivalent to the query *"Which patients are candidates for coronary angiography with atypical angina having age which is equal or more than 48, positive stress test and less than 2 risk factors?"*.

The behavior of algorithm **7.0** is such that preconditions, if there are any, are examined first at the classification layer of operational terms and subsequently at the term instances layer. This reflects the issue that classes of operations must be defined together with their expected behavior in terms of the nature and number of arguments expected. However, certain operations within the same class might show different or exceptional behavior in terms of more specific categories of arguments they need to operate meaningfully.

BEGIN Algorithm 7.0

q_c is the current query state
e.g., {Patient, Coronary angiography, atypical angina, age, stress test}
t_s is the activated term by the end-user
e.g., t'_s = {Patient} or t''_s = {Age} or t'''_s = {48} the activated term

get the *roles/classes* of t_s, r_{ts}
e.g., {Entity, Concept} when t'_s,
{Single Property, Categorical Variable} when t''_s,
{Domain Value, Atomic Value, Quantitative Value} when t'''_s
- in the following, only t'_s will be considered - get the *roles/classes* R_{tf} =
$\cup r_{tf}$ of operational
terms associated with r_{ts}
e.g., R_{tf}= {Interrelationship Function, Bivariate Function, Multivariate Function}
through {Entity} → {Interrelationship Function}
and through {Interrelationship Function} → {Bivariate Function, Multivariate Function}

set R_{tf} to list of candidate operational term classes
while there are operational term classes r_{tf} in R_{tf}
    if there is an assigned set of preconditions P_T to R_{tf} **then**
    e.g., P'_T = {p_1 = (Single Property), p_2 = (Single Property)} →
    {Bivariate Function}
    i.e. expecting two variable as input
    e.g., P''_T = {p_1 = (Single Property), p_2 = (Single Property),
    p_2 = (Single Property)} → {Multivariate Function}, i.e. expecting three variables as input

> **if** P_T is satisfied due to definition 2.7
>   **then**
>   *e.g., only P'$_T$ is satisfied*
>   **add** r_{tf} to list of inferred (suggested) operational term classes
>   *e.g., Bivariate*
>   **else**
>   ignore r_{tf}
>   *e.g., ignore {Multivariate}*
>   **else**
>   **add** to list of inferred (suggested) operational term classes
> **end of while**

Start examining preconditions at the instance layer of operational terms.

> **for** each inferred (suggested) operational term classes
>   **set** the operational term instances $t_f \in R_{tf}$ to list of candidate operational terms
>   *e.g., {OP100, OP101, OP102, OP103...}*
>   **while** there are operational term instances t_f in the list of candidate operational terms
>   **if** there is an assigned set of preconditions P_t to t_f **then**
>   *e.g., P$_t$ = {p$_1$ = (Single Property ∧ Numerical Variable),*
>   *p$_2$ = (single Property ∧ Numerical Variable)} → {OP101, OP102, OP103}*
>   **if** P_t is satisfied by definition then 2.7 **then**
>   **add** to list of inferred (suggested) operational terms
>   **else**
>   ignore t_f
>   *e.g., ignore {OP101, OP102, OP103}*
>
> **else**
>   **add** to list of inferred (suggested) operational terms
> **end of while**

suggest list of inferred (suggested) operational terms
e.g., {OP100}
< *SELECT* > operational term
e.g., {OP100}

END Algorithm 7.0

Satisfaction of preconditions for operational terms has been considered, up to now, as a "black box". The following algorithm incorporates the reasoning underlying the satisfaction of a set of preconditions for operational terms, whether they are posed to classes or instances of operational terms. The algorithm is based on constraints over operational the set of preconditions are examined against *contexts of roles* as the power set of roles (classes) within which a particular domain term is a member.

In other words, the algorithm sets the truth value of the set of preconditions PT to *true* only if each precondition as an assembly of classes/roles is set to *true* based on the assumption that all classes/roles (literals) of a precondition also appear within a particular context of roles. Since the number of preconditions indicates the number of expected arguments, PT is also turned to *false*, if it happens that not enough contexts of roles CTq have been considered—less arguments than expected—or more contexts of roles CTq than preconditions pt ∈ PT-more arguments than expected.

BEGIN Algorithm 8.0

q_c is a give query state
get activated term t_s
construct context (set) of roles C_{Tq} for t_s
if {Entity} or {Complex Property} ∈ C_{Tq} **then**
      apply recursively until {Single Property} ∈ C_{Tq}
              visit connected node (term) t'_s to t_s in partial graph q_c
              construct context (set) of roles C'_{Tq} for t'_s
              if {Single Property} C'_{Tq}
                  insert C'_{Tq} into set of context of roles ∪ C_{Tq}
              else ignore C'_{Tq}
else
        insert C_{Tq} into set of context of roles ∪ C_{Tq}
Start examining set of preconditions (circumstances) against set of contexts of roles ∪ C_{Tq}.

get set of preconditions P_T
while there is a precondition $p_t ∈ P_T$
      while there are any *unvisited* C_{Tq} in ∪C_{Tq} and $C_{Tq} ≠$ {}
            get next p_t
            get next *unvisited* C_{Tq}
            if it holds that literals $q_j ⊆ C_{Tq}, j ≥ 0$ **then**

$$\textbf{mark } C_{T_q} \text{ as } visited$$
$$\textbf{set } p_t \to true$$
else

$$\textbf{get } \text{next } unvisited \ CTq$$

end of while

while

if it holds that $\forall p_t,\ p_t \to true$ and
there are no *unvisited* C_{T_q} in UC_{T_q} **then**
(exclusion of case where more arguments than the expected ones
are in a given query state)

$$\textbf{set } P_T \to true$$

else

$$\textbf{set } P_T \to false$$

END Algorithm 8.0

7.3 Summary and Major Conclusions

In order to meet the goal of *high usability*, the driving force behind the presented querying approach has been *intelligence* in terms of query context recognition, system-aided formulation of queries and generation of answers. The key issue has been the *computer understanding of a user's world* and not *user's understanding of a computer world*.

Furthermore, it has been shown that a key factor for increasing *computer understanding of a user's world* can be *computational semantics*, an often neglected issue in computer science. Perhaps this is so because of the difficulties and/or complexity of injecting *semantics* into a *world of symbols*. However, without representation of meaning and semantics, no software could ever infer and anticipate or assess worlds, if these worlds are not known to it. Moreover, without representation of meaning and semantics, knowledge exchange becomes difficult, especially, when different domain scientists are involved in a case study or research assisting information system. Even within the same scientific domain, communication among experts depends on common understanding is enabled or supported by the system.

The representation model of meaning of a query alphabet, we have presented, resulted from the lack of adequate conceptual/data modelling tools or languages to cope with concrete value domains of attributes and multi-lingual expressions, through preconditions holding between entities and attributes or between attributes and value domains. Moreover, the capability of expressing preconditions, which turn any statement of value or attribute assignment to be *relative* and context dependent was a major contribution to the inference logic for meaningful suggestions. On the other hand, a major difficulty has been the consideration of *variables* within literals of preconditions.

To sum up, "smart" inferences are not only bound to connectionism models but also to "constraints" and context recognition from a systemic and knowledge representation point of view. The latter adds complexity to the inference engine but is considered crucial. If we would like to talk in terms of "intelligent" inferences, from a human-oriented point of view, we must also be able to talk about "context based reasoning". The latter might be a key issue for addressing "meaningful information" in the "age" of massively producing and disseminating information and knowledge through the Internet.

CHAPTER 8

Multi-Lingual Querying and Parametric Theory

8.1 Introduction

The World Wide Web has evolved into a tremendous source of knowledge and it continues to grow at an exponential rate. According to Global Reach, 68% of Web content is in English. While a vast amount of information is available on the internet, much of it is inaccessible because of the language barrier. In order to make this knowledge resource available to non-English speaking communities, search applications that use native language interfaces are needed. Users should be able to specify search queries in their own language in order to retrieve documents and useful information from the internet.

In order to provide cross-lingual content to end users, a number of research questions have to be answered. For example, how do we make the approach flexible and scalable so that multiple languages can be accommodated? How do we interface with existing commercial search engines such as Google and summarize the relevant information sought by the user in his or her language? Good solutions to these problems are needed in order to tap the vast internet resource. While some of the existing search engines support limited translation capabilities, the potential to tap multilingual Web resources has not been realized. Current search engines are geared primarily towards English and have several limitations. In addition to the language constraint, the search engines use different syntax and rely on keyword based techniques, which do not comply with the actual needs and requirements of users as reported by (Amer-Yahia et al. 2005).

A similar set of issues is being addressed by the Knowledge Management and Decision Support communities. In the current global economy, multinational organizations have become commonplace and many companies have acquired or partnered other companies in different countries. This invariably results in the scattering of data, information and knowledge across multiple divisions in different languages. For example, when Daimler Benz and Chrysler merged to become Daimler Chrysler, it started an organization wide knowledge management initiative in order to restructure the knowledge assets. Managing this knowledge poses considerable problems particularly in accessing and retrieving relevant knowledge from disparate repositories because of language barrier. Developing appropriate search tools that can obtain pertinent information and documents from multilingual sources is very essential for supporting the decision making process at the organizational level.

A few techniques have been reported in the Multi-Lingual Information Retrieval (MLIR) literature that facilitates searching for documents in more than one language. Typically, two approaches are used in MLIR—translating the query into all target document languages or translating the document information to match the query language. As pointed out in (Zhou et al. 2005), the output quality of these techniques is still not very satisfying because they rely on automated machine translation and do not consider contextual information to perform word sense disambiguation correctly. In addition, these techniques have been developed, based on standard TREC collections, which are somewhat homogeneous and structured. However, the Web documents are highly heterogeneous in structure and format. Hence, multilingual querying on the Web is more complicated and requires a systematic approach. Consequently, a concept based query language enables querying through the description of concepts rather than simply using terms as part of a keywords based search task. This description might include associations with other concepts, i.e., querying by associations, either through taxonomies, e.g., engines as part-of cars, or interactions among concepts or instances, e.g., cars manufactured in Detroit.

The overall objective of this chapter is to discuss an indicative integration of the salient features of Web search technology with distributed database querying and information retrieval, with respect to multi-lingual aspects of querying, as an exemplary approach for enabling multi-lingual querying systems. In particular, the chapter discusses an indicative methodology for concept based multi-lingual querying on the Web, as well as architecture and a proof-of-concept prototype that implements the methodology.

8.2 Baseline Methods and Algorithms

8.2.1 Background theory

Universal grammar and parametric theory in linguistics: The idea of a universal grammar underlying all natural languages is as old as Chomsky's minimalist program and his school of grammar known collectively as transformational-generative grammar. This idea relies on principles and parameters by which children learn a language innately and which has even been taken to reduce grammar to a genetic mechanism for acquiring a language. To this extent, it also triggered controversial philosophical discussions on how close the universal grammar and a potential Language Acquisition Device are supposed to be.

Despite all controversial discussions and apart from any biological implications they embarked on, the idea of a universal grammar based on principles and parameters is motivated by the attempt to illuminate two puzzles: how adults can use language creatively and how it is possible for children to learn their first language. The term "creative" means the inherent capability for speakers to utter and understand any of an infinite set of possible sentences, many of which are created anew and on an appropriate occasion.

Given that the set is infinite and the utterances may contain recognizable mistakes, possible sentences cannot be memorized simply. They should be the result of a set of mentally represented rules, which appear to be surprisingly complex and similar from language to language. The complexity of this knowledge, however, gave rise to the problem of how we can learn a language with a few years at our disposal. An answer has been given in terms of a language faculty, which must result from the interplay of the innate and the learned, leaving aside the question of what is precisely innate.

The framework known as "principles and parameters" (Baker 2001) gives some explanation and provides knowledge of how languages can differ from each other. Parameters are conceived like switches, which can be set one way or the other on the basis of a simple evidence of the language we are dealing with. Parametric Theory is considered the unification theory in Linguistics in a role similar to that of Mendeleyev's Periodic Table in Chemistry or the role of the "grand theory and unification formula" in Physics. To this extent, a combination of certain parameters determines the grammatical pattern of a natural language and, therefore, its semantics and syntax, even for those natural languages, which are unknown or undiscovered as yet.

The products of the language faculty, however, such as representations of the sounds and meaning of sentences, derive many of their properties from the interfaces with which they have to interact: systems of perception and production, on the one hand, and systems of thought and memory on the other. A crucial goal has always been to determine what is unique to language and humans. In our approach, we are concerned with what is unique to language rather than to humans. Our approach parallels the products of the language faculty, however, from a mechanistic point of view. The proposed system of perception (parsing) and production (translation) of multi-lingual queries is equipped with systems of thought and memory (partially constructed queries, contextual knowledge, linguistic ontology such as WordNet, linguistic parameters, etc.), which are common to all natural languages.

It is worth noting that the interactive aspects of the proposed approach is an important means of deriving the meaning of queries either for query completion or translation. The main reason is that it goes beyond "yet another human-computer interface" into the very principles of a Turing Machine as a theoretical (mathematical) foundation of the computing discipline.

Automata theory: All classes of languages can be specified formally by an automaton, which lies within the range of Finite State Automaton (FSA) and Turing Machine (TM), if acceptance of a sequence of words is emphasized. The specification of all kinds of automata, however, underlie the same philosophy having, for example, some finite set S_q of states, some alphabet Σ called the *input alphabet,* a transition function δ as a mapping $S \times \Sigma \rightarrow \Sigma$, some start state s_0 and some set of accepting or final states $F \subseteq S$. In other words, M can be roughly defined as $M=(S_q=\{q_1,...,q_n\}, \Sigma, \delta, s_0, F)$. An underlying assumption is that the specification elements of the language need to be known *a priori* and bound to a particular language.

Alternatively, a kind of grammar in terms of

- a finite set N of *non-terminal symbols*
- a finite alphabet Σ of *terminal symbols*, disjoint from N
- a start symbol S, and
- a finite set of some set P of productions of the form $\xi \to \omega$

is usually specified when the emphasis is put on *generation* rather than *acceptance of sequences of words*. To this extent, a language $L(G)$ can be defined by a grammar $G = \{N, \Sigma, S, P\}$ meaning that all words can be derived by S by repeated applications of the productions P.

Generation might be turned into an *acceptance* mechanism when *parsing*. Therefore, we might formally define the parser of a language $L(M)$ in terms of an automaton M, which is equivalent to some grammar G, meaning that L consists of all those words in Σ^*, with Σ^* the set of all words, which can be constructed out of elements of Σ and be accepted by some automaton M.

From a knowledge representation point of view, either the whole set of production rules P or a particular set of words in Σ^* as accepted by M can be represented by a tree, which is called *derivation* or *parse* tree. Normally, the vertices of a derivation tree are labeled with terminal or non-terminal (variable) symbols.

Based on these principles, we embark on, the formal specification of a *concept based, cross-lingual web* query language, which we will call *MDDQL* in the rest of the chapter. The holding premises are (a) being independent of a particular natural language, (b) being easily adaptable to any natural language. In order to meet these objectives, both *acceptance* and *translation/generation* are driven by the same conceptual structure, which is meant to be defined as a multi-dimensional space. The dimensions are set by universal features holding for all kinds of natural languages, in other words, some atoms of languages as expressed in terms of linguistic *parameters*. Example parameters are *word type order*, e.g., *Subject-Verb-Object (SVO), Subject-Object-Verb (SOV), head directionality*, etc.

Formally speaking, a cross-lingual *MDDQL* is defined as $MDDQL(M_L(V \subseteq P_1 \times ... \times P_n, A \subseteq L))$ or $MDDQL(G_L(V \subseteq P_1 \times ... \times P_{n_{1,1}} A \subseteq L))$ meaning that the underlying automaton M or grammar G is a *functional mapping* of an input alphabet (set of words) A as a subset of a lexicon L for a particular language and a set or subset V of the Cartesian product $P_1 \times ... \times P_n$ of values of a set of the applicable parameters $P = \{P_1, ... P_n\}$, which roughly characterize the given natural language L in which a concept based query has been constructed, to the specification elements of M or G.

The mathematical notation of the *MDDQL* automaton for cross-lingual, concept based querying takes the form $M = \{S_q(V), \Sigma(A), \delta(V), s_0(A) F(V)\}$ rather than the conventional $M = (S_q, \Sigma, \delta, s_0, F)$, reflecting the functional dependency of potential states, input alphabet, transitional function, start symbol and set of finite acceptable states, respectively, from the multi-dimensional space V, as formed by a set of linguistic parameters (universals) and a corresponding lexicon A. In terms of a grammar, it takes the form $G = \{N, \Sigma(A), S(A), P(V)\}$ rather than the conventional form $G = \{N, \Sigma, S, P\}$.

To this extent, there is no need to specify a new automaton or grammar each time a natural language is being added as a means of querying interaction. Each natural language needs to be positioned only within the n-dimensional parametric space. Therefore, besides lowering the overall system complexity, *flexibility* and *scalability* become an important issue too, even if natural languages might be addressed, the grammars of which are unknown.

On the other side, the same data structure, i.e., the n-dimensional parametric space is used for translating or generating a target natural language specific query, since the submitted query is always reflected by the MDDQL high level, conceptual query tree, which is a natural language independent conceptual structure. Formally speaking, the *derivation* or *parse* tree, as an outcome of a successfully parsed query, is different from conventional parse trees, where the vertices are labeled with *terminal* or *non-terminal* (*variable*) symbols. An *MDDQL* parse tree is actually a connected acyclic graph $G_{MDDQL} = \{V, E\}$ in which all vertices are connected by one edge.

The vertices of $G_{MDDQL} = \{V, E\}$, however, are not classified as representing exclusively *terminal* or *non-terminal* symbols. They represent words (terminal symbols) together with their syntax or semantic roles. To this extent, $G_{MDDQL} = \{V, E\}$ is independent of a particular set of production rules *P* and, therefore, natural language. It reflects compositional semantics of the query. For example, $G_{MDDQL}=\{(<v_1: car\ (noun,\ class),\ v_2:\ wheel\ (noun,\ class)>,\ e:\ "part\text{-}of")\}$ is the parse tree of both *"car wheels"* or *"wheels of cars"*, in a Japanese and English word type order, respectively.

8.2.2 An example

Let one consider, for example, that one is interested in parsing the following query *Q* as expressed in three different languages:

> *"big children chasing dogs of the neighbors"* (English)
> *"big children-SU the neighbors dogs-OB chasing"* (Japanese)[1]
> *"Dogs-OB big-SU chasing children-SU neighbors-OB"* (Walrpiri - Australian Dialect)

and submit it for search on the Web or a database. Deciding the correctness as well as identifying the meaning of each one of those queries in terms of subjects (classes or instances), objects (classes or instances), adjectives (properties), verbs (relationships), etc., we need to consider the parameter $P_1 = word\ type\ order$ as one parsing dimension, with dimensional points or values the patterns *subject-verb-object, subject-object-verb, no word type order*, which apply to *English, Japanese, Walrpiri*, respectively.

To this extent, if $M_{L=English}(v=\{P_1 = subject\text{-}verb\text{-}object\} \in V, A \subseteq L)$, then the starting symbol S_0 could be any *English* word from *A* standing for a noun or adjective, whereas the parser is expecting a verb phrase to follow the noun phrase. However, when $M_{L=Japanese}(v=\{P_1 = subject\text{-}object\text{-}verb\} \in V, A \subseteq L)$ holds, the starting symbol S_0 could be any *Japanese* word from *A* standing for a noun or adjective and the parser expecting another noun phrase, this time the *object* of the phrase, to be followed by a verb phrase. In the third case $M_{L=Walrpiri}$ $(v=\{P_1 = no\text{-}word\text{-}type\text{-}order\} \in V, A \subseteq L)$, where special markers are used in order to distinguish the semantic roles of words within a *Warlpiri* sentence, the parser should be guided according to the indications of those special markers.

---

[1]The suffixes *-SU* and *–OB* are used to indicate the position of subjects and objects within the query or sentence.

The *word order type* parameter might also have some further consequences for the grammatical structures of noun phrases, for example, where to locate the possessed noun and the possessing one. In languages, which follow the *Japanese* like pattern, the governing noun normally *follows* the possessing one, in contrast with the *English* like pattern of languages. Therefore, *"car its-wheels"* in Japanese is an acceptable *part-of* relationship, which is equivalent to *"wheel of a car"* in English, having *wheel* as the governing noun. There are some further parameters such as the *null subject,*[2] *verb attraction,*[3] etc., which might also have an impact on grammatical patterns to be determined dynamically.

It is out of the scope of this chapter to explain the impact of all these parameters on parsing of a concept (restricted natural language) based query language. It is, however, important to realize that given a set of all *potential* parameters S shaping a multi-dimensional space, the formal specification of *MDDQL* is a functional mapping of a subspace $P \subseteq S$ to all potential specification elements of the automaton or the grammar, with P as a multi-dimensional sub-space of potential parameters within which a particular natural language can be characterized.

Note that the dimensions of P are interdependent meaning that particular dimensional points are determined by other *<dimension, point>* pairs. For example, the *<null subject, yes>* pair is only possible if *<subject placement, high>* is given, which, in turn, is possible only if *<verb attraction, yes>* applies. Therefore, given that *Romanian* is the input query language, where the *null subject* parameter applies, i.e., it is partly indexed within P by *<null subject, yes>*, the grammatical structure can also be determined by the inferred pairs *<subject placement, high>* and *<verb attraction, yes>*, since these are constitutional indexing parts of *Romanian* as a natural language within P.

A tree based organizational structure might also reflect these interdependencies among parameters as determinants of patterns and, consequently, families of natural languages, however, a multi-

---

[2]Languages like *Spanish, Romanian* and *Italian* do not necessarily need a *subject* within a sentence.

[3]Subjects in languages such as *French* or *Italian* go with an auxiliary verb phrase, however, in languages such as *Welsh* and *Zapotec,* subjects go with a verb phrase.

dimensional space allows more flexibility, especially, in the case of a non-exhaustive analysis of all possible parameters from a linguistic theory point of view. Similarly, parsing the query Q in all three languages (*English, Japanese, Warlpiri*) results into a parse tree, which takes the common form: $G_{MDDQL}=\{(<v_1$: *children (noun, subject, class)*, v_2: *big (adjective, subject, property)>, e_1: ""), (<v_1$: *children (noun, subject, class)*, v_3: *chasing (event, verb)>, e_2: ""), (< v_3$: *chasing (event, verb)*, v_4: *dogs (noun, object, class)>, e_3: ""), (< v_4$: *dogs (noun, object, class)*, v_5: *neighbors (noun, object, class)>, e_4: "")\}$.

8.2.3 An indicative approach

The concept based multilingual web querying approach adapts the heuristics-based semantic query augmentation method discussed in (Burton-Jones et al 2003). The proposed approach consists of the following five phases: 1) MDDQL Query Parsing, 2) Query Expansion, 3) Query Formulation, 4) Search Knowledge Sources, and 5) Consolidate Results. A schematic representation of the approach as well the inputs and outputs to each of these phases are shown in Figure 1. All the phases are briefly described below.

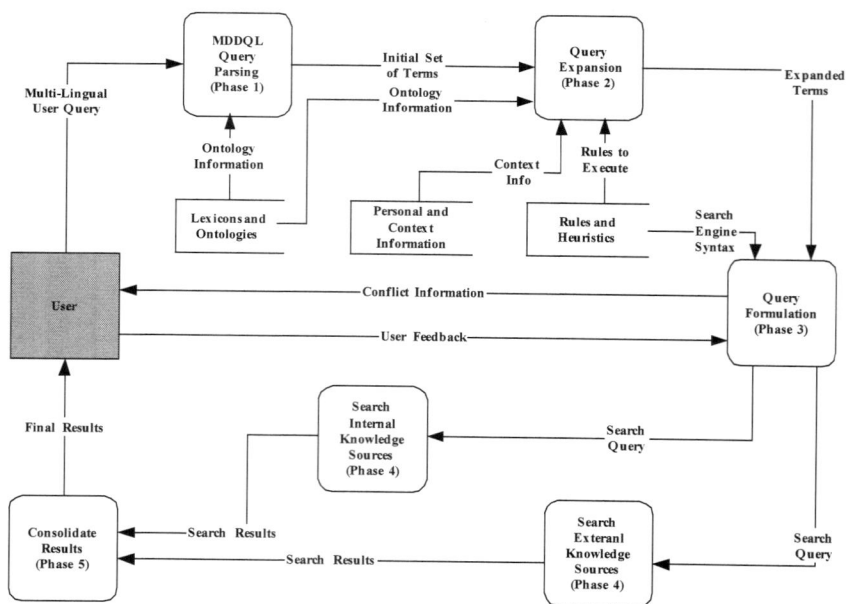

Figure 1: Multi-lingual Web Query Processing Approach

Phase 1—MDDQL Query Parsing: The first phase involves parsing the natural language query specified by the user. The query is segmented using an appropriate segmentation tool that is available for the language used to specify the query. The MDDQL parser parses the segments and creates the query graph. The parsing algorithm does not assume any underlying grammar, but adjusts its behavior based on the parameters specified. The vertices in the conceptual query graph created are eventually translated into English (or any other target language) to generate the initial query terms. The following algorithm exemplifies the transitional function and logic of this abstract machine with respect to a simple case, where the query[4] q: "*big children-SU the neighbors dogs-OB chasing*", in *Japanese*, and only the parameter *P: word-order-type* as a dimension of the space V, are considered.

Input: A query Q, an $A \subseteq L = Japanese$
Output: An *MDDQL* parse tree G_{MDDQL}

> ***SET*** *L=Japanese*
> ***SET*** *V with <P: word-order-type, subject-object-verb>*
> ***SET*** *input tape T to the query Q*
> ***SET*** *a set of auxiliary stacks* $\Delta = \{SU, OB, VP, PP, etc.\}$ *for the grammatical structures or sets-of-states to be determined dynamically*
> ***SET*** $G_{MDDQL} = \emptyset$
> ***WHILE*** *not the end of the tape T*
>> ***READ*** *word, e.g., "big"*
>> ***IDENTIFY syntactic/semantic*** *part of word as* $\Delta (V)=SU$
>> ***PUSH*** *word into stack* $\Delta (V)$
>> ***MOVE*** *head on T to the right*
> ***END OF WHILE***
> ***FOR ALL*** $\Delta \neq \emptyset$
>> ***CREATE vertices N(Δ,C) and edges E(Δ,C) for a sub-graph G***
> $\Delta \subseteq G_{MDDQL}$ *as a function of the nature of the stack and the ontological constraints C*
>> *e.g.,* G_{SU}: *(<v_1: children (noun, subject, class), v_2: big (adjective, subject, property)>, e_1: ""), **when*** $\Delta = SU$, *or*
>> G_{OB}: *(< v_4: dogs (noun, object, class), v_5: neighbors (noun, object, class)>, e_4: ""), **when*** $\Delta = OB$ *or*

---

[4]The suffixes -*SU* and −*OB* are used to indicate the position of subjects and objects within the query or sentence

$G_{VP:}$ < v_3: *chasing (event, verb)*> *when* $\Delta = VP$
END FOR ALL

FOR ALL different pairs <G_1,G_2> of sub-graphs G $\Delta \subseteq G_{MDDQL}$
   CREATE connecting edges E(<G_1,G_2>, C) *with respect to the* *ontological constraints C*
     e.g., < e_2: ""> *connecting* v_1: *children (noun, subject, class) with* v_3: *chasing (event, verb) and, therefore, connecting G_{SU} with G_{VP}*
     < e_3: ""> *connecting* v_3: *chasing (event, verb) with* v_4: *dogs (noun, object, class) and, therefore, connecting G_{VP} with G_{OB}*
   END FOR ALL

Phase2—Query Expansion: The output of the MDDQL parsing phase is a set of initial query terms which becomes the input to the query expansion phase. The query expansion process involves expanding the initial query using lexicons and ontologies. It also includes adding appropriate personal information as well as contextual information. For each query term, the first task is to identify the proper semantics of the term, given the user's context. To do so, the word senses from lexicon such as WordNet are used. For each term synonym sets are extracted. The appropriate word sense is determined based on the context and other query terms (may also need user input) and a synonym from that synset is added to the query. To ensure precise query results, it is important to take out pages that contain incorrect senses of each term. Thus, a synonym from the unselected synset with the highest frequency is added as negative knowledge to the query. Prior research has expanded queries with hypernyms and hyponyms to enhance *recall*. However, for web-based queries, *precision* is preferred over recall. Since ontologies contain domain specific concepts, appropriate hypernym(s) and hyponym(s) are added as mandatory terms to enhance precision. In this phase, personal information and preferences relevant to the query are also added. For example, the user may limit the search to certain geographical area or domain. Such information helps narrow the search space and improve precision.

Phase 3—Query Formulation: The output of the query expansion phase is the expanded set of query terms that includes the initial query terms, synonyms, negative knowledge, hypernyms, hyponyms, and information personal preference ragarding. This expanded set becomes the input to the query formulation phase. In this phase, the query is formulated according to the syntax of the search engine used. Appropriate boolean operators are used to construct the query depending upon the type of term added. For each query term, the synonym is added with an OR operator (e.g., query term OR synonym). Hypernym and hyponym are added using the AND operator (e.g., query term AND (hypernym OR hypernym)). Personal preference information is also added using the AND operator (e.g., query term AND preference). The negative knowledge is added using the NOT operator. The first synonym from the highest remaining synset not selected is included with the boolean NOT operator (e.g., query term NOT synonym). The following algorithm is used in constructing the expanded query:

> // *Expand each base query term with additional information*
> **let** *query_expn* = NULL
> **let** *WS* be a word senses set, $WS = \{WS_1, WS_2, .. WS_n\}$
> **let** each word sense set WS_i contain one or more synonyms, $WS_i = \{Syn_1, Syn_2, .. Syn_n\}$
> > **for** each term TN_i
> > > **let** the word senses from WordNet, $WS_{TNi} = \{WS_1, WS_2, .. WS_n\}$
> > > **let** the selected word sense for TN_i be WS_k
> > > **for** each synonym Syn_k in WS_k
> > > **if** appropriate,
> > > > $query\_expn \mathrel{+}= (TN_i \text{ OR } Syn_k)$
> > > **for** each hypernym $Hype_r$ and hyponym $Hypo_s$
> > > **if** appropriate
> > > > $query\_expn \mathrel{+}= \text{AND } (Hype_r \text{ OR } Hypo_s)$
> > > **if** personal info Pe appropriate
> > > $query\_expn \mathrel{+}= \text{AND } Pe$
> > > **get** the highest remaining word sense set WSN_n
> > > **get** the first synonym $SynN_1$ from this synset WSN_n
> > > $query\_expn \mathrel{+}= \text{AND NOT } SynN_1$
> > **end for**

Phase 4—Search Knowledge Sources: This phase submits the query to one or more web search engines (in their required syntax) for processing using the Application Programming Interface (API) provided by them. Our query construction heuristics work with most search engines. For example, AltaVista allows queries to use a NEAR constraint, but since other search engines such as Google and AlltheWeb do not, it is not used. Likewise, query expansion techniques in traditional information retrieval systems can add up to 800 terms to the query with varying weights. This approach is not used in our methodology because web search engines limit the number of query terms (e.g., Google has a limit of ten terms). The search query can also be submitted to internal knowledge sources within an organization that may contain relevant documents. Of course, this requires that the knowledge repositories provide a well defined set of APIs for various applications to use.

Phase 5—Consolidate Results: In the final phase, the results from the search engine Uniform Resource Location (URLs) and 'snippets' provided from the web pages are retrieved and presented to the user. The user can either accept the results or rewrite and resubmit the query to get more relevant results. This phase also integrates the search results from multiple language sources and takes care of the differences between the languages as well as the format. The results are organized, something that is based on the knowledge sources used, target language, or domain.

8.2.4 An indicative system architecture and implementation

The architecture of the proposed system is shown in Figure 2. It consists of the following components: a) MDDQL Query Parser, b) Query Expansion Module, c) Query Constructor, d) Query API Manager, and e) Results Integration Module. A proof-of-concept prototype has been developed using J2EE technologies. The user specifies search queries in natural language. The server contains Java application code and the WordNet database. The prototype will provide an interface with commercial search engines such as Google (www.google.com) and AlltheWeb (www.alltheweb. com). The MDDQL parser component is implemented using the Java programming language, where abstract data types have been defined for the construction and manipulation of the MDDQL (Kapetanios et al. 2005, 2006) query trees. A repository (knowledge base) for representing the parameters (dimensions), as well as a classification mechanism of a natural language into this parametric space is provided.

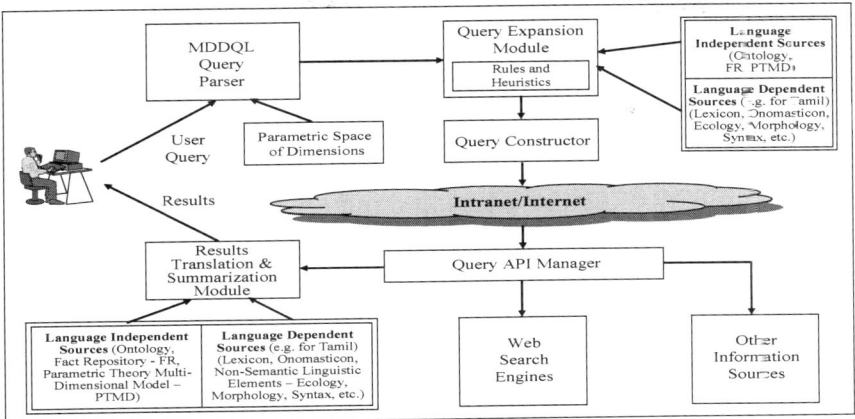

Figure 2: System Architecture for Cross-Lingual Web Querying.

The query API manager component uses URL encoding to communicate with the knowledge sources and the API provided by the search engines. The results integration module returns the first twenty pages from the search engines in the format they provide. The results contain the title of the page, the snippet, and the URL for each page. Each of the components and their functionalities are briefly described below. In order to verify the validity of our assumptions, and clarify the internal workings of the system as presented in Figure 2, one will describe in detail how a query is parsed, passed to the query expansion module, then to the query constructor, which will transform the query to be submitted to the query API manager.

Example classes used in the prototype are shown in Figure 3, some of which are briefly described below. The MDDQL conceptual query tree is defined as a generic QueryTermNode class. The instance variables TUI, Word, etc. represent the query term as an object rather than a single string value. In order to distinguish among the ontological role of nodes and, therefore, the included query terms on the tree (instance variable: Word), subclasses such as PropertyTermNode, RelationTermNode, ValueTermNode can be specified as inheriting from the generic QueryTermNode class, which indicate that a particular term is being considered as a Property, Association, Value, respectively. Of course, various other subclasses can be added, if needed.

The class QueryTreeContainer provides all the methods for the construction and manipulation of the MDDQL query tree based on its generic type QueryTermNode together with the initialization of some context for a particular query tree. The QtagQueryParser class currently receives as input the tokenized query as performed by the Part-of-Speech (POS) probabilistic tagger (QtagTokenFactory) and provides the natural language settings for the n-dimensional parametric space. The class QueryTreeMapping implements the conversion of the tokenized

All Classes
EntityTermNode
PropertyTermNode
QtagQueryParser
QtagTokenFactory
QueryTermNode
QueryTreeContainer
QueryTreeMapping
RelationTermNode
ValueTermNode

Package Class **Tree Deprecated Index Help**

PREV PACKAGE   NEXT PACKAGE   FRAMES   NO FRAMES   All Classes

Package mlq.parser

Class Summary

EntityTermNode	This class provides the structure of an entity term as o node of the MDDQL query tree.
PropertyTermNode	This class provides the structure of the terms standing for properties.
QtagQueryParser	
QtagTokenFactory	
QueryTermNode	This class provides the basic structure of the MDDQL query alphabet terms as query tree nodes.
QueryTreeContainer	This class implements the container for an MDDQL query tree.
QueryTreeMapping	
RelationTermNode	This class provides the structure of a relation term as a node of the MDDQL query tree.
ValueTermNode	This class implements the structure for value terms as objects (nodes) of the MDDQL query tree.

Package Class **Tree Deprecated Index Help**

PREV PACKAGE   NEXT PACKAGE   FRAMES   NO FRAMES   All Classes

Figure 3: Sample classes from prototype implementation.

query based on the POS list of tags and the natural language settings of the n-dimensional parametric space, into an MDDQL conceptual query tree. It takes into consideration what terms have been put into the various stacks, which refer mainly to the semantic role of clauses, i.e., a subject, an association, an object, within the query. Therefore, the sequence of query terms as they appear within a query does not have any impact on the order of appearance of the terms, i.e., query term nodes on the query tree.

In a querying session, the user is asked to define the natural language in which the query terms should be defined, for example, Japanese. For the sake of understanding, the query "cars Detroit manufacturing" is formulated with English terms, however, it follows the *Subject-Object-Verb* word type order as typical for Japanese. After having parsed the query, a conceptual query tree is constructed where it holds that the semantic relationship *cars – manufacturing – Detroit* is extracted. Even if the user had set up *English* as a querying interaction means, the same conceptual query tree would have been derived.

The MDDQL conceptual query tree is a kind of a conceptual graph. What is represented on the query tree is natural language independent, a kind of '*language of thought*'. For the query translation process, however, the query tree needs to be cloned in order to generate a copy as many times as the number of target languages. In the example, if the query is supposed to be targeted against an English and a Japanese corpus, there will be two cloned conceptual query trees. However, the terms on the nodes will be translated into the target language. From each natural language specific query tree, one finally, generates the query in the target language with the help of the n-dimensional parametric space. All one needs to do is simply traverse the conceptual query tree and generate a sequence of terms, which fit the parametric pattern of the target language.

For instance, the generated queries having Japanese and English as target natural languages would be "cars Detroit manufacturing" and "cars manufacturing Detroit", respectively. Doing so with a stable and NL independent representation of the query meaning does pay off in terms of simplifying the query translation process, which has always been an issue in all *Cross-Lingual Information Retrieval* approaches. In a sentence-by-sentence translation, one needs to specify rules for creating queries as mini-phrases out of possibly mutated query trees. This may add tremendously to complexity, which could be avoided, since one does have the guiding atoms for the generation of the corresponding query in a particular target language in one's n-dimensional parametric space and, therefore, one still works with the same flexible and scalable data structure regardless of the natural language concerned. From the query tree, the initial query phrase is created in the target language. In this example, let it be assumed that the user is interested in executing the following query: "cars Detroit manufacturing". The noun phrases are extracted and displayed and the user can uncheck nouns that are not relevant or appropriate. The selected terms form the initial query.

The initial query is represented as a semantic net with the nodes representing the terms and the arcs representing the relationships between them.

The initial query is then expanded by the "Query Expansion Module" by adding additional nodes and arcs corresponding to related information. First, the query expansion module retrieves word senses from WordNet for each query term and displays them to the user who selects the appropriate sense to use. If none is selected, the module uses the user's original query term. The term "car", for instance, may have five senses. After selecting the appropriate one, the user initiates the query refinement process. This module gathers the user selections and executes additional tasks to expand the underlying semantic net. Some of the tasks performed are: a) mark exclusion of a word-sense by excluding the first synonym from the highest-ordered synset not selected by the user, b) refine the context by including hypernyms and hyponyms from WordNet and an ontology, and c) resolve inconsistencies by checking if the ontology terms to be added, exist among the terms in the unwanted WordNet synsets. For example, a synonym could be identified and added to the semantic net, i.e., "Car" is expanded with the first term in the user's selected synset (i.e., car OR auto). A synonym of car is also added as negative knowledge from the next highest synset that has a synonym to exclude unrelated hits (i.e., chair AND NOT railcar). For each term, the hypernyms and hyponyms (superclass and subclass) corresponding to the selected word sense are retrieved from WordNet and the ontology and displayed to the user. The user-selected hypernyms and hyponyms are added to the semantic net.

The Query Constructor module transforms the semantic net into a query that follows the syntax required by search engines. This task automates query construction with Boolean operators to compensate for the possible lack of expertise on the part of users. The Boolean constraints depend upon the type of term, the connection to the node in the semantic net and the rationale provided in the earlier tasks. For instance, the first synonym from the intended synset is added with the OR operator such as: query term OR synonym; the hypernyms or hyponyms are added with the AND operator such as: query term AND (WordNet hypernym OR Ontology hypernym) AND (WordNet hyponym OR Ontology hyponym); and the first synonym from un-intended synset is added with the negation operator such as: query term AND NOT synonym. This transformation attempts to improve query precision, i.e., adding synonyms in combination with a hypernym or hyponym increases precision (unlike adding synonyms only, which increase recall at the cost of precision).

Once the expanded query is constructed, it is executed on publicly available search engines. The prototype interfaces with Google and AlltheWeb search engines. The query submitted to the search engine follows the required syntax. The results from the search engine are gathered through the Query API Manager and summarized and presented to the user through the Results Summarization module. Figure 5 shows the expanded query submitted to Google and the top twenty hits. The total number of hits for the query is also listed. The original query "car detroit manufacturing" returned 3,210,000 hits whereas the refined query returned 209,000 hits. The initial results indicate that the percentage of relevant documents in the first twenty hits is higher for the expanded query compared to the original query, which shows the usefulness of the query refinement process.

8.3 Summary and Major Conclusions

While a few retrieval techniques exist for searching in different languages, they are not adequate for performing multilingual querying on the Web because of their inability to handle the heterogeneity of the format and structure of Web documents. Simplification of query processing and translation is still an open issue for CLIR approaches. In this paper, a concept based methodology has been presented for multilingual querying on the Web. The MDDQL-based approach is language independent and hence the approach is scaleable and able to integrate even unknown grammars and/or natural languages.

Another distinguishing characteristic of one's approach is its flexibility and easy maintenance of patterns of NL grammars. It also uses contextual and semantic information in query refinement to improve precision. The architecture of a system that implements the methodology has been discussed. A proof-of-concept prototype is currently under development to demonstrate its feasibility. Advances in multilingual Web querying will move us a step closer to making the Semantic Web a reality.

PART IV

Knowledge Extraction and Engineering for NLP

CHAPTER 9

Word Sense Disambiguation

9.1 Introduction

This chapter is a short survey of Word Sense Disambiguation (WSD), this problem being the central one in Natural Language Processing (NLP). It is very difficult to write these days about WSD, since beginning with the *state of the art* work of N. Ide and J. Veronis in *Computational Linguistics* in 1998 (Ide and Veronis 1998), up to the extended survey of Roberto Navigli in *ACM Comput. Survey* in 2009 (Navigli 2009), many articles have focused and many books have earmarked one of their chapters on this subject: only two of them are mentioned here, the excellent books (Manning and Shütze 1999) or (Jurafsky and Martin 2000). Moreover, a whole content devoted to WSD is the comprehensive book (Agirre and Edmonds 2006). The present Chapter doesn't go to the details the way the mentioned works do, but reflects the theme at an appropriate proportion to the ensemble of this book. An attempt was made to approach WSD from an algorithmic point of view, describing algorithms in a way easy to understand, regardless of whether they represent some fundamental algorithms in WSD or not.

WSD is the process of identifying the correct meanings of words in particular contexts. Applications for which WSD is a potential improving stage are all language understanding applications (Ide and Veronis 1998): IR, Text mining, QA systems, Dialogue systems, Machine Translation, etc. WSD could provide an important contribution in the treatment of large scale amount of data contained in Semantic Web (Berners-Lee et al. 2001). However, "the success of WSD in real-word applications is still to be shown" (Navigli 2009) and it is currently considered that WSD is still an open research area.

In the last twenty years there has been a very large tendency to use statistical learning methods in computational linguistics (also called corpus-based methods). This large application of statistical methods has its origin in the growing availability of big machine-readable corpora. As for any statistical approach, the goal is to be able to treat uncertain and incomplete linguistic information successfully (however in the limits of some approximation). In this chapter a survey is made of statistical (or corpus based) methods for WSD following a *machine learning* approach categorization of WSD methods as: supervised, bootstrapping and unsupervised. All these methods are classified in this chapter into two broad categories: vectorial methods and non-vectorial methods. Another categorization of WSD methods could be made after the Part of Speach (POS) tag of disambiguated word: it is well known that the polysemy of different categories of words is decreasingly ordered as verb, adjective, adverb, noun. The remark is based on the number of verbs, adjectives, adverbs, nouns and the number of their senses in electronic on-line dictionary WordNet (Fellbaum ed. 1998, Miller 1995), the most well-developed and widely used lexical database in English.

One can remark finally that (full) Word Sense Disambiguation is the task of assigning sense labels to occurrences of one (or more) ambiguous words. This problem can be divided into two subproblems (Schütze 1998): sense discrimination and sense labeling. Word sense discrimination is, of course, easier than full disambiguation since one only needs to determine which occurrences have the same meaning and not what the meaning actually is. In the following WSD usually refers to both discrimination and sense labeling of ambiguous words.

9.1.1 Meaning and context

The systems in the *supervised learning* are trained to learn a classifier that can be used to assign a yet unseen example (examples) of a *target* word to one of a fixed number of senses. That means one has a training corpus, where the system learns the classifier and a test corpus which the system must annotate. So, supervised learning can be considered as a classification task, while unsupervised learning can be viewed as a clustering task.

The precise definition of a sense is a matter of considerable debate within the WSD community. However there are some rules very useful for all statistical methods:

- The context of a word is a source of information to identify the meaning of a polysemous word w;
- One would expect the words closest to the target polysemous word w to be of greater semantical importance than the other words in the text;
- If some words occur frequently in similar contexts one may assume that they have similar meanings.

Regarding the context, this is used in two ways: a) As a bag of words, without consideration for relationships to the target word in terms of distance, grammatical relations, etc. b) With relational information.

A context could be considered as a sentence containing w (in this case both approaches of a context a) and b) are possible) or a window centered on w (in this case only the approach a) is possible). For example the window context centered on w, c_i, could be represented as a list:

$$c_i = w_1, w_2, \ldots, w_t, w, w_{t+1}, \ldots, w_z$$

where $w_1, w_2, \ldots, w_t, w_{t+1}, \ldots, w_z$ are words from a set $\{v_1, \ldots, v_N\}$ of given features and t and z are selected by the user (usually $z = 2t$).

The *bag of words* approach works better for nouns than verbs (Ide and Veronis 1998) but is less effective than methods that take the other relations in consideration: compare the human act of understanding a sentence and understanding a mix of the same words in an arbitrary order. Of course, the sentence (in general, a text) is more than just a bag of words. On the other hand, verbs tend to have a larger number of senses, which explains why verbs are more difficult to approach in any WSD method. Regarding case b), studies about syntactic relations determined some interesting conclusions (Ide and Veronis 1998): verbs derive more disambiguation information from their objects than from their subjects, adjectives derive almost all disambiguation information from the nouns they modify and nouns are best disambiguated by directly placing adjectives or nouns adjacently.

Let also it be remarked that some linguistic preprocessing of a text, as stemming or part of speech tagging, could improve the precision of disambiguation. Stemming can be replaced by the simpler method of truncation. The idea is to approximate word stems by deleting the end of words after a certain predefined number of letters. An experiment in (Sahlgren 2001) has been reported as the best precision of WSD at around 63% for truncation, while with word stemming the precision was of 72 % for a 3+3 sized context window.

9.2 Methods and Algorithms: Vectorial Methods in WSD

9.2.1 Associating vectors to the contexts

In the following the vectorial representation of contexts will be set using the notations in (Manning and Schutze 1999): a context c is represented as a vector $\vec{c}$ of some features. The definition and the numbers of these features depend on the method selected. A common denominator between the methods is that they use co-occurrence and collocation statistics. The famous Wittgenstein's dictum: *"meaning is the use"* (Wittgenstein 1953) means that to understand the meaning of a word one has to consider its *use* in the frame of a concrete *context*. For vectorial methods, the *context* is the vector, and the *use* is represented by the set of selected features. These features could carry linguistic information (most of them) or non-linguistic information. Context size can vary from one word at each side of the focus word to more words *window* or even a complete sentence.

Let it be emphasized that one works in this survey with contexts (the vectors are associated with the contexts) but a vectorial approach where the vectors are associated with words (Navigli 2009) is also possible.

The notations used in (Manning and Schutze 1999) are:

- w is the word to be disambiguated (the target word);
- $s_1, \ldots, s_{N_s}$ are the possible senses for w;
- $c_1, \ldots, c_{N_c}$ are the contexts for w;
- $v_1, \ldots, v_{N_f}$ are the features (terms) selected for representing a context.

Even in these notations it is obvious that one's attention is focused on disambiguation of a target word(s), more than disambiguation of all (open class) words. However, some of the algorithms presented in this chapter could realize this desiderate (one asks in this aspect where the case is). Another aspect which is settled by the notations is that one uses an enumerative approach of senses.

In general, a list of the most frequently used words are selected for features $v_1, \ldots, v_{Nf}$. When these words have a specific position located to the left and/or the right of the target word w they are *collocation* features, when their exact position is ignored, they are called *co-occurrence* features. For a method of this kind, a set of *stop words*, as articles, pronouns, prepositions or conjunctions, are usually removed from the set of features. Another way of selection of features $v_1, \ldots, v_{Nf}$ is given by some statistical tests of feature relevance (as for example χ^2).

Once the list of words is selected as features for the target word w, one can associate a context c with the vector $\vec{c}$ as follows:

- $\vec{c} = (w_1, \ldots, w_{Nf})$ where w_i is the number of times the word v_i occurs in context c;
- $\vec{c} = (w_1, \ldots, w_{Nf})$ where w_i is 1 if the word v_i occurs in context c, or 0 otherwise;
- $\vec{c} = (\ldots w_{i-1}, w_{i+1} \ldots)$ where w_{i-1} (w_{i+1}) is 1 if the word v_i occurs in context c at the left (right) of the word w or 0 otherwise;
- $\vec{c} = (\ldots w_{i-k}, w_{i-(k-1)}, \ldots, w_{i-1}, w_{i+1}, \ldots, w_{i+k} \ldots)$ where w_{i-j} (w_{i+j}) is 1 if the word v_i occurs in context c at the left (right) of the word w at distance j or 0 otherwise. In this case the *window* has the type -j +j;
- $\vec{c} = (w_1, \ldots, w_{|W|})$ where w_i is 1 if the word v_i occurs in context c, or 0 otherwise, where v_i is a word from the entire text (or all the contexts) with the length $| W |$. In this last example the features are all the words which are not *stop words*.
- $\vec{c} = (w_1, \ldots, w_{Nf})$ where w_i is 1 if the verb v_i could have as complement the noun w and 0 otherwise (in this case w is a noun and the set $v_1, \ldots, v_{Nf}$ is a set of verbs) or w_i is 1 if the noun v_i could be a complement for the verb w and 0 otherwise (in this case w is a verb and $v_1, \ldots, v_{Nf}$ is a set of nouns).

In all the above examples the number w_i is the weight of the word (feature) v_i. This can be the frequency f_i of the feature v_i (term frequency or *tf*) as in the first case above. On the base of feature relevance principle, the features can be weighed to reflect the distance of the words to the focus word. For example, in a $-3 +3$ window the weights for the 6 features could be: 0.25, 0.5, 1, 1, 0.5, 0.25

Another method to establish the weight w_i is to capture the fashion of distribution of v_i in all the set of contexts on principle (Jurafsky and Martin 2000): features that are limited to a small number of contexts are useful for discriminating those contexts; features that occur frequently across the entire set of contexts are less useful in this discrimination. In this case it is used as a new weight for a feature, called *inverse document frequency*, denoted by *idf* and defined as below:

Definition

Let it be considered that the number of contexts is Nc and the number of contexts in which the feature v_i occurs is n_i. The inverse document frequency of feature v_i is:

$$idf_i = \frac{Nc}{n_i} \ or \ idf_i = \log(\frac{Nc}{n_i})$$

Combining *tf* with *idf* one obtains *tf.idf* weighting. In this case: $\vec{c} = (w_1, \ldots, w_{Nf})$, where $w_i = f_i \times idf_i$.

9.2.2 Measures of similarity

Almost all methods of WSD by vectorial approach are based on the similarity of contexts, expressed by the similarity of associated vectors. The most used similarity between two contexts c_a, c_b (of the same word or different words) is the *normalised cosine* between the vectors $\vec{c}_a$ and $\vec{c}_b$:

$$cos(\vec{c}_a, \vec{c}_b) = \frac{\sum_{j=1}^{m} w_{a,j} \times w_{b,j}}{\sqrt{\sum_{j=1}^{m} w_{a,j}^2 \times \sum_{j=1}^{m} w_{b,j}^2}}$$

and $sim(\vec{c}_a, \vec{c}_b) = cos(\vec{c}_a, \vec{c}_b)$.

Also, the similarity could be different vectorial measures as, for example, $sim(\vec{c}_a, \vec{c}_b) = 1/(|\vec{c}_a - \vec{c}_b|)$

When the vector of context is binary, the other measure used could be, where $|X \cap Y| = \Sigma_{i=1}^{n}(x_i \cap y_i)$ and $|X \cup Y| = \Sigma_{i=1}^{n}(x_i \cup y_i)$ and $\cap$, $\cup$ are logical operations *and, or*:

Matching coefficient: $| X \cap Y |$

Dice coefficient : $\dfrac{|X \cap Y|}{|X|+|Y|}$

Jaccard coefficient : $\dfrac{|X \cap Y|}{|X \cup Y|}$

Overlap coefficient : $\dfrac{|X \cap Y|}{min(|X|,|Y|)}$

In (Schütze 1998) the author introduces the concept of *word vector* as opposite to the *context vector* type. The word vector for a word w is $\vec{w} = (w_1, \ldots, w_{Nf})$ where w_i is the number of times the word v_i co-occurs with w in the whole set of contexts. The *context vector* for a context of an ambiguous word is obtained by summing the *word vectors* of all words in the context. Therefore two contexts are similar if the words in these contexts occur with similar words (the second-order co-occurrence). This is known as *strong contextual hypothesis*. The second order co-occurrence representation is more robust than the first order representation (Schütze 1998).

9.2.3 Supervised learning of WSD by vectorial methods

In supervised learning, a system is presented with a training set consisting of a set of input contexts labeled with their appropriate senses (disambiguated corpus). The task is to build a classifier which classifies new cases correctly based on their context. The most well known vectorial method is k-NN.

k-NN or memory based learning

At the time of training, a k-NN model memorizes all the contexts in the training set by their associated features. Later, when a new context c_{new} is provided, the classifier first selects k contexts in the training set that are closest to c_{new}, then picks a sense for c_{new}.

The features $v_1, \ldots, v_{Nf}$ could be selected as in the previous section and the vector of a context is $\vec{c}$. This supervised algorithm is adapted from (Jackson and Moulinier 2002) and calculates the sense s' of a target word w in the new context c_{new}:

- TRAINING Calculate $\vec{c}$ for each context c.
- TEST Calculate:

$$A = \{\vec{c} \mid sim(\vec{c}_{new}, \vec{c}) \text{ is maxim}, \mid A \mid = k\}$$

(*A* is the set of the *k* nearest neighbors contexts of $\vec{c}_{new}$).

for each sense s_j **do**

Calculate:

$$Score(c_{new}, s_j) = \sum_{c_i \in A} (sim(\vec{c}_{new}, \vec{c}_i) \times a_{ij})$$

(where a_{ij} is 1 if $\vec{c}_i$ has the sense s_j and a_{ij} is 0 otherwise)
endfor

Calculate

$$s' = argmax_j Score(c_{new}, s_j)$$

9.2.4 Unsupervised approach. Clustering contexts by vectorial method

Unsupervised approach of WSD does not use sense tagged data (training data). Strictly speaking, the task of unsupervised disambiguations is identical with sense discrimination. In this case, vector representations of unlabeled contexts are grouped into clusters, according to a similarity measure. One cluster is considered as representing a sense and a new context c_{new} is classified as having the sense of the cluster to which it is closest according to the similarity measure. In disambiguation an advantage of unsupervised methods is that the granularity of sense distinction is an adjustable parameter: for instance, a number of 20 clusters induces more fine-grained sense distinction than a number of 5 clusters.

As clustering methods we can use (agglomerative or divisive) hierarchical algorithms or non-hierarchical (flat) clustering algorithms (Manning and Schütze 1999, Tatar and Serban 2001). In the case of agglomerative hierarchical clustering, each of the unlabeled contexts is initially assigned to its own cluster. New clusters are then formed in bottom-up fashion by a successive fusion of two clusters that are most similar. This process continues until either a specified number of clusters is obtained or some condition about the similarity measure between the clusters is accomplished. In the case of divisive hierarchical clustering, one starts with all contexts in a unique cluster and by successive division the most dissimilar cluster is processed. In general, a good clustering method is defined as one that maximizes the within cluster similarity and minimizes the between cluster similarity.

A non-hierarchical algorithm starts out with a partition based on randomly selected seeds (one seed per cluster) and then refines this initial partition. The algorithm stops when a measure of cluster quality is accomplished.

Surveys about clustering are in (Manning and Schütze 1999, Tatar and Serban 2003).

A. Agglomerative algorithm for hierarchical clustering

Input The set of vectors of contexts to be clustered, the similarity function.

Output The set of hierarchical clusters C.

At each step the pair of clusters (C_{u*}, C_{v*}) to be unified is identified as:

$$(C_{u*},C_{v*}) := argmax_{(C_u,C_v)}sim^x(C_u,C_v)$$

The similarity $sim^x(C_u,C_v)$ between clusters, where $x = a, s, c$, could be group average similarity, a:

$$sim^a(C_u,C_v) = \frac{\Sigma_{a_i \in C_u}\Sigma_{b_j \in C_v} sim(a_i, b_j)}{|C_u| \times |C_v|}$$

single link, s, when the similarity of two clusters is equal to the similarity of the most similar elements from C_u and C_v:

$$sim^s(C_u,C_v) = max_{a_i \in C_u a_j \in C_v} sim(a_i,a_j)$$

complete link, c, when the similarity of two clusters is equal to the similarity of the least similar elements from C_u and C_v:

$$sim^c(C_u,C_v) = min_{a_i \in C_u a_j \in C_v} sim(a_i, a_j)$$

B. Non-hierarhical clustering algorithm: k-means algorithm (reproduced from (Manning and Schütze 1999))

Input. The set of vector contexts to be clustered, the similarity measure, a function μ for computing the mean.

Output. The set of clusters.
begin

      Select k initial centers $\{\vec{f}_1, \vec{f}_2, \ldots, \vec{f}_k\}$
      while *condition of stopping unaccomplished* **do**
          for *all clusters* C_j **do**

$$C_j = \{\vec{x}_i \mid \forall \vec{f}_l \; sim(\vec{x}_i, \vec{f}_j) > sim(\vec{x}_i, \vec{f}_l)\}$$
          endfor
          for *all clusters* C_j **do**
$$\vec{f}_j = \vec{\mu}(C_j)$$
          endfor
      endwile
end

As center we can calculate:

$$\vec{\mu}(C_j) = \frac{1}{|C_j|} \sum_{\vec{x} \in C_j} \vec{x}$$

9.3 Methods and Algorithms: Non-vectorial Methods in WSD

9.3.1 Naive Bayes classifier approach to WSD

In this case the context of a word w is treated as a *bag of words* without structure. What one wants to find is the best sense s' for a new input context c_{new} of an ambiguous word w. This is obtained using Bayes' rule as:

$$s' = argmax_{s_k} P(s_k \mid c_{new}) = argmax_{s_k} \frac{P(c_{new} \mid s_k) \times P(s_k)}{P(c_{new})}$$

$$= argmax_{s_k} P(c_{new} \mid s_k) \times P(s_k)$$

The independence assumption (naive Bayes assumption) is, where v_i could be some frequently occurring words in training corpus (the set of contexts of target word w, annotated with senses) or even all the words in the corpus:

$$P(c_{new} \mid s_k) = P(\{v_i \mid v_i \in c_{new}\} \mid s_k) = \prod_{v_i \in c_{new}} P(v_i \mid s_k)$$

This assumption (often referred to as a *bag of words* model) has two consequences:

- the structure and order of words in the context is ignored;
- the presence of one word in the context doesn't depend on the presence of another.

Despite this coarse approximation in the naive Bayes assumption, it is well known that NBC algorithm works well.

Finally, $s' = argmax_{s_k} P(s_k) \times \Pi_{v_i \in c_{new}} P(v_i \mid s_k)$.
Thus the supervised algorithm is (Manning and Schutze 1999):

- TRAINING Calculate:

for all senses s_k of w **do**

$$P(s_k) = \frac{N(s_k)}{N(w)}$$

   for all words v_j in corpus **do**

$$P(v_j \mid s_k) = \frac{N(v_j, s_k)}{N(s_k)}$$

   endfor

endfor

where $N(s_k)$ is the number of contexts annotated with the sense s_k in the training corpus, $N(v_j, s_k)$ is the number of occurrences of term v_j in the contexts annotated with the sense s_k and $N(w)$ is the number of total occurrences of word w.

- DISAMBIGUATION STEP Calculate for a new context c_{new} the appropriate sense:

$$s' = argmax_{s_k} P(s_k \mid c_{new}) = argmax_{s_k} P(s_k) \times \prod_{v_i \in c_{new}} P(v_i \mid s_k).$$

In (Manning and Schutze 1999) some very high precisions are reported for disambiguation systems based on this algorithm.

9.4 Methods and Algorithms: Bootstrapping Approach of WSD

A major problem with supervised approaches is the need for a large sense tagged training set. The bootstrapping methods (also called semisupervised) use a small number of contexts labeled with senses having a high degree of confidence.

In (Yarowsky 1999) the author observed that there are constraints between different occurrences of contextual features that can be used for disambiguation. Two such constraints are a) *one sense per discourse* and b) *one sense per collocation*. The constraint a) means that the sense of a target word is highly consistent within a given discourse (as (Gale et al. 1992) observed, "it appeared to be extremely unusual to find two or more senses of a polysemous word in the same discourse)." The constraint b) provides the idea that the contextual features (near words) provide strong clues to the sense of a target word.

A bootstrapping method begins by hand tagging with senses the contexts of an ambiguous word w for which the sense of w is clear because some *seed collocations* (Yarowsky 1995, Yarowsky 1999) occur in these contexts.

These labeled contexts are further used to train an initial classifier. This classifier is then used to extract a larger training set from the remaining untagged contexts. By repeating this process, the number of training contexts grows and the number of untagged contexts reduces. One will stop when the remaining unannotated corpus is empty or any new context cannot be annotated.

The bootstrapping approach is situated between the supervised and unsupervised approach of WSD.

For the word *bass* for example, one might begin with *fish* as a reasonable sense for sense $bass^A$ (*bass as fish*), as presented in WordNet and *play* as a reasonable sense for $bass^B$ (*bass as music*). A small number of contexts can be labeled with the sense A and B. These labeled contexts are used to extract a larger set of labeled contexts (as in (Yarowsky 1995)).

(Tatar and Serban 2001) presents an original bootstrapping algorithm that uses an NBC classifier. Let it be remarked that it is provided for separate contexts, which do not constitute a discourse, so the principle of *one sense per discourse* is not used. This algorithm will be described below (Tatar and Serban 2001):

Let it be considered that the words $V = \{v^1, \ldots, v^l\} \subset \{v_1, \ldots, v_{N^p}\}$ where l is small (for the above example, in (Yarowsky 1995) l is 2) are surely associated with the senses for w, so that the occurrence of v^i in the context of w determines the choice for w of a sense s_j (*one sense per collocation*).

These rules can be made as a decision list:

if v^i occurs in a context of w then s_j (R)

The set of v^i in all rules (R) forms the set V.

So, some contexts could be solved using on-line dictionaries or thesaurus querying for collocations. In above example these contexts with A or B are marked. In the following algorithm $C_{resolved}$ denotes the set of contexts already disambiguated, C_{un} indicates the set of context rested unsolved and C denotes the set of all contexts. A context of the word w, is denoted as $(w_1, \ldots, w_t, w, w_{t+1}, \ldots w_z)$. Here $w_1, \ldots,$ w_z are words surrounding w.

(note the difference with the weights in vectorial representation of contexts).

Bootstrapping Algorithm

$C_{resolved} = \varnothing$; Determine the set $V = \{v^1, \ldots, v^l\}$
for each context $c \in C$ **do**
   if $\exists v^i \in V$ such that $v^i \in c$ and the rule (R) is valid
      then the sense of c is s_j

$C_{resolved} = C_{resolved} \cup \{c\}$ and $C_{un} = C\backslash\{c\}$
endfor
while $C_{un} \neq \varnothing$ or $C_{resolved}$ *changes* **do**
   Determine a set V^* of words with the frequency bigger
   than a threshold in $C_{resolved}$.
   Define $V = V \cup V^* = \Sigma_{j=1}^l V_{s_j}$ where V_{s_j} is the set of words associated

   with the sense s_j (If $v \in V$, v occurs in the context c and the context c has been solved with the sense s_j, then $v \in V_{s_j}$, according with the principle "one sense per collocation")
for each $c_i \in C_{un}$ **do**
   apply the BNC algorithm :
   $s^* = argmax_{s_j} P(s_j \mid c_i) =$
   $argmax_{s_j} P(c_i \mid s_j) \ P(s_j)$
   where $P(c_i \mid s_j) = P(w_1 \mid s_j) \ldots P(w_t \mid s_j) P(w_{t+1} \mid s_j) \ldots P(w_z \mid s_j)$
   and $P(w_k \mid s_j) = 1$ *if* $w_k \in V_{s_j}$ and $\dfrac{nr.of\ occ\ of\ w_k}{nr.total\ of\ words}$ else

   If $P(s^* \mid c_i) >$ fixed threshold then c_i is solved with the sense s^*,
      $C_{resolved} = C_{resolved} \cup \{c_i\}$ and $C_{un} = C\backslash\{c_i\}$
  endfor
  endwhile

Concerning the set V^* and the probabilities $P(w \mid s)$, they are re-estimated until all the contexts are solved or $C_{resolved}$ is unchanged.

In the paper (Tatar and Serban 2001) is obtained with this algorithm an accuracy of around 60% for disambiguation of Romanian texts.

9.5 Methods and Algorithms: Dictionary-based Disambiguation

Exactly as in the case when an annotated corpus could help WSD process, the use of dictionaries, thesaurus, ontologies, etc. could play the same role of additional knowledge. Work in WSD reached a turning point in the 1980s when large-scale lexical resources such as dictionaries, became widely available. The machine readable dictionaries (MRD) developed a lot in those days. This section describes disambiguation methods that rely on the definitions of senses of a word in dictionaries (also called knowledge-based methods). The first author who used a method in this category is Lesk (Lesk 1986) and we will describe his algorithm (formalized like in (Manning and Schutze 1999)) below.

9.5.1 Lesk's algorithms

Lesk starts from the idea that a word's dictionary definition (or a gloss) is a good indicator for its different connotations. He used the definitions in the Oxford Advanced Learner's Dictionary (of English) directly.

Suppose that a polysemous word w has in a dictionary Ns senses $s_1, s_2, \ldots, s_{Ns}$ given by an equal number of definitions $D_1, D_2, \ldots, D_{Ns}$ (here $D_1, D_2, \ldots, D_{Ns}$ are sets of stemmed words corresponding to the dictionary definitions). In the original algorithm of Lesk it is considered that E_{v_j} represents the union of all possible dictionary definitions of a polysemous word v_j occurring in the context c_{new} of w. If v_j has the senses $\{s_{j_1}, s_{j_2}, \ldots, s_{j_m}\}$ then $E_{v_j} = \cup_{i=1}^{m} D_{j_i}$. In this case the score of each senses is calculated as:

> **begin**
>> **for** $k = 1, \ldots, Ns$ **do**
>>> $score(s_k) = |\ D_k \cap (\cup_{v_j \in c_{new}} E_{v_j})\ |$
>> **endfor**
>> *Calculate* $s' = argmax_k score(s_k)$.
> **end**

The idea of Lesk's algorithm was simplied or extended in numerous ways. Thus, a reduced form is the following:

for $k = 1, \ldots, Ns$ **do**

$$score(s_k) = |\ D_k \cap (\cup_{v_j \in c_{new}} \{v_j\})\ |$$

endfor

Calculate $s' = argmax_k score(s_k)$

Here v_j are words occurring in the context c_{new} of w. The score of a sense is the number of words that are shared by the definition of the sense and the context.

A generalization of Lesk's algorithm, presented for example in (Sidorov and Ghelbukh 2001) and (Pedersen et al. 2004) is to find overlaps between the glosses or the definitions of words (as above) as well as glosses of the synonym words.

The same role as a dictionary could be played by a thesaurus. (Walker 1987) and (Yarowsky 1992) used the semantic categorization provided by Roget's thesaurus and calculated the sense s_k for a new context c_{new} as the sense of the maximally joined semantic category of the words in context c_{new}.

9.5.2 Yarowsky's bootstrapping algorithm

Yarowsky (Yarowsky 1999) observes that the sense of a target word is highly consistent within any given document or discourse. This is the content of *one sense per discourse* principle. For example, if a document is about biological life, then each occurrence of the ambiguous word *plant* is more probably linked with the sense of "living being". If the document is about industrial aspects, then *plant* is more probably linked with the sense *factory*. Of course, the definition of discourse is central for this principle.

On the other hand, the sense of a target word is strongly correlated with certain other words in the same phrasal unit, named collocation features. By a collocation we mean usually first/second/third/etc. word to the left/right of the target word. In fact, there are words that collocate with the target word w with a high probability. Such words are considered the strongest in the disambiguation process (*one sense per collocation* principle). The algorithm proposed by Yarowsky combines both types of constraints. It is presented below, as adapted from (Manning and Schutze 1999). It iterates building two sets , F_k and E_k for each sense s_k: F_k contains characteristic collocations for the sense s_k, E_k is the set of contexts of the target word w that are assigned to the sense s_k.

INPUT: w, s_1, s_2, . . . , s_{Ns}, c_1, c_2, . . . , c_{Nc}, the contexts of w to be disambiguated. The contexts are given in Nd documents, d_1, d_2, . . . , d_{Nd}.

OUTPUT: The set c_1, c_2, . . . , c_{Nc} tagged with senses in each document.

begin
    Initialization
    for *all sense s_k of w* **do**
        F_k = {*the set of collocations v_i, in the dictionary definition of s_k sense of w*}
        $E_k = \varnothing$

    endfor
    One sense per collocation
    while *at least one E_k changed in the last iteration* **do**
        for *all sense s_k of w* **do**
            $E_k = \{c_i \mid c_i \cap F_k \neq \Phi\}$
        endfor
        for *all sense s_k of w* **do**

$$F_k = \{v_m \mid \forall n \neq k, \frac{P(s_k \mid v_m)}{P(s_n \mid v_m)} > \alpha \; (usually \; \alpha = 1)\}$$

        endfor
    endwhile
    One sense per discourse
    for *all documents d_i* **do**
        determine the majority sense of w in d_i
        assign all occurrences of w in d_i to this sense.
    endfor
end

9.5.3 WordNet-based methods

We cite in a separately defined section the WordNet (WN) based methods because, until now, WordNet (Miller 1995) has been the most complete and used free machine readable dictionary of English. WN was created by hand in 1990s and maintained at Princeton

University. It includes definitions for individual senses of words, as in a dictionary, also defines groups of synonymous words representing the same lexical concept (synset), and organizes them into a conceptual hierarchy. WordNet 3.0 contains about 155,000 words organized in over 117,000 synsets (Miller 1995, Navigli 2009).

The algorithm of Banarjee and Pedersen.

The algorithm of Lesk was successfully developed in (Banerjee and Pedersen 2002) by using WordNet as a dictionary. Their work improves the original method of Lesk by using WordNet definitions (glosses) augmented with non-gloss information (synonyms, examples) and glosses of semantic related words (hypernyms, hyponyms). Also, the authors introduced a novel overlap measure between glosses that favors multi-word matching. The algorithm introduced in (Banerjee and Pedersen 2002) takes as input a context with a single *target word* and outputs a WN sense for it based on information provided by a few immediately surrounding words that can be derived from WN. Each surrounding word w_i has one or more possible sense, in a number of $|w_i|$. The algorithm evaluates each possible combination of senses, and selects for the *target word* the sense obtained from maximum score combination. Namely, when a pair of senses is treated in a combination, the score is the length of the longest sequence of consecutive words that occurs in both glosses (the *overlap* (Banerjee and Pedersen 2003)). The score of a combination of senses is the sum of scores of every pair of two senses. The combination with the highest score is the winner, and the *target word* is assigned to the sense given in that combination. Thus, the winner combination also provides sense tags for the other words in the context. In (Banerjee and Pedersen 2002) an overall accuracy of 31.7% is reported for a set of 4328 contexts.

CHAD algorithm for WSD

In the paper (Tatar et al. 2007a) an all-word (vs. *target word*) disambiguation algorithm called **chain algorithm** of disambiguation is proposed, CHAD, which presents elements of both points of view about a context:

- because this algorithm is *order sensitive* it belongs to the class of algorithms which depend of relational information;
- because it doesn't require syntactic analysis and syntactic parsing it regards the context as a "bag of word".

CHAD is based on the disambiguation of triplets of words and realizes disambiguation of all words in a text with any length, without increasing the computational complexity.

The disambiguation of a single triplet of words $w_1 w_2 w_3$ using Dice measure consists of the following algorithm, where $D_{w_i}^{s_j}$ is the WordNet gloss of the word w_i with the sense s_j (actually $D_{w_i}^{s_j}$ is considered as a set of stemmed words occurring in the WordNet gloss):

begin
   for *each sense* $s_{w_1}^i$ **do**
     for *each sense* $s_{w_2}^j$ **do**
       for *each sense* $s_{w_3}^k$ **do**

$$score(i,j,k) = \frac{|D_{w_1}^{s_i} \cap D_{w_2}^{s_j} \cap D_{w_3}^{s_k}|}{|D_{w_1}^{s_i}| + |D_{w_2}^{s_j}| + |D_{w_3}^{s_k}|}$$

      endfor
     endfor
   endfor

$(i^*, j^*, k^*) = argmax_{(i,j,k)} score(i, j, k)$

Comment: found sense of w_1 is $s_{w_1}^{i^*}$, found sense of w_2 is $s_{w_2}^{j^*}$, found sense of w_3 is $s_{w_3}^{k^*}$. Further, a found sense is denoted by $*$.

end

For the Overlap measure the score is calculated as:

$$score(i,j,k) = \frac{|D_{w_1}^{s_i} \cap D_{w_2}^{s_j} \cap D_{w_3}^{s_k}|}{min(|D_{w_1}^{s_i}|, |D_{w_2}^{s_j}|, |D_{w_3}^{s_k}|)}$$

For the Jaccard measure the score is calculates as:

$$score(i,j,k) = \frac{|D_{w_1}^{s_i} \cap D_{w_2}^{s_j} \cap D_{w_3}^{s_k}|}{|D_{w_1}^{s_i} \cup D_{w_2}^{s_j} \cup D_{w_3}^{s_k}|}$$

CHAD begins with the disambiguation of a triplet $w_1 w_2 w_3$ and then adds to the right the next word to be disambiguated. Hence, it disambiguates a new triplet at a time, where the first two words are already associated with the best senses and the disambiguation of the third word depends on these disambiguations. The whole algorithm for Dice measure is:

begin
Disambiguate triplet $w_1w_2w_3$
$i = 4$
while $i \le N$ do
    for *each sense* s_j *of* w_i **do**

$$score(s_j) = \frac{|\ D^*_{w_{i-2}} \cap D^*_{w_{i-1}} \cap D^{s_j}_{w_i}\ |}{|D^*_{w_{i-2}}\ | + |\ D^*_{w_{i-1}}\ | + |\ D^{s_j}_{w_i}\ |}$$

    endfor
    $s_i^* := argmax_{s_j}\ score(s_j)$
    $i := i + 1$
endwhile
end

Remark: Due to the brevity of definitions in WN, many values of $|\ D^*_{w_{i-2}} \cap D^*_{w_{i-1}} \cap D^{s_j}_{w_i}\ |$ are 0. It is assigned the first sense in WN for s_i^* in these cases.

Application of CHAD algorithm supposes preprocessing the glosses in WN and replacing inflected words with their stems using http:// snowball. tartarus.org/ algorithms/ porter/ diffs.txt. We tested CHAD on 10 files of Brown corpus, which are POS tagged, treating the following cases: a) disambiguate nouns, b) disambiguate verbs, c)disambiguate nouns, verbs, adjective and adverbs (case noted as All POS). The correctly disambiguated words in our statistics mean for us the same disambiguation as in SemCor. SemCor is a free on-line corpus formed with about 25% of Brown corpus files (that means 352 texts), where the content words have been manually annotated with POS and word senses from WordNet inventory, thus constituting the largest sense-tagged corpus. The comparison of the results of CHAD with Semcor are given in the following tables, for nouns and All POS. CHAD was ran separately for nouns, verbs, and All POS also for Senseval-2 corpus. The precision results obtained for Senseval-2 corpus are slightly lower than those obtained for Brown corpus: the explanation is probably that the order of senses in WordNet corresponds to words frequency in Brown corpus and CHAD is an WordNet based algorithm.

As CHAD algorithm is dependent on the length of glosses, and as nouns have the longest glosses, the highest precision is obtained for nouns.

Table 1.1: Precision for Nouns, sorted descending by the precision of Overlap measure.

File	Words	Dice	Jaccard	Overlap	WN1
Bra01	486	0.758	0.758	0.767	0.800
Bra02	479	0.735	0.731	0.758	0.808
Bra14	401	0.736	0.736	0.754	0.769
Bra11	413	0.724	0.726	0.746	0.773
Brb20	394	0.740	0.740	0.743	0.751
Bra13	399	0.734	0.734	0.739	0.746
Brb13	467	0.708	0.708	0.717	0.732
Bra12	433	0.696	0.696	0.710	0.781
Bra15	354	0.677	0.674	0.682	0.725
Brc01	434	0.653	0.653	0.661	0.728

Mihalcea and Moldovan's algorithm of disambiguation, using Word- Net

In (Mihalcea and Moldovan 2000) the authors report a very efficient algorithm of WSD, which disambiguated a part of 55% nouns and verbs with 92% precision using WordNet and SemCor corpus. There are 8 Procedures in the main algorithm which are called in an iterative manner: after the Procedure i is ended, the Procedure $i + 1$

Table 1.2: Precision for All POS, sorted descending by the precision of Overlap measure.

File	Words	Dice	Jaccard	Overlap	WN1
Bra14	931	0.699	0.701	0.711	0.742
Bra02	959	0.637	0.685	0.697	0.753
Brb20	930	0.672	0.674	0.693	0.731
Bra15	1071	0.653	0.651	0.684	0.732
Bra13	924	0.667	0.673	0.682	0.735
Bra01	1033	0.650	0.648	0.674	0.714
Brb13	947	0.649	0.650	0.674	0.722
Bra12	1163	0.626	0.622	0.649	0.717
Bra11	1043	0.634	0.639	0.648	0.708
Brc01	1100	0.625	0.627	0.638	0.688

is called. The goal of the algorithm is to disambiguate each occurrence of nouns and verbs in a text (corpus) T, and the descriptions of each Procedure are:

- Procedure 1: Recognize the named entities in the text. From now they are disambiguated words.
- Procedure 2: Identify the words in T which are monosemous (one sense in WordNet) and disambiguate them with the single sense.
- Procedure 3: If a target word (to be disambiguated) is in position i in the text, denote it by W_i. Form the pairs $W_{i-1}W_i$ and W_iW_{i+1} and extract from the tagged corpus SemCor all the tagged correspondent pairs. If in these tagged occurrences the word W_i has the sense #k in a number which is larger than a given threshold, then disambiguate the word W_i as having the sense #k.
- Procedure 4 (only for nouns): For each sense j of a noun N determine a list of hypernyms and a list of nouns which appears in given length windows of the each tagged noun N#j in Semcor. These lists, for each sense, represent the noun-contexts of N, C_j. For a given occurrence of N as a target word (to be disambiguate) select the sense which maximize *clearly* the cardinal of $C_j \cap Win$, where Win is a window of a given dimension of the occurrence of N. The term *clearly* means that the cardinal of $C_j \cap Win$ for the sense j is much bigger than the other senses.
- Procedure 5. Find all the pairs of words in the text T which are semantically connected (belong to the same synset in WordNet) but one of them is already a disambiguated word. Tag the not yet disambiguated word with the correspondent tag.
- Procedure 6. Find all the pairs of words in the text T which are semantically connected (belong to the same synset in WordNet) but none of them is an already disambiguated word. Tag the pairs with the senses in the synset (let us remark that the procedure is very expensive, but Procedure 6 is called after Procedure 1 to 5 which has already tagged a number of words).
- Procedure 7. Find all the pairs of words in the text T which are semantically connected at the distance 1 (belong to an hypernymy/hyponymy relation in WordNet) but one of them is already a disambiguated word. Tag the not yet disambiguated word with the correspondent tag.

- Procedure 8. Find all the pairs of words in the text T which are semantically connected at the distance 1 but none of them is an already disambiguated word (similarly with the Procedure 6). Tag the pairs with the senses in the relation.

The sequence of procedures 1 to 8 is applied (in this order) on T such that, after each procedure, the number of disambiguated words grows. Of course, there are nouns and verbs that remain undisambiguated. But for those disambiguated, the precision is very good (Mihalcea and Moldovan 2000).

Disambiguation with WordNet and a bilingual dictionary

In (Brun 2000) the author presents an algorithm for disambiguation using an English-French dictionary and WordNet. From the description of the target word in the dictionary some rules for disambiguation are constructed. Defining a distance measure between rules and applying these rules in a decreasing order of this distance, the translation for the target ambiguous word is performed.

The on-line dictionary uses some SGML tags, as for example:

- $< S1 >, \ldots </S1 >$ for different POS;
- specification of POS between $< PS >$ and $< /PS >$;
- definitions of different senses between $< S2 >, \ldots < /S2 >$;
- different translations between $< TR >, \ldots < /TR >$;
- collocations of the target word between $< CO >, \ldots < /CO >$;
- examples between $< LE >, \ldots < /LE >$;
- idioms examples between $< LI >, \ldots < /LI >$

In (Brun 2000) the algorithm is illustrated for the target word *seize*. The reduced description in a SGML form dictionary (where we emphasized the word *seize* by quotes "and") is as below. Let us remark that there are 4 senses of *seize* as intransitive verb (0.I.1,0.I.2,0.I.3,0.I.4) and one sense as transitive verb (0.II.1).

<SE>

<HW>"seize"</HW>

    <S1> <PS> vtr </PS>

0.I.1 <S2>

    <TR> saisir <CO> person, object</CO> </TR>

    <TR> <LE> to "seize" somebody around the waist </LE>

saisir par la taille </TR>

    <TR><LI> to "seize" hold of </LI> se saisir de
                        <CO> person </CO></TR>

    <TR> s'empare de <CO> object</CO> </TR>

    <TR> sauter sur <CO> idea</CO> </TR>

    </S2>

0.I.2 <S2>

    <TR> saisir<CO> opportunity, moment</CO></TR>

    <TR> prendre <CO> initiative </CO> </TR>

    <TR><LI> to be "seized" by </LI> etre pris de
                        <CO> emotion</CO> </TR>

    </S2>

0.I.3 <S2>

    <TR> s'emparer de <CO>territory, hostage,
prisoner,installation</CO></TR>

      <TR> prendre<CO> control </CO></TR>

    </S2>

0.I.4 <S2>

      <TR> saisir <CO> arms, drugs, property</CO></TF>

      <TR> apprehender <CO> person </CO></TR>

    </S2>

    </S1>

0.II.1<S1> <PS> vi</PS>

    <TR> <CO> engine, mechanisme</CO> se gripper </TR>

    </S1>

</SE>

The rules (called at the word level) obtained from this analysis, for example for the sense 0.II.1, are:

$$SUBJ(engine, seize) \rightarrow 0.II.1$$
$$SUBJ(mechanism, seize) \rightarrow 0.II.1$$

Consider that the phrase to be translated is (Brun 2000):

The police seized a man employed by the X branch of Y Society on approximately 1985.

The relations obtained by the syntactic analyze of this phrase are:

SUBJ(police, seize)

DIR – OBJ(seize, man)

VERB – MODIF(seize, about, 1985)

VERB – MODIF(seize, of, society)

VERB – MODIF(seize, by, branch)

If the left size of a rule from syntactic analyzer unifies with the left size of a rule at the word level obtained from the dictionary, the disambiguation is made up with the right size of a sense rule (for our phrase this is not the case). Otherwise, there are formed new rules using WordNet, for example:

a. *DIR – OBJ(seize,* [18]) → 0.*I*.3, obtained from

DIR – OBJ(seize, *prisoner*) → 0.*I*.3 and

b. *DIR – OBJ(seize,* [18]) → 0.*I*.1, obtained from

DIR – OBJ(seize, somebody) → 0.*I*.1.

Here [18] is the semantic class of *noun.animate* in WordNet (the semantic class for *prisoner* and *somebody*). The rules have been obtained from occurrence of the words *prisoner* in the sense 0.I.3 and *somebody* in the sense 0.I.1. Let denote they by semantic class rules. A distance between the semantic class rules R1 and R2 is defined in (Brun 2000) as:

$$d = \frac{card(L_1 \cup L_2) - card(L_1 \cap L_2)}{card(L_1 \cup L_2)}$$

where L_1 is the list of semantical classes from the semantic class rule R1, and L2 is the list of semantical classes from the semantic class rule R2, as obtained from WordNet.

Defining a distance measure between rules permits us for applying rules in a decreasing order of distance.

In the above example the rules with the least distance from

DIR – OBJ(seize, man) are two:

a. *DIR – OBJ(seize,* [18]) → 0.*I*.3,

b. *DIR – OBJ(seize,* [18]) → 0.*I*.1.

When the rules have the same distance (as in our case, d=1), some rules of preference are established. For example, if a rule is obtained using collocations, it is preferred to a rule obtained from compound example. In our example *DIR–OBJ(seize, prisoner)* → *0.I.3* is obtained from a collocation and *DIR–OBJ(seize, somebody)* → *0.I.1*, is obtained from an example, so the sense is 0.I.3. The translation of *The police seize a man....* will thus be *Police s'empare d'un homme...*.

The system based on the algorithm was evaluated for English on the 34 words used in Senseval competitions '98 and '99 as well as on the Senseval corpus Hector. For the test set of around 8500 sentences the precisions obtained were the following: for nouns 83.7%, for adjectives 81.3 % and for verbs 75% (Brun 2000).

Exploring parallel texts

Let on mention as a final remark that parallel texts, beside bilingual dictionaries, could be useful for WSD. Research in (Ng et al. 2003) and others has shown how parallel corpora in English and Chinese are used for training WSD examples. Namely, after the corpora are sentence-aligned and word-aligned, some possible translations of each sense of an English word w are assigned. From the word alignment those occurrences of w which were aligned to one of the Chinese translation chosen are selected. The English contexts of these occurrences serve for training data of w. The method was used to automatically acquire sense tagged training data, used further for disambiguating with high score the nouns in the Senseval-2 contest (in the coarse-grained English all-words task).

9.6 Evaluation of WSD Task

The battle in WSD is continuously done in the domain of the efficiency (precision and recall) of different methods. A short picture of the evaluation frameworks is given in this section.

Given the variety in the domain, it is very difficult to compare one method with another. Producing a gold standard annotated corpus is both expansive (many person-months of effort) and subjective (different individuals will often assign different senses to the same word-in-context). This led to some conferences dedicated to a comparison of different methods in WSD. In April 1997 a workshop of ACL included for the first time a session of WSD evaluation (Resnik and Yarowski 1998). Beginning with 1998 (then in Sussex, England) was opened a series of some WSD evaluation workshops, named SENSEVAL (since 2007 the name is SEMEVAL). In the first edition 25 systems and 23 teams participated and most of the research has been done in English. The test set contained 8400 instances of 35 target words and the best systems performed with between 74 % and 78% accuracy (Navigli 2009). Senseval 2, which took place in Toulouse (France) in 2001, had 12 languages, 93 systems and 34 participating teams. The performance was lower than in Senseval 1, probably due to a much larger number of target words (Navigli 2009). In 2004 (Senseval 3, Barcelona) 160 systems from 55 teams participated for English, French, Spanish and Italian languages. In that edition, was a task for Romanian language organized for the first time, with seven participating teams. Our participation at Senseval 3 is presented in (Serban and Tatar 2004). The very hight performance for English task ranged between 70.9% and 72.9% (Navigli 2009). The fourth edition of Senseval, now renamed Semeval, took place in 2007 in Prague (Czeck Republic) and contained some tasks of semantic analysis. For the English task the contest was divided into fine-grained and coarse-grained disambiguation. For coarse-grained exercise the best participating system attained an 82.5% accuracy. Beginning with Semeval 2007, the calendar and the panel of tasks was drastically enlarged.

A general remark could be made on Senseval (Semeval) competitions: almost half of the systems used supervised training methods, memory-based learning and SVM method. In fact, these approaches proved to be among the best systems in several competitions.

The evaluation involves a comparison of the output of each system using as measures the *precision* and the *recall*. The *precision* is defined as the number of correct answers provided over the total number of answers provided. The *recall* is the number of correct answers provided over the total number of answers to be provided. In WSD literature *recall* is also referred to as *accuracy*.

The upper bound for accuracy of a WSD system is usually human performance. This is between 97% and 99 % (Manning and Schutze 1999). The lower bound is the performance of the simplest algorithm, baseline, usually consisting of the assignment of all contexts to the most frequent sense. Another method of evaluation of WSD is the so called pseudo-word testing. A pseudo-word is a new word formed by two distinct words, which is considered a single word with two senses. Replacing all instances of both words by the pseudo-word, a good algorithm will identify the instances with the first sense in the contexts with the first word, and with the second sense in others. The performance of pseudo-words is better than that of real words.

9.6.1 The benefits of WSD

Disambiguation and Information Retrieval

It is clear that using a WSD system is beneficial especially when this is integrated in another NLP application, where WSD is only an intermediate task. The problem is how good the performance of WSD system must be for a real apport to this application. A special case is that of Information Retrieval (IR) where the problem of finding whether a particular sense is connected with an instance of a word is likely the IR task of finding whether a document is relevant to a query. The search engines today do not use explicit methods to eliminate documents which are not relevant to a query. An accurate disambiguation of documents and queries may allow one to eliminate documents containing the same words with different meanings (increasing the precision) and retrieve documents with the same meaning but with different word (increasing the recall) (Navigli 2009). In fact, (Sanderson 1994) asserted that the benefit in IR performance could be observed only for WSD system performing with at least 90%. More exactly, he concluded that IR is very sensitive to erroneous disambiguation, especially for short queries.

The statement is attenuated in (Stokoe et al. 2003) where it is proved that with WSD accuracy of 62% an improvement of 1.73% can still be obtained. As IR is used by millions of users, small average of improvement could be seen as very significant. More exactly, the idea was that benefits from disambiguation may not be found from the overall WSD accuracy but rather how successful a WSD system is at disambiguating the rare case where a word is used in an infrequent sense.

In (Ide and Wilks 2006) it is observed that the standard fine-grained division of senses may not be an appropriate goal for IR and that the level of sense-discrimination needed corresponds roughly to homographs. That means, even the WordNet has become a standard in sense inventories in WSD, the resulting distinction is very finely grained. In fact IR, as oppositely to other applications (for example Machine Translation), demands only shallow WSD.

In (Schütze 1998) the author makes a proposal that is beneficial for IR. In this proposal the features in the definition of vectors are senses and not words: a feature in a context has a nonzero value if the context contains a word assigned to the sense represented by the feature. This method increased performance by 7,4 % compared to "features equal words" case.

Another recent work advocating the improve of IR by WSD is (Zhong and Ng 2012). It proposes the idea to annotate the senses for short queries, to incorporate word senses into the language modeling approach of IR, and utilize sense synonym relations to further improve the performance of IR. Using one's own language model, the probability of a query term in a document is enlarged by the synonyms of its sense. Consequently, documents with more synonyms senses of the query terms will get higher retrieval rankings.

Disambiguation and Machine Translation

Statistical machine translation (SMT) is consistently improved incorporating the predictions of a WSD system (Carpuat and Wu 2007). The results are better for phrasal-based SMT and multi-word phrasal WSD. The WSD task directly disambiguates between all phrasal translation candidates seen during SMT training. In this order, building WSD systems for the specific purpose of translation seems to be a good solution. (Carpuat and Wu 2007) shows that integrating the WSD module, the Chinese-English translation of 3 different available test sets is improved, accordingly to some different standard measures.

9.7 Conclusions and Recent Research

To conclude this Chapter let us remark that the problem of WSD is AI complete and that it is one of the most important open problems in NLP (Ide and Veronis 1998, Navigli 2009).

Despite the high performance, the supervised WSD systems have an important drawback, namely their limitation to the words for which sense tagged data is available. The battle in this time is to generate sense-tagged data which can be used to construct accurate sense disambiguating tools. The newest approach for this goal is the use of Wikipedia, the biggest free online encyclopedia (Mihalcea 2007). The basic entry in Wikipedia is an article (or page) and consists of a document with hyperlinks to other articles within or outside Wikipedia. Each article is uniquely referenced by an identifier, and hyperlinks are created using these unique identifiers. Starting with a given ambiguous word, the sense tagged corpus is derived following three main steps (Mihalcea 2007): the first step consists in extracting all the paragraphs in Wikipedia that contain an occurrence of the ambiguous word as part of a link. The second step involves collecting all the possible labels by extracting the leftmost component of the hyperlinks. Finally, the labels are manually mapped to their corresponding WordNet sense. To evaluate the quality of the sense tagged corpus generated using the algorithm from Wikipedia, 30 ambiguous nouns used during Senseval 2 and Senseval 3 were disambiguated, and the average of precision was 84.65% (when the average of examples was of 316 for each ambiguous noun). Beginning with this seminal paper, many works approach the problem of WSD in connection with Wikipedia, as for example (Fogarolli 2011).

Another recent direction is WSD with multilingual feature representation. In (Banea and Mihalcea 2011) a new supervised WSD method is investigated that can handle more than two languages at a time, so that the accuracy of the disambiguation increases with the number of languages used. The method consists of the translation of each context from English in French, German, and Spanish (with Google Translate) and the construction of a multilingual vector. The method allows the vector space model to capture information pertaining to both the target word and its translation in the other languages. It has been applied to two datasets and the highest accuracy was achieved when all languages are taken into consideration: 86.02% for first dataset and 83.36% for the second.

A paper addressed also to multilingual processing is (Navigli and Ponzetto 2012). This paper observes that the growing amounts of text in different languages has been a major cause of growing research on multilingual processing. Recent organization of SemEval 2010 tasks on cross-lingual WSD and cross-lingual lexical substitution proves this. However, the lack of resources is still a problem in the development of effective multilingual approaches to WSD. (Navigli and Ponzetto 2012) provides a contribution in this domain consisting of an API for efficiently accessing the information available in BabelNet, a very large knowledge base with concept lexicalization in

6 languages (at the word sense, named entities and conceptual levels). The disambiguation API takes as input a sentence, a knowledge base (BabelNet, WordNet) and a graph connectivity measure. It creates a graph for the input words, then scores their senses and disambiguate the sample sentence. BabelNet and its API are available for download at http://lcl.uniroma1.it/babelnet.

CHAPTER 10

Text Entailment

10.1 Introduction

Text Entailment relation between two texts: T (the text) and H (the hypothesis) represents a fundamental phenomenon of natural language. It is denoted by $T \to H$ and means that the meaning of H can be inferred from the meaning of T (or T *entails* H). Even this definition suggests that the Text Entailment (TE) problem concerns a semantic aspect (the word *meaning*) and a logical aspect (the word *inference*). Indeed, the recognition of textual entailment is one of the most complex tasks in natural language processing and the progress on this task is the key to many applications such as Question Answering, Information Extraction, Information Retrieval, Text summarization, and others. For example, a Question Answering (QA) system has to identify texts that entail the expected answer. Given a question (Q), and turning this into an correct answer (Q'), the text found as answer (A) entails the expected answer (Q'), ($A \to Q'$). Similarly, in Information Retrieval (IR) the concept denoted by a query expression (Q) should be entailed by relevant documents retrieved (R), ($R \to Q$). In text automated summarization or (TS) a redundant sentence or expression (R), to be omitted from the summary (S), should be entailed by other sentences in the summary ($S \to R$). In Information Extraction (IE) entailment holds between different text variants that express the same target extraction. In Machine Translation (MT) a correct translation should be semantically equivalent to the standard translation, and thus both translations have to entail each other. Let it be remarked that the paraphrasing problem is in fact related with TE, as a bidirectional entailment: T_1 is a paraphrase of T_2 if T_1 entails T_2 and T_2 entails T_1. Thus, in a similar way with Word Sense Disambiguation which is recognized as a generic task, solving textual entailment may consolidate the performance of a number of different tasks of NLP (QA, IR, MT, IE, TS).

The situation of TE is opposite to the state of affairs in syntactic processing, where clear application-independent tasks and communities have matured. The efforts in TE field can promote the development of entailment recognition engines which may provide useful generic modules across applications (Dagan et al. 2009).

Although the problem of text entailment is not new, most of the automatic approaches have been proposed only recently within the framework of the Pascal Textual Entailment Challenges RTE-1 (2005), RTE-2 (2006) and RTE-3 (2007). The Proceedings of these first conferences could be found at the address http://pascallin.ecs. soton.ac.uk/Challenges/RTE/. The Fourth PASCAL Recognizing Textual Entailment Challenge was organized in 2008 (http:// www. nist.gov/tac/), and the further challenges in 2009 (http://www.nist. gov/tac/2009/RTE/), 2010 (http:// www.nist.gov/tac/2010/RTE/) and 2011 (http:// www.nist.gov/tac/2011/RTE). In 2012 the CLTE (Cross-Lingual Text Entailment) task from the SemEval 2012 contained an important participation of forces: 10 teams, and 92 runs. Consequently, beginning with 2005 and until now Text Entailment gets a continue attention from the top of NLP researchers. One will analyze only RTE-1 and RTE-2 challenges shortly.

10.2 Methods and Algorithms: A Survey of RTE-1 and RTE-2

A common feature for both contests RTE-1 and RTE-2 was the launch before the fixed dates of the contest of a Dataset formed by Development dataset (for training) and Test dataset (for test). The features of Dataset are:

- They are formed by 576 pairs and 800 pairs for training and test respectively: a pair is of the form (T,H) and for Development data set the entailment relation between T and H is specified (for RTE-2 and the following contests, Development dataset was formed by 800 pairs);
- the pairs are balanced between *True* (50%) and *False* (50%);
- the pairs are of different levels of difficulty;
- the examples have been collected from different domains: QA, IR, MT, IE, TS in an equal proportion;

- each *T* and *H* texts are formed by only one sentence (beginning with RTE-3 they were enable to contain more than one sentence).

The evaluation measures were the following:

- Accuracy or Precision: $Precision = \dfrac{nr.of.correct.responses}{total.number.of.pairs}$
- Confidence weighted score (CWS):

$$CWS = \frac{1}{N} \sum_{i=1}^{N} \frac{number.of.correct.resp.up.to.rank.i}{i}$$

It is also possible to evaluate TE recognition methods indirectly, by measuring their impact on the performance of larger natural language processing systems. For instance, one could measure the difference in the performance of a QA system with and without a TE recognition module, or the degree to which the redundancy of a generated summary is reduced when using textual entailment recognizers (Androutsopoulos and Malakasiotis 2010).

At the RTE-1 contest 16 teams participated and the accuracy results were between 49.5% and 58.6% while at the RTE-2 contest 23 teams participated, with the results between 52.5% and 75.38% accuracy. The best results were thus considerably higher than in the previous year. In order to make the challenge data more accessible, some pre-processing for the examples was provided, including sentence splitting and dependency parsing. A table with the names (in alphabetical order), short descriptions of systems, and the performances regarding RTE-1 could be found in (Dagan et al. 2005). A similar table for RTE-2 is presented in (Bar-Haim et al.2006). The RTE systems results demonstrate general improvement with time, with overall accuracy levels ranging from 49% to 80% on RTE-3 (26 submissions) and from 45% to 74% on RTE-4 (26 submissions, three-way task) (Dagan et al. 2009).

The RTE challenges provide a significant thrust to recognition entailment work and they have helped establish benchmarks and attract more researchers (Androutsopoulos and Malakasiotis 2010). Some of the different ideas of the participants will be surveyed shortly further. A first remark is that almost all teams used supervised machine learning techniques, with a variety of features, including lexical-syntactic and semantic features, document co-occurrence counts, first-order logic, syntactic rules. All these features train a classifier on Dataset offered by the challenge. Once trained, the classifier can classify new pairs as correct or incorrect textual entailment pairs by examining their features.

Most of the approaches require intensive processing to be carried out over the (T,H) pairs, and the final results often depend crucially on the quality of the tools and resources which are used. The most exploited of such resources is WordNet (Fellbaum 1998), with its extensions (e.g., EuroWordNet, eXtended WordNet). Also verb-oriented resources such as VerbNet and FrameNet have been used by some systems. A new tendency in considering the web as a resource arises in the successful use of Wikipedia by some participating systems, in order to extract entailment rules, named entities and background knowledge. Furthermore, various text collections are exploited as sources of information (such as the Reuters corpus and gazetteers), to extract features based on documents co-occurence counts and draw lexical similarity judgements (Dagan et al. 2009).

As a matter of fact, the ability to manage and combine good linguistic tools and resources has proven to be one of the key factors enabling high performance of some systems. Let the case of *Ilickl (LCC)* team, ranked on first place at RTE-2 with a highest accuracy of 75.38% be cited. In this case many preprocessing and original features are used in addition to syntactic parsing: POS tagging; stemming; Named Entity annotation and their co-reference resolution; dates, times and numeric values identification; semantic parsing and predicate argument annotation; paragraphs matching verification; negation and polarity features. Finally, after features extraction, a classifier is trained to select a *True* or *False* decision for a pair (T,H). No explicit logical aspects are treated in the frame of this approach. However, these are implicitly used in paraphrase acquisition, as background knowledge.

10.2.1 Logical aspect of TE

Thinking of textual entailment in terms of logical entailment allows us to borrow notions and methods from logic. A frame for this parallelism is the well known fact that a linguistic text can be represented by a set of logical formulas, called logic forms. Various methods were given for associating logical formulas with a text: (Thayse 1990, Harabagiu and Moldovan 1999, Rus 2001, Bos and Markert 2006). From the logical point of view, if each sentence is represented as a logical formula, proving a textual inference consists in showing that a logical formula is deducible or inferable from a set of others formulas. The problem is a classical (semidecidable) one. In the last few years, when text mining was very important in many AI applications, text inference from both points of view, theorem proving and linguistics perspective, has been a very active field of research.

If one regards the relation of entailment as logical implication $T \to H$ then one must distinguish between two kinds of logical proof:

- directly;
- by refutation.

A *direct* proof of logical implication could be made as $\vdash T \to H$ ($T \to H$ is a theorem) or by using deduction theorem $T \vdash H$ (*H* is deductible from *T*). For example, in the second case one must construct a sequence of formulas, $U_1, \ldots, U_n$, where $U_n = H$, such that:

a) U_i is *T* or
b) U_i is an axiom or
c) takes place: from $U_j, U_j \to U_i$, by *modus ponens* rule, U_i is obtained.

To prove $T \vdash H$ by *refutation* means to prove that the set of disjunctive clauses obtained from *T* and $\neg H$ (the set denoted by $C_{\{T,\neg H\}}$) is a contradiction. By disjunctive clause one means a disjunction of literals (negated or not negated atoms). A set of disjunctive clauses is a contradiction if, by Robinson's theorem, the empty clause denoted by [ ], is obtained by the repeated application of *resolution* rule to the set of disjunctive clauses. The resolution is defined as follows:

Definition

Two (disjunctive) clauses c_i and c_j provide by *resolution* the (disjunctive) clause c_k written as

$$c_i, c_j \models_{resolution} c_k$$

if $c_i = l \lor c'_i$, $c_j = \neg l' \lor c'_j$, *l* and *l'* are unifiable with the unificator σ. The resulting clause is $c_k = \sigma(c'_i) \lor \sigma(c'_j)$.

In this case, to prove $T \vdash H$ by *refutation* it is enough to prove

$$C_{\{T,\neg H\}} \models^*_{resolution} [\,]$$

Many systems participating at RTE-1 and RTE-2 used logical methods and some "classical" theorem provers, many available on web.

10.2.2 Logical approaches in RTE-1 and RTE-2

One will begin with RTE-1 contest. The first (alphabetically) team, *Akhmatova (Macquarie)* obtaining the accuracy of 51.9 used the automated theorem prover OTTER, where the texts T and H have been translated in logic forms. Synonyms in WordNet are translated in equivalent logic forms. Another team approaching the entailment relation by logic is *Bos(Edinburgh and Leeds)* with 59.3% accuracy. They used a system of axioms (*BK*) obtained fromWordNet (ex: if w_2 is an hypernym of w_1, or w_1 "*is–a*" w_2 then the following formula is generated: $(\forall x)(w_1(x) \rightarrow w_2(x))$. These formulas are the inputs for the theorem prover Vampire and the following implication is checked: $BK \wedge T \rightarrow H$. The team also used an overlapping condition, with the priority on the result of theorem prover (the overlap is checked only if the prover fails). In spite of sophisticated tools, the system does not correctly solve some pairs (obtaining false positive) where the overlap is big but the prover doesn't state the entailment. However, with the same method, they obtained at RTE-2 a better result of 61.6%. A similar approach is made by *Fowler (LCC)* team (56%) regarding the logical aspect. The automated theorem prover by refutation is now COGEX (Rus 2001). A second logical way is used by *Raina (Stanford)* (62.1%) which obtains scored axioms from WordNet and constructs scored proofs, such that the minimum score (below a certain threshold) is searched (see (Raina et al. 2005)). The team *Tatu (LCC)*, ranked second at RTE-2 with an accuracy of 73.75%, transformed the texts T and H into logical forms and used COGEX as a refutation theorem prover. They rely on some logical properties of implication, as for example the cases when formulas of the form $\exists \rightarrow \exists$, $\forall \rightarrow \forall$ or $\forall \rightarrow \exists$ are valid. The cases are pointed out by a system of penalizing some modifiers of T and H.

10.2.3 The directional character of the entailment relation and some directional methods in RTE-1 and RTE-2

Only a few authors exploited the *directional* character of the entailment relation, which means that if $T \rightarrow H$ holds, it is unlikely that the reverse $H \rightarrow T$ also holds. From a logical point of view, the entailment relation is alike to the logic implication which, contrary to the equivalence, is not symmetric.

The most notable directional method used in RTE-1 Challenge was that of *Glickman(Bar Ilan)* (Glickman et al. 2005). He uses as criterion: *T entails H iff P(H | T) > P(H)*. The probabilities are calculated on the base of Web. Namely, if $T = (t_1, \ldots, t_n)$ and $H = (h_1, \ldots, h_m)$ then:

$$P(H \mid T) = \prod_{i=1}^{m} P(h_i \mid T) = \prod_{i=1}^{m} max_{t_j \in T} P(h_i \mid t_j)$$

The probabilities $P(h_i \mid t_j)$ are calculated using the Web. The accuracy of the system is the best for RTE-1 (58,5%). Let it be remarked that the system does not rely on syntactic analysis or parsing.

Another directional method is that of *Kouylekov (IRST)* (Kouylekov and Magnini 2005), who use the criterion: *T entails H* if there is a sequence of transformations applied to the dependency graph of *T* such that the dependency graph of *H* is obtained with a total cost below of a certain threshold. The output of a dependency grammar parser is a graph (usually a tree) whose nodes are the words of the sentence and whose (labeled) edges correspond to syntactic dependencies between words (for example the dependency between a verb and the head noun of its subject noun phrase, or the dependency between a noun and an adjective that modifies it (Androutsopoulos and Malakasiotis 2010)). The following transformations are allowed:

- Insertion: insert a node from the dependency tree of *H* into the dependency tree of *T*.
- Deletion: delete a note from the dependency tree of *T*. When a node is deleted all its children are attached to its parent.
- Substitution: change a node of *T* into a node of *H*.

Each transformation has a cost and the cost of edit distance between *T* and *H*, denoted by *ed(T, H)*, is the sum of costs of all applied transformations. The entailment score of a given pair is calculated as:

$$score(T, H) = \frac{ed(T, H)}{ed(\,, H)}$$

where $ed(\ ,H)$ is the cost of inserting the entire tree H. If this score is below a learned threshold, the relation $T \to H$ holds.

Another team, *Punyakanok (UIUC)* (accuracy 56.9%) uses a definition which in terms of representation of knowledge as feature structures could be formulated as: *T entails H if H subsumes T* (Salvo Braz et al. 2005).

In all the above mentioned directional methods, the following criterion formulated in (Monz and Rijke 2001) could be verified: *T entails H if H is not informative in respect to T*. One will revert to this last property when one introduces other directional methods.

10.2.4 Text entailment recognition by similarities between words and texts

There are many measures that exploit WordNet (or similar resources) and compute the semantic similarity between two words or two texts (see (Budanitsky and Hirst 2006) for some measures evaluations). Some directional methods used in RTE-1 and RTE-2 contests use different kinds of similarity between texts. For example, in RTE-1 challenge, the team *Jijcoun (Amsterdam)* (accuracy 55.2%) calculates for each pair (T,H) a similarity $sim(T,H)$. They consider T and H as a bag of words: $T = \{T_1, \ldots, T_n\}$ and $H = \{H_1, \ldots, H_m\}$ and use as $sim(T_i,H_j)$ the WordNet similarity between the words T_i and H_j. $sim(T,H)$ is the sum of the maximal similarity between each H_j and all T_i over a weight depending on $idf(H_j)$. If the entity $sim(T,H)$ is bigger than a threshold, then the answer is *True* (which means $T \to H$). As a remark, $sim(T_i,H_j)$ is a symmetrical relation but $sim(T,H)$ is not, because of the different role of T and H in the calculus. This is not surprising, since the entailment relation is not a symmetrical relation.

The syntactic matching between T and H also measures a kind a similarity between these texts. The group *Raina (Stanford)* (accuracy 56.3%) (Dagan et al. 2005) represented each sentence as a dependency graph (G_T and G_H) and stated that, if a matching M exists from the nodes of G_H to the nodes of G_T and from the edges of G_H to the edges of G_T, so that the cost of this matching is low, then $T \to H$. The cost of matching two nodes (words) is equal to the similarity between words and the cost of matching two edges depends on the dependency relation (it is 0 for identity relation). Of course, the more similar is the text T to H, the smaller the cost of matching from G_T to G_H, denoted by *MatchCost(T,H)*, is. It may again be noted that *MatchCost(T,H)* and *MatchCost(H,T)* are not equal.

The following directional approach of the teams participating at RTE-1 is that of *Zanzotto (Rome-Milan)* (accuracy 52.4%). They explicitly introduce a measure of text similarity using the concept of anchor. An anchor is a pair of words, (T_i, H_j), T_i is a word of T and H_j is a word of H, so that, for a fixed word T_i, H_j is the most similar to T_i. When the set of anchors is denoted by A, the similarity between T and H is defined as:

$$sim_{(T,H)} = \frac{\sum_{(T_i,H_j)\in A} sim(T_i,H_j) \times idf(H_j)}{\sum_{H_j} idf(H_j)}$$

The similarity between words, $sim(T_i,H_j)$, is calculated here as WN similarity. If $sim_{(T,H)}$ is bigger than a given threshold, then T entails H.

This measure is very close to the (Corley and Mihalcea 2005) one, introduced in the next section. Let us remark again that $sim_{(T,H)} \neq sim_{(H,T)}$ correlates with the directional character of entailment relation.

As regarding RTE-2, the contribution of *Adam (Dallas)* (accuracy of 62.6%), relies on a similarity measure formed by two cumulated components: one is a WordNet similarity, the second is a probability. Thus, if $T = \{T_1, \ldots, T_n\}$ and $H = \{H_1, \ldots, H_m\}$ then $sim(T_i,H_j)$ is defined as:

$$sim(T_i,H_j) = sim_{WN}(T_i,H_j) + \alpha \times P(H_i \mid T_j).$$

where $sim_{WN}(T_i,H_j) = 1 - 0.1 \times length\ of\ shortest WN\ lexical\ chain(T_i,H_j)$. In a lexical chain all Word Net relations could occur (hypernyms, hyponyms, cause, etc). The calculus of $sim(T_i,H_j)$ is used for establishing a map M between the words of H and of T: $M(H_i) = T_j$ if T_j has the highest similarity score with H_i. Finally, a base score of entailment $T \rightarrow H$ is calculated as:

$$score_{BASE} = \prod_{(H_i,T_j)\in M} sim(T_i, H_j).$$

The criterion is: $T \rightarrow H$ if $score_{BASE} \geq \alpha$ where α is a learned threshold. The further similarity based contribution is that of *Ferrandez (Alicante)*, with an accuracy of 55.63%. The system is composed of two components: the first performs derivation of logic forms of T and H using eXtended WN (Rus 2001, Harabagiu' and Moldovan 1999) and the second calculates the similarity between the obtained logic forms. This similarity accounts only for verbs and again fixes a verb in H and searches the most similar verb in T. The similarity between verbs is calculated in two ways: the weight of lexical chains connecting them and the Word Net similarity.

Another RTE-2 matching-based contribution is *Rus (Memphis)* with the accuracy of 59.0%. This relies on the parse trees obtained using Charniak parser, on a mapping between the nodes and the edges of the T parse tree and H parse tree. An entailment score (*entscore*) is calculated from the matching scores between nodes, $match(v_H, v_T)$, and the matching scores between edges, $match(e_H, e_T)$. A score of $match(v_H, v_T)$ is 1 if nodes v_H and v_T are equal, is 0.5 if the nodes are synonym words and is 0 otherwise. A similar score is calculated for $match(e_H, e_T)$. If the sets of nodes are V_T and V_H and the sets of edges are E_T and E_H, the entailment score $entscore(T,H)$ is defined as:

$$entscore(T,H) = (\alpha \frac{\sum_{v_H \in V_H} max_{v_T \in V_T} match(v_H, v_T)}{|V_H|} +$$

$$\beta \frac{\sum_{e_H \in E_H} max_{e_T \in E_T} match(e_H, e_T)}{|E_H|} + \gamma) \times \frac{1 + (-1)^{nr. of\ .negations\ .in\ T \cup H}}{2}$$

Here α, β, γ are from the interval [0,1]. The criterion is:

$T \rightarrow H$ if $entscore(T,H) \geq \tau$ where τ is a given threshold. Let it be remarked again that $entscore(T,H) \neq entscore(H,T)$.

The last method based on the similarity mentioned here is the one used by *Schilder (Thomson & Minnesota)*, which reached an accuracy of 54.37% (Bar-Haim et al. 2006). They used the directional similarity of the texts introduced in (Corley and Mihalcea 2005) to calculate $sim(T,H)_T$ (see the next section) and stated that if $sim(T,H)_T \geq \tau$, where τ is a given threshold, then $T \rightarrow H$. Again, here $sim(T,H)_T \neq sim(T,H)_H$.

10.2.5 A few words about RTE-3 and the last RTE challenges

RTE-3 (Giampiccolo et al. 2007) followed the same basic structure as RTE-1 and RTE-2, in order to facilitate the participation of newcomers and allow "veterans" to assess the improvements of their systems in a comparable setting. The main novelty of RTE-3 was that part of the pairs contained longer texts (up to one paragraph), encouraging participants to move towards discourse-level inference (Dagan et al. 2009). As stated in (Mirkin et al. 2010), the experiments suggest that discourse provides useful information, which improves entailment inference significantly and should be better addressed by future entailment systems. This conclusion is emphasized with the RTE-5, where entailing sentences are situated within documents and depend on other sentences for their correct interpretation. Given a topic and a hypothesis, entailment systems are required to identify all sentences in the corpus inside the topic, that entail the hypothesis. In this setting, texts need to be interpreted based on their entire discourse (Bentivogli et al. 2009a), hence attending to discourse issues becomes essential.

More specific tools address anaphora resolution (relevant for long text, introduced in RTE-3) and word sense disambiguation. For anaphora resolution all documents are parsed and processed with standard tools for named entity recognition and coreference resolution and then the substitution of coreferring terms is performed.

The major innovation of the fourth challenge RTE-4 was a three-way classification of entailment pairs. In three-way judgement, non-entailment cases are split between contradiction, where the negation of the hypothesis is entailed from the text, and unknown, where the truth of the hypothesis cannot be determined based on the text.

10.3 Proposal for Direct Comparison Criterion

In all the above mentioned methods, based on similarity between T and H, this similarity is compared with a given (learned) threshold. If this similarity is higher than this threshold, then $T \rightarrow H$ is considered as a *True* relation, and otherwise as a *False* relation. On the other hand, the similarity between T and H is a *directional* one: this means that, regardless of how $sim(T,H)$ is calculated, $sim(T,H) \neq sim(H,T)$. This remark, correlated with the following: *T entails H if H is not informative in respect* to T led us to establish a relation not between $sim(T,H)$ and a threshold (which must be determined!) but between $sim(T,H)$ and $sim(H,T)$. This criterion of textual entailment is operational and doesn't depend on a set of trained dataset.

In (Tatar et al. 2009b) this criterion is proven using a lexical extension of Robinson's classical principle of refutation. We will summarize *lexical refutation* in the next section.

10.3.1 Lexical refutation

(Tatar and Frentiu 2006) proposed a refutation method to solve the problem of text entailment. The method is called *lexical refutation* and the modified unification *lexical unification,* and it is presented below. Let us analyze the method introduced in (Rus 2001) for obtaining the logic forms, where each *open-class* word in a sentence (that means: noun, verb, adjective, adverb) is transformed into a logic predicate (atom). The method is applied to texts (which are part of speech tagged and syntactic analyzed) following the steps:

- A (logic) predicate is generated for every noun, verb, adjective and adverb (possibly even for prepositions and conjunctions). The name of the predicate is obtained from the morpheme of the word.
- If the word is a noun, the corresponding predicate will have as argument a variable, as individual object.
- If the word is an intransitive verb, the corresponding predicate will have as first argument an argument for the event (or action denoted by the verb) and a second argument for the subject. If the verb is transitive it will have as arguments three variables: one for the event, one for the subject and one for the direct complement. If the verb is ditransitive it will have as arguments four variables: two for the event and the subject and two for the direct complement and the indirect complement.
- The arguments of verb predicates are always placed in the order: event, subject, direct object, indirect object (the condition of ordering is not necessary for modified unification).
- If the word is an adjective (adverb) it will introduce a predicate with the same argument as the predicate introduced for modified noun (verb).

The *lexical unification* method of two atoms proposed in (Tatar and Frentiu 2006) supposes the use of a lexical knowledge base (as, for example, WordNet) where the similarity between two words is quantified. In the algorithm of *lexical unification* it is considered that $sim(p, p')$ between two words p, p' is that obtained, for example, by the Word::similarity interface (Pedersen et al. 2004). This similarity between two words is used to calculate a score for the unifiability of two atoms.

In the algorithm of *lexical unification* the input and the output are:

INPUT: Two atoms $a = p(t_1, \ldots, t_n)$ and $a' = p'(t_1', \ldots, t_m')$, $n \leq m$, where names p and p' are words in WordNet and $t_1, \ldots, t_n$ and $t_1', \ldots, t_n'$ are variables or constants (in the last case also words in WordNet), and a threshold τ.

OUTPUT: Decision: The atoms are

  a) *lexical unifiable* with a calculated score τ' and the unificator is σ, OR
  b) they are not unifiable (the score τ' of unification is less than τ).

The score τ' is the sum of all similarities between p, $t_1, \ldots, t_n$ and p', $t_1', \ldots, t_m'$ during the process of unification. As the score is expected to be large, these similarities are needed to be significant.

Let us observe that two terms t_i and t_j' are unifiable in the following two cases.

  1. The first one refers to the regular cases in First Order Predicate Calculus:
     • terms are equal constants;
     • one is a variable, the other is a constant;
     • both are variables.
  2. In the second case, if t_i and t_j' are two different constants, as they are words, then they are unifiable if $sim(t_i, t_j')$ is big enough. The similarity between two different constants (words) is maximal when they are words from the same synset in WordNet.

Let us observe that in the method of obtaining logic form, which we use, the arguments of predicate are only variable or constants.

The other "classical" to "lexical" extensions are:

Definition

Two (disjunctive) clauses c_i and c_j provide by *lexical resolution rule* (l_r) the (disjunctive) clause c_k with the score τ', written as

$$c_i,\, c_j \models l_r\, c_k$$

if $c_i = l \vee c_i'$, $c_j = \neg l' \vee c_j'$, l and l' are lexical unifiable with the score τ' and the unificator σ. The resulting clause is $c_k = \sigma(c_i') \vee \sigma(c_j')$.

Let it be denoted by $\models^* l_r$ the transitive and reflexive closure of $\models l_r$. The following definition is a translation of Robinson's property concerning a set of disjunctive clauses which are contradictory. As *lexical resolution* is the rule used, we denote this property as *lexical contradictoryness*:

Definition

A set of disjunctive clauses C (obtained from formulas associated with sentences from a text) is *lexical contradictory* with the score of refutation τ' if the empty clause [ ] is obtained from the set of formulas C by repeatedly applying the *lexical resolution rule*:

$$C \models^* l_r\, [\,]$$

and the sum of all scores of *lexical resolution rule* applications is τ'

The test of relation $T \rightarrow H$ is that the score of refutation (the sum of scores of all *lexical unificators* needed in resolutions) is larger than a threshold.

The steps of demonstrating by *lexical refutation* that a text T entails the text H for the threshold τ consist in:

- translating T in a set of logical formulas T' and H in H';
- considering the set of formulas $T' \cup negH'$, where by $negH'$ we mean the logical negation of all formulas in H';
- finding the set C of disjunctive clauses of formulas T' and $negH'$;
- verifying if the set C is *lexical contradictory* with the score of refutation τ'. If $\tau' \geq \tau$ then the text T entails the text H.

Let us remark that the *lexical refutation* is a directional method: to demonstrate $T \rightarrow H$, the set of clauses is obtained from formulas T' and $negH'$ which is different, of course, from the set of clauses considered if $H \rightarrow T$ is to be demonstrated.

10.3.2 Directional similarity of texts and the comparison criterion

In (Corley and Mihalcea 2005), the authors define the directional similarity between the texts T_i and T_j with respect to T_i as:

$$Sim(T_i, T_j)_{T_i} = \frac{\sum_{pos} \left(\sum_{w_k \in WS^{T_i}_{pos}} \left(maxSim(w_k) \times idf_{w_k} \right) \right)}{\sum_{pos} \sum_{w_k \in WS^{T_i}_{pos}} idf_{w_k}} \qquad (1)$$

Here the sets of open-class words (*pos* is noun, verb, adjective and adverb) in each text T_i and T_j are denoted by $WS^{T_i}_{pos}$ and $WS^{T_j}_{pos}$. For a word w_k with a given *pos* in T_i, the highest similarity for all words with the same *pos* in the other text T_j is denoted by $maxSim(w_k)$.

Starting with this text-to-text similarity metric, we derive (Tatar et al. 2007b) a textual entailment recognition *criterion* by applying the *lexical refutation* theory presented above: $T \rightarrow H$ iff the following relation will take place:

$$sim(T,H)_T < sim(T,H)_H \qquad (2)$$

Indeed, to prove $T \rightarrow H$ it is necessary to prove that the set of formulas {T, *negH*} is *lexical contradictory* (we denote also by T and *negH* the sets of disjunctive clauses of T and *negH*). That means the empty clause must be obtained with a big enough score from this set of clauses. As *negH* is the support set of clauses, the clauses in *negH* are to be preferred in the refutation. The clauses in *negH* are preferred in the refutation if the score of unifications of atoms in H (firstly) with atoms in T is bigger than the score of unifications of atoms in T (firstly) with atoms in H. So, the following relation holds:

the sum of maximum similarities between atoms of H with atoms of T > the sum of maximum similarities between atoms of T with atoms of H. As atoms are provided by words, this is exactly the relation (2), with $T_i = T$ and $T_j = H$ and ignoring $idf(w)$.

The criterion obtained from (2) has been applied for the development dataset of RTE-1 and the obtained accuracy was above 55% (Tatar et al. 2009b). In (Perini and Tatar 2009) are being used, beside the similarity relation, the other relations provided by WordNet (see Fellbaum 1998), such that $maxRel(w_k)$ is defined as the highest "relatedness" between w_k and words from the other text having the same part of speech as w_k. A system based on the directional criterion (2), where $maxSim(w_k)$ is replaced by $maxRel(w_k)$, obtained at Semeval-2012 (Task 8, crosslingual TE) some notable results (Perini 2012).

10.3.3 Two more examples of the comparison criterion

A cosine directional criterion

In (Tatar et al. 2009b) two cosine measures considering the words of $T = \{t_1, t_2, \ldots, t_m\}$ and of $H = \{h_1, h_2, \ldots, h_n\}$ are defined. Since usually $n \neq m$, let denote by $\overrightarrow{T}$ and $\overrightarrow{H}$ two vectors so that they have an equal number of components:

- The two vectors for calculating $cos_T (T, H)$ are : $\overrightarrow{T} = (1, 1, \ldots 1)$ (an m-dimensional vector) and $\overrightarrow{H}$, where $\overrightarrow{H}_i = 1$, if t_i is a word in the sentence H and $\overrightarrow{H}_i = 0$ otherwise.
- The two vectors for calculating $cos_H(T, H)$ are : $\overrightarrow{H} = (1, 1, \ldots 1)$ (an n-dimensional vector) and $\overrightarrow{T}_i = 1$, if h_i is a word in the sentence T and $\overrightarrow{T}_i = 0$ otherwise.
- For $cos_{H \cup T} (T, H)$ the first vector is obtained from the words of T contained in $T \cup H$ and the second, from the words of H contained in $T \cup H$.

We will denote in the following the vectors obtained from T and H with the same symbols (T and H). The above criterion (2) is expressed as:

$$cos_H(T, H) \geq cos_T(T, H).$$

Moreover, some relations are learned from a trained corpus. The similarities between T and H calculated with respect to T and to $H \cup T$ must be very close. In the same way, the similarities between T and H calculated in respect to H and to $H \cup T$ must be very close. Also, all these three similarities must be bigger than an appropriate threshold. To sum up, the conditions of entailment are:

- 1. $|cos_T (T, H) - cos_{H \cup T}(T, H)| \leq \tau_1$
- 2. $|cos_H(T, H) - cos_{H \cup T}(T, H)| \leq \tau_2$
- 3. $max\{cos_T(T, H), cos_H(T, H), cos_{H \cup T}(T, H)\} \geq \tau_3$

The thresholds founded by a learning method were: $\tau_1 = 0.095$, $\tau_2 = 0.15$ and $\tau_3 = 0.7$.

The accuracy of the method for *True* pairs is 68.92 and for *False* pairs is 46.36. The global accuracy is 57.62.

A modified Levenshtein distance as a directional criterion

Let us consider that for two words w_1 and w_2 a *modified* Levenshtein distance (Tatar et al. 2009b) is denoted by $LD(w_1, w_2)$. This is defined as the cost of the minimal number of transformations (deletions, insertions and substitutions) so that w_1 *is transformed in* w_2 and this cost is, in a way, the quantity of information of w_2 with respect to w_1. It is denoted by T_{word} the "word" obtained from the sentence T by considering the empty space between words as a new letter, and by concatenating all words of T. Similarly a "word" H_{word} is obtained. $LD(T_{word}, H_{word})$ represents the quantity of information of H with respect to T. Let it be remarked that the modified Levenshtein distance $LD(w_1, w_2)$ is not a distance in its usual sense, such that $LD(w_1, w_2) \neq LD(w_2, w_1)$.

As *T entails H iff H is not informative with respect to T* the following relation must hold:

$$LD(T_{word}, H_{word}) < LD(H_{word}, T_{word})$$

One checked the criterion on a set of 800 pairs of RTE-1 development dataset and obtained the accuracy of 53.19 (Tatar et al. 2009b).

The costs of transformations *from the word w_1 to the word w_2* are as following: the cost for substitution with a synonym is 1, the cost for deletions is 3, the cost for insertions is 3, the cost for substitution (with an arbitrary word) is 5.

10.4 Conclusions and Recent Research

Predicting entailment relation between two given texts is an important stage in numerous NLP task. Consequently, a big interest is dedicated to this problem in the last years.

Although many logic-based methods have been developed, most methods operate at the lexical, syntactic and semantic levels. The most recent papers model tree-edit operations on dependency parse trees (Wang and Manning 2010, Stern and Dagan 2012). In (Wang and Manning 2010) the edit operation sequences over parse tree are represented as state transitions in a Finite State Machine. Each edit operation (delete, insert, substitution) is mapped to a unique state, and an edit sequence is mapped into a transition sequence among states. The probability of an edit sequence between two parse trees is then calculated. The higher probability determines the binary judgement about entailment relation. The method is tested on RTE-2 and RTE-3 dataset and obtained 63 % accuracy (the second result for a set of 5 systems) and, respectively, 61.1 % accuracy (the best result for a set of 4 systems).

(Stern and Dagan 2012) presents a modular open-source system (BIUTEE) which, given a pair (T, H), applies a search algorithm to find a proof that transforms T into H, and finds the lowest cost proof. A proof is defined as a sequence of operations between a set of parse trees that were constructed from T and a set of trees embedding the parse tree of H. The applied operations are lexical, syntactic, coreference-substitution etc. and the visual tracing tool of the system presents detailed information on each proof step. BIUTEE outperforms all systems at RTE-6 (Stern and Dagan 2012).

Another recent orientation of research in TE is on semantic inference over texts written in different languages (Negri et al. 2012).

Let it be mentioned at the end that the whole website, Textual Entailment Resource Pool hosted by ACL, provides a repository of linguistic tools and resources for textual entailment (http://aclweb. org/aclwiki/index.php). The excellent tutorial on Recognizing Textual Entailment presented at NAACL 2010 by Mark Sammons, Idan Szpektor and V.G. Vinod Vydiswaran is stored here.

CHAPTER 11

Text Segmentation

11.1 Introduction

The problem of segmenting text, words and sentences has been of paramount importance in natural language processing. Text segmentation is defined as the process of dividing written text into meaningful units such as words, sentences, or topics. The term applies both to mental processes used by humans when reading a text as well as automated processes implemented in computers.

Strongly related to text segmentation is *word segmentation*, which is defined as the problem of dividing a string of written language into its component words. In English, as well as in many other languages, using some form of the Greek or Latin alphabet, the space between words is a good approximation of a word delimiter.

The problem is exacerbated by the fact that *sentence delimitation* is bound with further challenges. For instance, in English and some other languages, using punctuation, particularly the full stop, is a reasonable approximation. However, even in English this problem is not trivial due to the use of the full stop for abbreviations, which may or may not also terminate a sentence. The usual way to work around this problem has been to set up tables of abbreviations that contain periods. This can help to prevent incorrect assignment of sentence boundaries. When punctuation and similar clues are not consistently available, the segmentation task often requires fairly non-trivial techniques, such as statistical decision making, large dictionaries, as well as consideration of syntactic and semantic constraints.

In this chapter the *topic* segmentation will be addressed. This is often a valuable stage in many natural language applications and the prominent role of segmentation has continued to grow due to increased access in creating and consuming information. For example, the identification of a passage in Information Retrieval is more useful than whole-document retrieval. Thus segmenting into distinct topics is useful as Information Retrieval needs to find relevant portions of text that match with a given query. An experiment with French documents (Prince and Labadie 2007) shows how matching the query content with segments of the retrieved text could improve the results of Information Retrieval. In *Anaphora Resolution*, dividing texts into segments, with each segment limited to a single topic, would result in a minimal usage of pronominal reference to entities outside a given segment (Reynar 1998). Segmentation is an important task in speech processing as well, as for example in *Topic Detection and Tracking* (TDT) (Allan et al. 1998, Stokes 2004). The goal of TDT initiative is to find out and follow new events in a stream of news stories broadcast.

However, the most important application of segmentation is, Text summarization. A clear separation of the text into segments helps in summarization tasks, as argued in (Boguraev and Neff 2000, Orasan 2006). In Chapter 12 this aspect will be approached in more details.

11.1.1 Topic segmentation

Literature on discourse structure shows that there are several definitions of topics and, consequently, of topic segmentation. Generally speaking, discourse segments are always defined with respect to something (Alonso 2005). Thus one may admit that the *topic* of a segment is *what it is talking about* and topic segmentation is dividing a text into segments, each sentence of which talks about the same subject (Labadie and Prince 2008b).

The purpose of topic text segmentation is to obtain groups of successive sentences which are linked with each other on the basis of their topic. In this respect a segment is a contiguous piece of text that is linked internally but disconnected from the adjacent text because of the topic. To realize these desiderata, most methods rely on two internal properties of discourse: cohesion and coherence. As stated in (Morris and Hirst 1991), cohesion relations are relations among elements in a text (references, conjunctions, lexical elements) and coherence relations are relations between text units (clauses and/or sentences). Moreover, after (Morris and Hirst 1991, Stokes et al. 2004, Barzilay and Elhadad 1997), the cohesion is the textual quality making the sentences seem to hang together, while coherence refers to the fact that there is a sense in the text as a whole. Let one remark that cohesion is a surface relationship more accessible for automatic identification than coherence: lexical cohesion relations is far easier to be established than other cohesion relations (this is the reason for a separate section on segmenting by lexical chains in this book).

(Harabagiu 1999) is an attempt to connect the cohesion structure of a text to its coherence structure. The construct used is the semantic (lexical) path between two words w_1 and w_2, defined as:

$$LP(w_1, w_2) = c_1 r_1 c_2 r_2 ... r_{n-1} c_n$$

in which c_1 is a concept representing word w_1 while the concept c_n represents w_2, and r_i is a lexical relation between the concepts c_i and c_{i+1}. As concepts, the WordNet *synsets*, and as lexical relations, the semantic WordNet relations are used. The author shows that semantic paths between text units, combined with discourse cues (markers), are likely to enable the recognition of some coherence relations between text units (as for example *resemblance* relations, *cause-effect* relations and *contiguity* relations).

11.2 Methods and Algorithms

11.2.1 Discourse structure and hierarchical segmentation

The first hypothesis in most of the following theories is that different levels of normalization of text are supposed to be fulfilled, as for example tokenization, lemmatization, and removal of stop-words. The second hypothesis is that the discourse (text) is coherent. Moreover, it could be expected that the more coherent the text, the more structure it has (Yaari 1998). Much of the work in discourse structure makes use of hierarchical models, which we survey in this section.

One of the first attempts to structure a discourse was provided by the research of Hobbs (Hobbs 1985), who stated that between some units of a text a coherence relation is present. Displaying the coherence relations between text units, the structure of discourse appears like a binary tree where the nodes are relations and the leafs are text units (sentences or clauses). A hierarchical segmentation of the discourse is provided by this tree. As in all other cases of tree-like representations of a discourse, proximity to the root implies some kind of relevance or centrality to the topic of discourse. This is the reason that the tree-like representations are useful in the summarization stage (see Chapter Text Summarization).

In (Hobbs 1985) it is stated that the coherence relations could be established by describing the situation between a speaker and a listener: the speaker wants to convey a message; the message is in the service of some goal; the speaker must link what he says to what the listener already knows; the speaker should ease the listeners difficulties in comprehension. Corresponding to these cases there are four classes of coherence relations: each coherence relation in the class is motivated by the requirements of the discourse situation. The first class contains *occasion* with the special cases *cause* and *enablement*. The second class results from the need to relate what has been said to some goal of the conversation and is called *evaluation*. The coherence relations in the third class are those directed toward relating a segment of discourse to the listeners' prior knowledge. The two relations in this class are the *background* relation and the *explanation* relation. The final class of coherence relations is the largest and contains : *elaboration, parallel, exemplification (generalization), contrast.*

The following idea in (Hobbs 1985) is conclusive both for the definition of a segment and the hierarchical structure of a discourse: *a clause is a segment of discourse, and when two segments of discourse are discovered to be linked by some coherence relation, the two together thereby constitute a single segment of the discourse.*

Regarding the resources annotated with coherence relations, let it be mentioned that the Penn Discourse TreeBank is currently the largest resource manually annotated for discourse connectives, their arguments, and the senses they convey.

Another kind of hierarchical structure of a discourse is provided by the Rhetorical Structure Theory (RST) of Mann and Thompson (Mann and Thompson 1988). The principle of structural pattern in RST is that two spans of text units (clauses or sentences) are related so that one of them has a specific role relative to the other (or a rhetorical relation, *RR*, takes place between these). This principle also says that one of the spans (the nucleus) is more important to the text than the other (the satellite). The order of spans is not constrained, but there are more likely and less likely orders for all of the *RR* relations.

A rhetorical tree is one where the nodes are spans and the edges are rhetorical relations RR between two spans: an edge between spans SP1 and SP2, oriented from SP1 to SP2, means that a rhetorical relation RR between a sentence (or a clause) of SP1 (as satellite) and one sentence (or a clause) of SP2 (as nucleus) takes place. For a given set of RRs (as *justification, evidence, concession*, etc.) there are more than one rhetorical tree which could be constructed. The problem of construction of the *optimal* rhetorical tree for a given text was successfully solved in (Marcu 1997a) as a constraint satisfaction problem. He also resolved the problem of establishing the *optimal* RR between two text spans using a corpus annotated with RRs and an NBC learning model. However, in (Power at al. 2003) the authors observe that a concordance does not always exist between rhetorical structure and document structure (the organization of a document into graphical constituents like sections, subsections, paragraphs, sentences).

In both these theories (Hobbs's and RST) the problem is that of the high subjectivity appearing in the resulting analysis, as for example the identified coherence relations and rhetorical relations.

At the end of the series of hierarchical segmenting methods, let the stack algorithm of J. Allen (Allen 1995) be mentioned. The sentences (clauses) are analyzed sequentially and the following rules for PUSH and POP stack operations are applied:

- When a new segment begins, it is set above the current segment (PUSH operation).
- When a segment (already present in the stack) extends (continues), all the segments above it must be deleted from the stack (POP operation).
- A sentence (clause) begins a new segment when a temporal shift or a cue phrase is encountered in this sentence (clause).
- A segment continues when the constraints of extending a segment are fulfilled. The constraints of extending a segment are: tense constraint (a text unit can extend a segment only if its tense is identical to the tense used in the segment so far), reference constraint (a unit can extend a segment only if its anaphoric components can be linked to the entities of this segment or in a parent segment).
- A continuation is preferred over beginning a new segment, in the case of ambiguity.

The example in (Allen 1995), page 504, has a number of 12 units (sentences or clauses) and also two PUSH operations and two POP operations. The procedure results in 3 segments:

- Segment 1 contains the units 1,2,6,7,12;
- Segment 2 contains the units 3,4,5;
- Segment 3 contains 8,9,10,11.

The root of the discourse tree is Segment 1, with two descendents, Segment 2 and Segment 3. It is obvious that this text segmentation is not linear: Segment 1 is not a sequence of units. Segment 2 begins as a new segment by a PUSH operation, because unit 3 contains a temporal shift. Unit 6 continues with Segment 1 by a POP operation because it contains a temporal shift to the tense of Segment 1. A similar situation takes place at the beginning and end of Segment 3.

Another stack algorithm is introduced in (Grosz and Sidner 1986), where the elements of the stack are data structures called focus spaces, which contain lists of available entities introduced by text units. The stack operations POP and PUSH are determined by attentional and intentional features addressed to the focus spaces (see the Preferred center algorithm below).

As always, in a such theories there remain open questions as for example: how consistent individual observers' structure discourse analysis is and how deep their subjectivity is.

A natural question could arise about the connection between hierarchical clustering of text units and hierarchical segmentation of a text. The answer was first given in (Yaari 1997) where a method of hierarchical segmentation arises naturally in the agglomerative hierarchical clustering of the text units (here sentences). Here a vectorial representation is used for sentences, and a measure of similarity (*cosine*) is introduced. For a given similarity, the hierarchical segmentation corresponds to the hierarchical clusters partitioning the sentences. Combining the agglomerative hierarchical clustering with lexical cohesion, (Yaari 1998) provides *Texplore*, a method of exploring expository texts and dynamic presentation of their contents, where the user controls the level of details.

11.2.2 Linear segmentation

All theories considered above describe how a hierarchical structuring of the discourse results in an *hierarchical* segmentation. *Linear* segmentation is the opposite case, when the segments form a linear sequence. In spite of a large class of hierarchical structure theories, the majority of algorithmic approaches belong to the linear segmentation. The explanation could be the easier comparison with the linear structures in the evaluation process apart from a lower complexity.

Among linear segmentation methods, a large class relies on lexical chains. Therefore, lexical chain methods are surveyed separately in the next section.

The best known and largely implemented method of (topic) linear segmentation is TextTiling (Hearst 1997, Reynar 1993, Stokes 2004). A first remark about TextTiling language is the working hypothesis: a discourse is considered having a single "topic" and the segments bear different "subtopics". This is somehow different from the common view, that the discourse has some topics, one for each segment. In one's view, the concepts of "topic" and "subtopic" are used in a similar way. The seminal article (Hearst 1997) describes the algorithm for subdividing expository texts into contiguous, non-overlapping subtopic segments (from here the name TextTiling of the method derives). As the author explains, instead of undertaking the difficult task of attempting to define what a (sub)topic is, one should concentrate on describing what one recognizes as (sub)topic shift. TextTiling assumes that a set of lexical items is in use during the course of a given subtopic discussion, and when that subtopic changes, a significant proportion of the vocabulary changes as well (Manning and Schuze 1999). The central idea of TextTiling is dividing the text artificially into fixed length blocks and comparing adjacent blocks. The more words the blocks have in common, the higher the lexical score at the gap (the boundary candidate) between them. If a low lexical score is preceded by and followed by high lexical scores, it is assumed to indicate a shift in vocabulary corresponding to a subtopic change (a boundary) (Hearst 1997). The lexical score of a gap is calculated as the *cosine* of vectors associated with the adjacent block texts, where a vector contains the number of times each lexical item occurs in the corresponding text. The heuristics to avoid a sequence of many small segments is to use a boundary selector: each boundary candidate has a *depth score* measuring how low its lexical score is, compared to that of the previous and next candidates. Boundary selector estimates average μ and standard deviation σ of *depth scores* and selects only the gaps which have a *depth score* higher than $\mu - c\sigma$ as boundaries, for some constant c (Manning and Schuze 1999).

Many variants of the standard method (Hearst 1997) have been developed. Kaufmann's *VecTiling* system (Kaufmann 1999) augments TextTiling algorithm with a more sophisticated approach to determine block similarity. (Reynar 1998) describes an algorithm (*Dotplots*) similar to TextTiling with a difference in the way in which the blocks of adjacent regions are chosen. The algorithm generates a matrix (*a dotplot*) in which cells (x, y) are set to 1 when word number x and word number y are the same or have the same root and then identifies boundaries which maximize the density within segments that lie diagonally to the matrix. The crucial difference between *Dotplot* and TextTiling algorithm is that the last identifies boundaries by comparing neighboring regions only, while the first compares each region to all other regions (Reynar 1998).

Another popular method in linear text segmentation is *C*99, developed in (Choi 2000). The method is based on Reynar's maximization algorithm (Reynar 1998) and uses a sentences pair (*cosine*) similarity matrix, where each element in the matrix represents the proportion of neighboring elements that have lower similarity values. The process begins by considering the entire document placed in a set B as a single coherent text segment. Each step of the process splits one of the segments in B in such a manner that the split point is a potential boundary which maximizes the inside density of B. For a document with b potential boundaries, b steps of divisive clustering are performed, from which the optimal split point is selected.

As stated in (Labadie and Prince 2008a), most of the segmentation methods neglect the syntactic text structure. Thus, they propose a new method, *Transeg*, which combines both the semantic and the syntactic information of a sentence, the last obtained as an output of a parser providing constituents and dependencies analysis. *Transeg* tries to identify the last sentence of the previous segment and the first sentence of the beginning of the new segment. In this respect, the likelihood (score) of being the first sentence and of being the last sentence of a segment is calculated separately. For each sentence, the first score is multiplied with the second score of the previous sentence. Higher score for a sentence indicates a higher probability of being the first sentence of a new segment.

A variant of hierarchical text segmentation based on clustering is that of clustering based linear segmentation. The method in (Lamprier et al. 2007), *ClassStruggle*, reaches the boundaries in two steps. In the first step, the sentences are vectorially represented and clustered by their similarity (*cosine* measure). In the second, a score for each sentence and cluster is calculated (using the similarities between clusters) and the sentences having the same best dominant cluster according to this score are grouped into new clusters. If the new clusters are different from the old ones, the algorithm restarts at the first step. Finally, between all adjacent sentences not belonging to the same cluster, a boundary is fixed.

The clustering method could solve the problem of subjective segmentation as case opposite to the general one in which the process of segmentation is seen as an objective method, which provides one clearly defined result. However, different users have quite different needs with regard to segmentation because they view the same text from completely different, subjective, perspectives. Segmenting a text must be associated with an explanation of why a given set of segments is produced. All these could be realized by viewing the process of segmentation as a clustering process of the sentences of a text (Tatar et al. 2010a). When the cluster $Cl = \{S_{i_1}, S_{i_2} ..., S_{i_m}\}$ is one of the set of obtained clusters, and $i_1 < i_2 < ... < i_m$, then the linear segmentation is: $[S_1, S_{i_1-1}][S_{i_1}, S_{i_2-1}]...[S_{i_m}, S_n]$. When two sentences are adjacent and one is in the cluster and the previous is not, a new segment starts. For each cluster a different segmentation could be generated. The clustering criterion and the center of the cluster Cl explain the reason of the segmentation.

As a suggestion of another general method of linear segmentation, it will be outlined how the study of the reference chains could help in linear segmentation. Consider that all the *chains of antecedents-anaphors* or *coreference chains* (Mitkov 2002) of a text are CHR_1, ..., CHR_m. A chain CHR_i contains the occurrences of the entities identified as antecedents for a given anaphor and also the occurrences of this anaphor. The principle for detecting a boundary between two segments could be the following: the most frequent pair (antecedent-anaphor) is changed at a boundary, and stays unchanged at the interior of a segment. So, if the most frequent pair (antecedent-anaphor) for the sentences $S_1, ... S_i$, denoted by $P(S_1, ... S_i)$ is different of $P(S_1, ... S_{i+1})$, then there is a boundary between the sentences S_i and S_{i+1}. Otherwise, S_i and S_{i+1} are in the same segment.

Input: Text= $\{S_1, S_2, \ldots, S_n\}$, Coreference Chains $CHR_1, \ldots, CHR_m$.
Output: Segments $Seg_1, \ldots, Seg_t$.

$\quad i = 1, j = 1, Seg_1 = \{S_1\}$

while $i < n$ **do**

$\quad\quad$ Calculate $P_{S_1}, \ldots, S_{i+1}$, the most frequent pair (antecedent-anaphor) in sentences $S_1, \ldots, S_{i+1}$;

$\quad\quad\quad$ **if** $P_{S_1}, \ldots, S_{i+1} = P_{S_1}, \ldots, S_i$

$\quad\quad\quad\quad$ **then**

$\quad\quad\quad\quad\quad\quad Seg_j := Seg_j \cup \{S_{i+1}\}$
$\quad\quad\quad\quad\quad\quad i := i+1$

$\quad\quad\quad\quad$ **else**

$\quad\quad\quad\quad\quad\quad j := j + 1, Seg_j := \{S_{i+1}\}$
$\quad\quad\quad\quad\quad\quad i := i+1$

$\quad\quad\quad$ **endif**

$\quad$ **endwhile**

Let a reverse relation between the anaphora resolution and text segmentation be observed: in a (fine-grained) segmented text, by restricting the candidates for antecedents to those that exist in the same segment with the anaphor, the quality of the anaphora resolution can be improved greatly.

Another similar method for linear segmentation is provided by the Centering Theory (Grosz et al. 1995). Let the *Forward looking* centers (the syntactically ranked entities introduced by a sentence) be calculated and the highest ranked center (the *Preferred* center) be established for each sentence S_i. The calculus of the *Forward looking* centers and *Preferred* centers supposes that anaphora resolution is fulfilled. The principle for detecting a boundary between two segments is the following: the *Preferred* center is changed at a boundary, and stays unchanged at the interior of a segment. Thus, if the *Preferred* center of S_i, denoted by $CP(S_i)$ is different to $CP(S_{i+1})$, then there is a boundary between the sentences S_i and S_{i+1}. Otherwise, S_i and S_{i+1} are in the same segment.

Input: Text= $\{S_1, S_2, \ldots, S_n\}$, Preferred centers $CP_{S_1}, \ldots, CP_{S_n}$.
Output: Segments $Seg_1, \ldots, Seg_t$.
    $i = 1; j = 1; Seg_1 = \{S_1\}$
    while $i < n$ **do**
        if $CP_{S_i} = CP_{S_{i+1}}$
        then
            $Seg_j := Seg_j \cup \{S_{i+1}\}$
            $i=i+1$
        else
            $j := j + 1, Seg_j := \{S_{i+1}\}$
            $i=i+1$
        endif
    endwhile

The last method enumerated here is proposed in (Tatar et al. 2008a), and it is called *Logical segmentation* because the score of a sentence is the number of sentences in the text that are *entailed* by it. The scores form a structure which indicates how the most important sentences alternate with less important ones and organizes the text according to its logical content. Due to some similarities with TextTiling algorithm this method is called *Logical TextTiling* (LTT). Similarly with the topic segmentation (Hearst 1997), the focus in logic segmentation is on describing what a shift is in information relevant for a discourse. To put it simply, a valley (a local minim) in the obtained graph (logical structure) is a boundary between two (logical) segments. This is in accordance with the definition of a boundary as a perceptible discontinuity in the text structure (Boguraev and Neff 2000), in this case a perceptible logical discontinuity in the connectedness between sentences. This idea will also be the basis of the methods of segmentation presented in the next sections.

A short summary of the algorithm is given below. The condition:

$$(score(S_{i-1}) + score(S_{i+1}) - 2 \times score(S_i)) \geq \tau$$

assures that the difference between $score(S_i)$ and $score(S_{i-1})$, on one hand, and $score(S_i)$ and $score(S_{i+1})$ on the other hand, is big enough to be considered a significant local minim. Thus τ is a threshold which determines the granularity of the segmentation (τ is small for fined grained segmentation):

Input: Text= $\{S_1, S_2, \ldots, S_n\}$
Output: Segments $Seg_1, \ldots, Seg_t$.
  $k = 1, t = 1, Seg_t = \{S_1\}$
    while $k \le n$ **do**
      Calculate $score(S_k)$= number of sentences entailed by
S_k
    endwhile
    Determine the local minima *Min* in the graph of function
score.
    for $i = 2$ **to** $n - 1$ **do**
      if $score(Si) \in Min$ and $(score(S_{i-1})+score(S_{i+1})) -2 \times score(S_i)) \ge \tau$
      then
        $t := t + 1, Seg_t := \{S_i\}$
      else
        $Seg_t := Seg_t \cup \{S_i\}$
      endif
    endfor

Another simple algorithm of segmentation, (Pure entailment), based on the entailment relation, and proposed in (Tatar et al. 2008a) as well is the following:

Input: Text= $\{S_1, S_2, \ldots S_n\}$
Output: Segments $Seg_1, \ldots, Seg_t$.
  $k = 2, t = 1, Seg_t = \{S_1\}$
  while $k < n$ **do**
    if $(Seg_t \rightarrow S_k)$
    then
      $Seg_t := Seg_t \cup \{S_k\}$
    else
      $t := t + 1, Seg_t := \{S_k\}$
    endif
    $k:=k+1$
  endwhile

Let it be remarked that a new segment begins when the relation $Seg_t \rightarrow S_k$ does not hold. When Seg_t is formed by a few sentences, the probability that these sentences entail the next sentence S_k is less and thus a new segment begins. This is the reason why the Pure Entailment method generates many segments. The LTT method corrects this inconvenience (Tatar et al. 2008a).

It can be mentioned here the older methods of linguistic cues (referential noun phrases, cue words) as universally acknowledged in the literature on discourse presented in (Passonneau and Litman 1997). Moreover, the authors show that when multiple sources of linguistic knowledge features are used concurrently, the performance of the algorithm improves.

One can imagine combinations of methods in the process of looking for boundaries in a discourse, because as (Hearst 1997) says: "There are certain points at which there may be radical changes in space, time, character configuration, event structure, or , even, world... At points where all of these change in a maximal way, an episode boundary is strongly present". As examples of combinations one could imagine pairs of methods enumerated above, for example *Preference-center* changes and most frequent pair *anaphor-antecedent* changes, etc.

Let it be said at the end of this section that linear segmentation and (extractive) summarization are two interdependent goals. Moreover, good segmentation of a text could improve the summarization (see Chapter Text Summarization). On the other hand, the method of extracting sentences from the segments is decisive for the quality of the summary. Some largely applied strategies (rules) are (Tatar et al. 2008a):

1. The first sentence of a segment is selected.
2. For each segment the sentence with a maximal score is considered the most important and hence it is selected (for example, for LTT method, the sentence with a maximal score for a segment is a sentence which entails a maximal number of other sentences and thus it is the most important one).
3. The third way of reasoning is that from each segment the most informative sentence (the least similar) relative to the previously selected sentences is picked up.

Thus, one can say that while determining segmentation of a text, a summary of that text can be obtained as well. Further connections between the segmentation and the summarization will be presented in Chapter Text summarization.

11.2.3 Linear segmentation by Lexical Chains

Lexical Chains (LCs) are sequences of words which are in a lexical cohesion relation with each other and they tend to indicate portions of a text that form semantic units (Okumura and Honda 1994, Stokes et al. 2004, Labadie and Prince 2008a). Lexical cohesion relationships between the words of LCs are established using an auxiliary knowledge source such us a dictionary or a thesaurus. LCs could further serve as a basis for text segmentation (and also for WSD and Text summarization, see Chapters WSD and Text summarization). The first paper which used LCs (manually built) to indicate the structure of a text was (Morris and Hirst 1991) and it relies on the hierarchical structure of Roget's Thesaurus to find semantic relations between words. Since the chains are used to structure the text according to the attentional/intentional theory of Grosz and Sidner theory, (Grosz and Sidner 1986), their algorithm divides texts into segments which form hierarchical structures (each segment is represented by the span of a LC).

As regards the automatic LCs determination, in (Barzilay and Elhadad 1997, Stokes 2004) a distinction between two types of lexical chaining is made: greedy chaining, where each addition of a term to a chain is based only on those words that occur before it in the text, and non-greedy chaining, which postpones assigning a term to a chain until it has seen all possible combinations of chains that could be generated from the text and then selects the appropriate chain. In (Barzilay and Elhadad 1997) the advantages of a non-greedy chaining approach are argued. Once all chains have been generated, only the strongest of them are retained, where the strength is derived from the length of the chain and the homogeneity index. In order to reduce the run time of this algorithm, some modifications have been introduced in (Silber and McCoy 2002) by creating a structure that stores all chain interpretations implicitly and then decides which chain is to be selected. However, the greedy chaining algorithm *LexNews* presented in (Stokes 2004), constructed for a Topic Detection and Tracking goal, which evaluates the quality of the chains with respect to the disambiguation accuracy, is as effective as a chaining algorithm in a non-greedy approach.

One might expect that there is only one possible set of lexical chains that could be generated for a specific text, i.e., the correct set. However, in reality, terms have multiple senses and can be added to a variety of different chains (Stokes 2004). Regarding the subjective interpretation of the lexical cohesion of a text, an empirical study shows approximately 40% individual differences between human annotations (Morris 2006).

The paper (Tatar et al. 2008b) proposes a top-down method of linear text segmentation based on lexical cohesion of a text. Namely, first a single chain of disambiguated words in a text is established, then the rips of this chain are considered. These rips are boundaries of the segments in the cohesion structure of the text. The chain of disambiguated words of a text is obtained by a Lesk type algorithm, CHAD, (see the Chapter Word Sense Disambiguation) based on Word Net (Tatar et al. 2007a). The CHAD algorithm shows what words in a sentence are semantically unrelated with the previous words and, consequently, receive a "forced" first WordNet sense. Of course, these words are regarded differently from the words which receive a "justified" first WordNet sense. Scoring each sentence of a text by the number of "forced" first WordNet sense words in this sentence, a representation of the lexical cohesive structure of the text will be provided. If for a sentence S_i this number is F_i, the local maxima in the graph representing the function $score(S_i) = F_i$ will represent the boundaries between segments spanned by lexical sub-chains. Thus, a segment is considered a piece of text where the disambiguation of contained words is "chained". Due to some similarities with TextTiling algorithm, the method is called *Cohesion TextTiling* (CTT).

Let it be remarked further that the segmentation and the (extractive) summarization are two interdependent goals: a good segmentation of a text can improve the summarization and from a summary of the text a segmentation of this text is available (see the Chapter Text summarization). The paper (Tatar et al. 2008b) compares segmentations by LTT and CTT by comparing the summaries obtained while applying different strategies. The conclusion obtained is that the quality of CTT summaries is better than the quality of the LTT summaries from the point of view of informativeness (for informativeness see (Orasan 2006)).

The last method of segmentation by lexical chains summarized here is the *Lexical Chains Distribution* (LCD) method (Tatar et al. 2009a) that relies on the number (or weight) of chains which end in a sentence, which begin in the following sentence and traverse two successive sentences. The method improves a scoring method of (Okumura and Honda 1994). Let the number of LCs which end in S_i be denoted by $input(S_i)$, the number of LCs which begin in S_i by $output(S_i)$ and the number of LCs which traverse S_i (and S_i is not the beginning for LC) and S_{i+1} (and S_{i+1} is not the end for LC) by $during$ (S_i, S_{i+1}). The score of S_i to be the last sentence of a segment is:

$$score(S_i) = \frac{input(S_i) + output(S_{i+1})}{during(S_i, S_{i+1})} \tag{1}$$

The bigger the $score(S_i)$ is, the bigger is the chance to have a boundary between S_i and S_{i+1}. In this way, the points of the local maxim in the graph $score(S_i)$ indicate the boundaries of the text $S_1, \ldots, S_n$: if $score(S_i)$ is a local maxim in the graph, S_i is the last sentence of a segment, and S_{i+1} is the first sentence in the next segment. The performance of the LCD method is experimentally proved as better than CTT in (Tatar et al. 2009a). An outline of LCD algorithm of segmentation is given below, where τ is a threshold selected according to the granularity of the segmentation (τ is small for fine grained segmentation):

> **Input:** Text= $\{S_1, S_2, \ldots, S_n\}$, τ
> **Output:** Segments $Seg_1, \ldots, Seg_t$.
> $i = 1$, $t = 1$, $Seg_t = \{S_1\}$
> **while** $i < n$ **do**
>   Calculate $score(S_i) = \dfrac{input(Si) + output(S_{i+1})}{during(S_i, S_{i+1})}$
> **endwhile**
> Determine the local maxima Max in the graph of function $score$.
> **for** $i = 2$ to $n - 1$ **do**
>   **if** $score(S_i) \in Max$ and $2 \times score(S_i) - (score(S_{i-1}) + score(S_{i+1})) \geq \tau$
>   **then**
> $t := t + 1$, $Seg_t := \{S_{i+1}\}$
>   **else**
> $Seg_t := Seg_t \cup \{S_i\}$
>   **endif**
> **endfor**

Let it be remarked that a recently presented paper (Marathe and Hirst 2010) and an older one (Stokes 2004) introduce a method very similar to the above LCD method. A score $score(S_i) =$

$input(S_i)+output(S_i)$ is calculated for each sentence S_i (paragraphs for (Stokes 2004)). However, as is proved in (Tatar et al 2009a), the influence of the denominator $during(S_i, S_{i+1})$ in the process of determination of segments is important: close by 4 %.

The performance of the LCD method (equation 1) is verified on lexical chains based on Roget's Thesaurus in (Tatar et al. 2012). We used the freely available 1911 version and also the 1987 edition of Penguins Roget's Thesaurus of English Words and Phrases from Pearson Education. The later has a wider coverage, since recent technical terms are missing from the 1911 version, however the 1911 version works very well.

Another feature of the LCD method presented in (Tatar et al. 2012) is the use of a weight for each LC. Let us remember that a lexical chain LC_k is represented as $[S_{i_k}, S_{j_k}]$ regardless of the number of other sentences which contain words from LC_k and are situated in the interior of this interval (the sentences situated after S_{i_k} and before S_{j_k}). To express this information about a lexical chain, the weight could be a solution, for example the ratio between $j_k - i_k$ and the number of interior sentences of the span $[S_{i_k}, S_{j_k}]$ containing elements of LC_k. Weights for lexical chains could also be their length, their density (the ratio between the number of total terms and the number of distinct terms in a chain), the types of cohesion relationships, etc. Using different schemes for the weights can help in the study of the influence of different features: it is known that the question of preferring long or short chains for the study of the cohesion is not yet solved.

The functions: $input(S_i)$, $output(S_i)$ and $during(S_i, S_{i+1})$ in weighted version are modified into $input(S_i)^p$ and $output(S_i)^p$ and $during(S_i, S_{i+1})^p$, as follows, where w_p^j is the weight of the lexical chain LC_j and p is the selected index of the weight from a set of possible weights:

$$input(S_i)^p = \sum_{LCj \ ends \ in \ S_i} w_p^j$$

$$output(S_i)^p = \sum_{LCj \ begins \ in \ Si} w_p^j$$

$$during(S_i, S_{i+1})^p = \sum_{LC_j \ strict \ traverses \ S_i \ and \ S_{i+1}} w_p^j$$

The above formula 1 for scoring a sentence is slightly modified to:

$$score(S_i)^p = \frac{input(S_i)^p + output(S_{i+1})^p}{during(S_i, S_{i+1})^p} \qquad (2)$$

In our study (Tatar et al. 2012) a single weight is used for each lexical chain which reflects the importance and the strength of the chain. It is obtained from Roget's Thesaurus based on the degree of similarity between the words in the chain, on how far apart they are in the structure of the thesaurus and on how many words are present in the chain. Firstly, we compared the LCD method with un-weighted LCs with the LCD method with weighted LCs and observed that for the second variant the results are better regarding the correctness (see the definition of measure of correctness in the next section). Secondly, the averages of correctness of a set of 20 documents from the CAST project corpus (http://www.clg.wlv.ac.uk/projects/CAST/corpus/listfiles.php) by the LCD method versus two other different methods of segmentations have been compared. The averages were better for the LCD method, where Roget based weighted LCs have been used, rather than the case where WordNet based un-weighted LCs have been used, along with the segmentation obtained by TextTiling method. The remark was repeated when 10 documents from DUC2002 corpus were used.

As a final remark, let it be mentioned that segmentation techniques differ not only with regard to the type of discourse structure (linear or hierarchical) but they also differ with respect to the level of granularity. Segmentation approaches are categorized into two levels of segmentation granularity: fine-grained segmentation and coarse-grained segmentation (Stokes 2004). This problem is strongly connected with the definition of what a unit of the text represents (clauses, sentences or paragraphs).

11.2.4 Linear segmentation by FCA

In the paper (Tatar et al. 2010a) the FCA theory is applied in order to obtain the segments of a text. We started from the Intentional structure principle (Grosz and Sidner 1986) which is based on the idea that *every discourse has an overall purpose, and every discourse segment has a purpose, specifying how it contributes to the overall purpose.* It is reasonable to consider that the *purpose* of a segment is maximally coherent (all the sentences contribute to this *purpose*) and that the *purposes* of two different segments are minimally similar. The corresponding FCA idea is to determine a set of concepts (where the objects are the sentences and the attributes are Lexical Chains) such that (see Chapter 5):

- the set of concepts extents has to cover the text;
- the set of concepts has to contain disjoint extents;
- the set of concepts has to present a maximal internal cohesion (semantical and lexical);
- the set of concepts has to display a minimal "in between" cohesion (cohesion between a concept and the other concepts).

All these four constraints applied to the set of concepts led one to identify the sentences of a text as objects, the Lexical Chains as attributes and the formal context $K = (G, M, I)$ as below:

- $G = \{S_1, S_2, \ldots, S_n\}$
- $M = \{L_1, L_2, \ldots, L_m\}$. For each Lexical Chain L_i the first sentence (where the chain begins) and the last sentence (where the chain ends) is given.
- $(S_i, L_j) \in I$ if the Lexical Chain L_j begins in a sentence S_b, with $b \leq i$ and ends in a sentence S_e, with $i \leq e$

The formal concepts of the context K are pairs formed by a set of sequential sentences and a set of Lexical Chains shared by these sentences. The needed set of formal concepts (which determine a segmentation) consists of a special constrained set of concepts $\{c_1, \ldots, c_l\}$ such that:

- $sup\{c_1, \ldots, c_l\} = 1_{\mathfrak{B}(K)};$
- $inf\{c_1, \ldots, c_l\} = 0_{\mathfrak{B}(K)};$

(as the lattice $\mathfrak{B}(K)$ is complete, the *sup* and *inf* does exist for each set of concepts)

- if the set of concepts with the above property is $\{c_1, \ldots, c_l\} = \{(A_1, B_1), \ldots (A_l, B_l)\}$, then $A_1 \cup \ldots A_l = \{S_1, S_2, \ldots S_n\}$ and $A_1 \cap \ldots A_l = \varnothing$.

In other words, the set of concepts $\{c_1, \ldots, c_l\}$ represents a kind of "complementing" concepts of the Concept lattice $\mathfrak{B}(K)$. As a remark, the set of concepts fulfilling the above constraints is, possibly, not unique, providing different segmentations for the same text.

The theorem below assures that the extents of concepts contained in a set of concepts are formed by adjoining sentences and form, indeed, a segmentation:

Theorem *If* $K := (G, M, I)$ *is a finite formal context,* $G := \{g_i \mid i = 1, \ldots, n\}$, $M := \{m_j \mid j = 1, \ldots, m\}$, *having the property that for each attribute* m_j, *the incidences are of the form* $(g_i, m_j) \in I$, $(g_{i+1}, m_j) \in I, \ldots, (g_{i+k}, m_j) \in I$ *(i.e., the objects are adjoining), then all formal concepts* (A, B) *have their extents formed by adjoining objects, too.*

Proof Let (A, B) be a formal concept of K. Its intent is defined to be $A' := \{m \in M \mid \forall g \in A, (g, m) \in I\}$. By the Basic Theorem of FCA, we have that $(A, B) = \wedge_{m \in B} \mu m$, i.e., it is the infimum of all attribute concepts of its intent. Hence, by the same Basic Theorem, $A = \cap_{m \in B} m'$ since every attribute concept is of the form $\mu m = (m', m'')$.

By the assumptions made, for every attribute $m_j \in M$, its attribute extent is formed by adjoining objects, $m'_j = \{g_i, g_{i+1}, \ldots, g_{i+k}\}$. Since every meet of sets consisting of adjoining objects has the same property, it follows that the extent A is formed by adjoining objects too.

The introduced method could be summarized in the following algorithm called **FCA segmentation** algorithm:

Input A text T formed by the sentences $\{S_1, S_2, \ldots, S_n\}$ and a set of Lexical Chains $\{L_1, L_2, \ldots, L_m\}$.

Output One or more linear segmentations of the text T corresponding to one or more sets of constrained concepts fulfilling the above properties.

Table 1: The incidence matrix for Example 1.

	L_1	L_2	L_3
S_1	X		
S_2	X		X
S_3	X		X
S_4	X		X
S_5		X	X
S_6		X	

Examples

Let us consider a first Example 1, where the incidence matrix is given as in Table 1. The Concept Lattice is given in Figure 1.

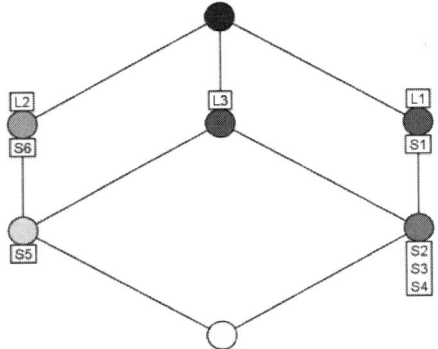

Figure 1: Concept lattice for Example 1.

The formal concepts in the Concept Lattice of the Example 1 are:

$C_1 = (\{S_1, S_2, S_3, S_4\}, \{L_1\})$

$C_2 = (\{S_5, S_6\}, \{L_2\})$

$C_3 = (\{S_2, S_3, S_4, S_5\}, \{L_3\})$

$C_4 = (\{S_2, S_3, S_4\}, \{L_1, L_3\})$

$C_5 = (\{S_5\}, \{L_2, L_3\})$

$C_6 = (\{S_1, S_2, S_3, S_4, S_5, S_6\}, \Phi)$

$C_7 = (\Phi, \{L_1, L_2, L_3\})$

Let it be remarked that the defined formal concepts fulfill the properties mentioned in the Formal Concept Analysis Chapter.

The single set of concepts such that $inf = C_7$, $sup = C_6$ and $A_{C_1} \cup A_{C_2} = \{S_1, S_2, \ldots, S_6\}$ and $A_{C_1} \cap A_{C_2} = \varnothing$ is the set $\{C_1, C_2\}$. The determined segmentation is: $[S_1, S_4][S_5, S_6]$. Let it be observed that the summary with the *first sentence of a segment* rule is $\{S_1, S_5\}$.

As a second example we display in the Figure 2 a short 9 sentences text from DUC2002 (AP880314). The manual Lexical Chains are presented in the Figure 3. The incidence matrix is established from the Lexical Chains as for example: $L1$ is an attribute for the objects $S1, \ldots, S7$, $L2$ is an attribute for the objects $S1, \ldots, S8$, etc.

Text:

S1. The communist world gets its first McDonald's next week, and some people here are wondering whether its American hamburgers will be as popular as the local fast-food treat, Pljeskavica.

S2. The long-awaited opening of the restaurant on one of Belgrade's main downtown squares will take place March 24, the Yugoslav news agency Tanjug reported, and it will offer Big Macs, fries and the other specialities familiar to McDonald's customers in the West.

S3. The Belgrade media have suggested that the success of the American restaurant depends on its acceptance by Yugoslavians who are long accustomed to the hamburger-like Pljeskavica.

S4. Pljeskavica is made of ground pork and onions, and it is served on bread and eaten with the hands.

S5. It is sold at fast-food restaurants across the country and costs about a dollar.

S6. "In fact, this is a clash between the Big Mac and Pljeskavica," said an official of Genex, Yugoslavia's largest state-run enterprise that will operate the McDonald's.

S7. John Onoda, a spokesman at McDonald's Oak Brook, Ill., headquarters, said it was the first of the chain's outlets in a communist country.

S8. The next East European McDonald's is scheduled to be opened in Budapest, Hungary, by the end of this year, said Vesna Milosevic, another Genex official.

S9. Negotiations have been going on for years for expanding the fast-food chain to the Soviet Union, but no agreement has been announced.

Figure 2: Text from DUC2002 (AP880314), Example 2

Manual Lexical Chains for Example 2:

L1: world (S1) , country (S5) (part/whole), country (S7) (repetition)

L2: McDonald's (S1), McDonald's (S2), McDonald's (S6), McDonald's (S7), McDonald's (S8) - repetition all cases

L3: people (S1), customers (S2) (specification)

L4: hamburgers (S1) (part/whole), fast-food (S1) (part/whole), fries (S2) (part/whole), specialties (S2) (generalization), hamburger (S3) (repetition), pork (S4) (part/whole), onions (S4) (part/whole), bread (S4) (part/whole), fast-food (S5) (repetition), fast-food (S9) (repetition)

L5: Pljeskavica (S1), Plesjkavica (S3), Plesjkavica (S4), Plesjkavica (S6)-repetition all cases

L6: week (S1) (part/whole), March (S2) (part/whole), year (S8), years (S9) (repetition)

L7: restaurant (S2), restaurant (S3) (repetition), restaurants (S5) (repetition)

L8: Belgrade (S2), Belgrade (S3) (repetition)

L9: Big Macs (S2), Big Macs (S6) (repetition)

L10: news (S2), agency (S2) (specification), media (S3) (generalization)

L11: success (S3), acceptance (S3) (part/whole), clash (S6) (opposition)

L12: headquarters (S7) (part/whole), chain (S7) (specification), outlets (S7) (part/whole), chain (S9) (repetition)

L13: official (S6), spokesman (S7) (specification), official (S8) (repetition)

L14: Genex (S6), Genex (S8) (repetition)

L15: American (S1), American(S3) (repetition)

Figure 3: Manual Lexical Chains extracted from the text in the Figure 2 (Example 2).

FCA tool *Toscana* was used to obtain the Concept Lattice (see Figure 4), and to segment the text of Example 2, Figure 2, with the FCA segmentation algorithm. Toscana is a software tool aimed for handling the tasks involved in the study of lattice theory, mainly formal concepts. There are several FCA tools available, a list may be found on the Formal Concept Analysis homepage: http://www.upriss.org.uk/fca/fca.html.

The extents of Formal concepts which satisfy the constraints of FCA segmentation algorithm are: $\{S_1, S_2\}, \{S_3, S_4, S_5, S_6\}, \{S_7, S_8, S_9\}$. The corresponding segmentation is [1,2][3,6][7,9]. The binary vector which signals the beginning and the end of a segment is 110001101. Let this result (called FCA) be compared with that obtained with Lexical Chains Distribution (LCD) method (Tatar et al. 2009a) using *WindowDiff* (see the Evaluation section in this Chapter). For the segmentation [1,6] [7,9], obtained with LCD method, corresponds the binary vector 100001101, and for the manual segmentation (Man), [1,3][4,6][7,9] corresponds the binary vector 101101101. When $k = 0$, *WindowDiff(Hyp, Ref)* is the ratio of the number of different elements in the binary vectors of the segmentation *Hyp* and the segmentation *Ref* over N (the length of the text). The obtained values for Example 2 (Figure 2) with $k = 0$ are:

$$WindowDiff(FCA, Man) = 2/9;$$

$$WindowDiff(LCD, Man) = 2/9.$$

Other extents which satisfy the constraints of FCA-segmentation algorithm are: $\{S_1, S_2, S_3, S_4, S_5, S_6\}, \{S_7, S_8, S_9\}$ and they determine the segmentation [1,6] [7,9], identical with LCD segmentation.

Thus, for this short text, the performances of the FCA algorithm and LCD algorithm (based on Lexical Chain Distribution) are equal.

Concept-oriented segmentation by clustering

The process of segmentation is seen as an objective method, which provides one a clearly defined result (for a given number of segments). However, different users have quite different needs with regard to a segmentation because they view the same text from a completely different, subjective, perspective. Segmenting a text must be associated with an explanation of why a given set of segments is produced.

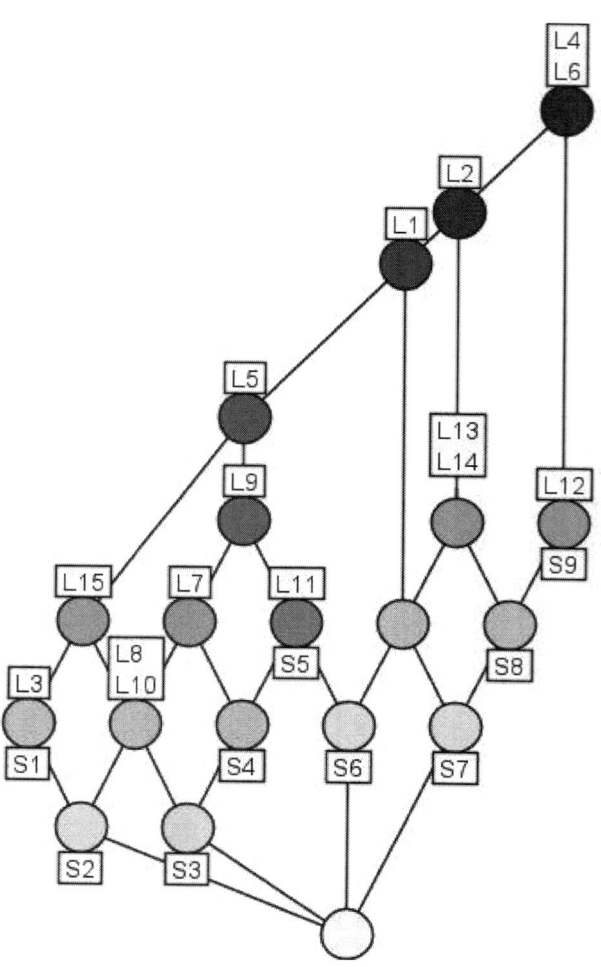

Figure 4: Concept lattice for Example 2.

We view the process of segmentation as connected to the clustering process of the sentences of a text. When the cluster $Cl = \{S_{i_1}, \ldots, S_{i_m}\}$ is one of the set of obtained clusters, where $i_1 < i_2 \ldots < i_m$, then the linear segmentation is: $[S_1, S_{i_1-1}][S_{i_1}, S_{i_2-1}], \ldots, [S_{i_m}, S_n]$.

The concept terms which are "specific" to this cluster Cl explain the reason of the segmentation.

Let it be remarked that clustering texts usually means selecting the most important (by frequency) words as features of clustering (Cimiano et al. 2005). In (Tatar et al. 2010b) transitive verbs and complement nouns are chosen as words which form the concepts in the FCA approach (Cimiano et al. 2004). In what follows these words are referred to as *concept terms*, namely *concept attribute terms*, M (transitive verbs), and *concept object terms*, G (complement nouns).

A sentence is represented as a vector of concept terms: an entry of each vector specifies the frequency that a concept term occurs in the text, including the frequency of subconcept terms. The frequency of subconcept terms is obtained from a taxonomy extracted from the text as in (Cimiano et al. 2005).

Concept-oriented segmentation by clustering algorithm

Input A text $T = \{S_1, \ldots, S_n\}$, the set of concept terms $G \cup M$ (M is formed by transitive verbs and G is formed by their complement nouns), the concept lattice $\mathfrak{B}(K)$ and the taxonomy H obtained from $\mathfrak{B}(K)$ (Cimiano et al. 2005).

Output Different segmentations of the text T, after different sets of concepts.

- Calculate the frequency of the concept term $t \in G \cup M$ in the sentence S_i denoted by $f(i, t)$;
- Calculate total frequency of the concept term t in the sentence S_i as $Total_S(i, t) = f(i, t) + \sum_{t' \text{ is a direct descendent of } t \text{ in the taxonomy } H} f(i, t')$;
- Calculate the total frequency of t for all sentences as $Total(t) = \sum_{i=1}^{n} Total_S(i, t)$;
- Choose the first m best supported *concept terms* (which maximize $Total(t)$), $t_1, \ldots, t_m$;
- Represent each sentence S_i by an *m-concept term* vector: $V(i) = (Total_S(i, t_1), \ldots, Total_S(i, t_m))$
- Cluster the set of sentences $T = \{S_1, \ldots, S_n\}$, where $sin(S_i, S_j) = cosine(V(i), V(j))$
- A cluster corresponds to a segmentation as above. The *concept terms* specific for this cluster explain the "view" of segmentation and help the user to understand the differences between clustering (segmentation) results.

11.3 Evaluation

There are several ways to evaluate a segmentation algorithm and the first challenge is to select appropriate measures. These ways include comparing the segments against that of some human judges, comparing the segments against other automated segmentation strategies and, finally, studying how well the results improve a computational task (Hearst 1997, Orasan 2006). In (Tatar et al. 2008a) all these ways of evaluation are used, including the study of how the segmentation method affects the outcome of summarization.

Since there are no large corpora with segment annotations, a frequent method to provide references is that of concatenated texts: the system takes a concatenated set of texts and returns segments consisting of single texts. This method presents the advantage that the subjectivity in boundary judgements is eliminated. However, (Labadie and Prince 2008b) differentiated between the task of finding text boundaries in a corpus of concatenated texts and the task of finding transitions between topics inside the same text. The reason of this separation is that a concatenation of texts is not a discourse as a whole. The conclusion in (Labadie and Prince 2008b) is that common segmentation methods get good results in finding concatenated text boundaries, but their performance drops when handling a single text topic boundaries, whereas their method works well in both cases.

Regarding the comparison with the human judge, a method is evaluated according to the IR style measures of precision and recall: precision is the percentage of the boundaries identified by the algorithm that are indeed true boundaries, the recall is the percentage of the true boundaries that are identified by the algorithm (Pevzner and Hearst 2002). A problem with the precision and the recall is that they are not sensitive to "near misses": if two algorithms fail to match any boundary precisely, but one is close to correct, whereas another is entirely off, both algorithms would receive scores 0 for precision and recall (Stockes 2004). This is the reason some works have tried to introduce a modified version for precision and recall. Namely, they count:

- how many of the same (or very close) boundaries with the human judge the method selects out (precision);
- how many true (or very close) boundaries are found out of the possible total (recall)

Thus, a margin of error of +n/-n is admitted for a correct boundary, with the precaution that each boundary may only be counted once as correct.

In (Tatar et al. 2008a) the evaluation of a segmentation *Method* against a *Manual* segmentations is made with the following *Precision* and *Recall* measures:

$$Precision_{Method,\ Manual} = \frac{Number\ of\ correct\ gaps}{Number\ of\ gaps\ of\ Method}$$

$$Recall_{Method,\ Manual} = \frac{Number\ of\ correct\ gaps}{Number\ of\ gaps\ of\ Manual}$$

For *F* measure is used the formula:

$$F_{Method,Manual} = \frac{2 \times Recall_{Method,Manual} \times Precision_{Method,Manual}}{Recall_{Method,Manual} + Precision_{Method,Manual}}$$

Here *Number of correct gaps* is the numbers of beginnings of segments found by *Method* which differ with -1,0,+1 to the beginnings of segments found by *Manual*. This definition for *Number of correct gaps* increases the sensitivity of the precision and the recall to "false negatives" when they are a "near-miss".

Another metric used is the Beeferman probabilistic metric P_k (Beeferman et al. 1997) which calculates segmentation accuracy in terms of "false positive", "false negative", and "near-misses" boundaries. More specifically, P_k is defined as the probability that a randomly chosen pair of sentences (or other text units) at the distance k is incorrectly classified by the hypothesized segmentation comparing with reference segmentation. Correctly classified means that if two sentences are in the same segment in the reference segmentation, or they are in different segments, they are in the same situation in the hypothesized segmentation.

The paper (Pevzner and Hearst 2002) evaluates several drawbacks of the metric P_k: the metric penalizes "false negatives" more heavily than "false positives", overpenalizes "near misses", and is sensitive to variations in segment size. As a solution, they propose a simple modification of P_k into a new metric, called *WindowDiff*, which penalizes "near misses" less than "false positives" and "false negatives". *WindowDiff* moves a fixed-sized window across the text and compares the number of boundaries assigned by the reference segmentation (*Ref*) with the number of boundaries assigned by the hypothesized segmentation (*Hyp*) in this window. The segmentation (*Hyp*) is penalized if the module of the difference of these numbers is strictly greater than 0.

The formula for this measure is:

$$WindowDiff(Hyp, Ref) = \frac{\sum_{i=1}^{N-k} |r(i, k) - h(i,k)|}{N-k}$$

Here $r(i, k)$ represents the number of boundaries of *Ref(erence)* segmentation contained between sentences i and $i + k$ and $h(i, k)$ represents the number of boundaries of *Hyp(othesis)* segmentation contained between the sentences i and $i + k$ (k is the size of the window, N is the number of sentences).

When $k = 0$, *WindowDiff(Hyp, Ref)*= (number of different elements in the binary vectors of the segmentation *Hyp* and the segmentation *Ref*) /N. The binary vector is simply associated with a segmentation as in the following example: consider that a segmentation is [S1,S3] [S4,S5][S6,S7]. The binary vector put 1 on the initial and final sentence in a segment, and 0 on an internal sentence, and thus the binary vector signals the beginning and the end of a segment. Thus, the binary vector for the segmentation example is [1011111]. Consider further that *Ref* segmentation of a text is [S1,S4][S5,S7], and *Hyp* segmentation is: [S1,S3][S4,S5][S6,S7]. The binary vector for *Ref* is [1001101] and the binary vector for *Hyp* is [1011111]. *WindowDiff(Hyp, Ref)* =2/7=0.285. The penalization is small, since the "false positives" (S4 and S6) in *Hyp* are very close to the boundary (S5) in *Ref*. If *Hyp* is [S1,S2][S3,S7], the binary vector is [1110001] and *WindowDiff(Hyp, Ref)* =4/7= 0.57. The penalization is bigger, since the "false positive" (S3) is more distant to the boundary (S5) in *Ref*.

A measure of correctness of a segmentation *Hyp* relative to a reference segmentation *Ref* in percents % is (Tatar et al. 2012):

$$Correctness(Hyp, Ref) = (1 - WindowDiff(Hyp, Ref)) \times 100$$

Correctness(Hyp, Ref) for the second above example is 43%.

The measure *WindowDiff* has been also analyzed in (Lamprier et al. 2007) where some improvements are proposed. Namely, the authors observe that whereas errors at the extremities of a text are as important as others, *WindowDiff* penalizes errors located between sentences 1 and k and between sentences $n-k+1$ and n less than others, since the window (of the length k) takes this error into account a fewer number of times. The problem can be solved, as the authors propose, by adding k fictive sentences at the begin and at the end of the text.

The paper (Lamprier et al. 2007) observes that the comparison of a hypothesized segmentation and reference segmentation raises some problems, among others the differences of sensitivity of the methods w.r.t to the granularity or to the specific point of view. They propose an evaluation method that does not need a segmentation of reference and is based on the definition of topic segments as parts of a text with strong internal relationships which are disconnected from other adjacent parts. These two criteria do not take into account the order of the sentences in the segments, therefore changes in the order of sentences inside the segments of a hypothesized segmentation should not modify the boundaries. On the contrary, the less accurate a segmentation method is, the fewer chances there are to retrieve the same boundaries, after sentences permutations inside each segment. Between the segmentations before and after permutations inside segments they propose to apply recall, precision and F-measure.

As a last method of evaluation, we mention here that the performance of the segmentation could be measured by the quality of the output results for a specified task. The paper (Tatar et al. 2008b) compares segmentations obtained by two different methods (LTT and CTT) by comparing the summaries obtained starting from these segmentations. The conclusion arrived at is that the quality of CTT segmentation is better than the quality of the LTT segmentation from the point of view of informativeness of the obtained summaries (the informativeness is the similarity, calculated as *cosine* between the summarized text and the summaries obtained with one of the above strategy). See in the next Chapter Text summarization starting from linear segmentation.

11.4 Conclusions and Recent Research

The segmentation depends on the view we take on the text, on the goal of a fine-grained segmentation or coarse-grained segmentation, or on the conception of the topic. All these can vary depending on the domain and the interests of application. A possible further solution might be to use an adaptive approach of segmentation. For example, for extending existing text segmentation approaches to produce more readable segments and increasing the effectiveness of QA systems one could introduce, as a second component of the similarity of two sentences, a measure of the amount of linkage between them (Ganguly et al. 2011).

In (Purver 2011), an excellent overview and survey of topic segmentation, it is concluded that the approaches in Text segmentation are based on changes in lexical similarity (the low cohesion areas), on high cohesion areas (determined by clustering), on cue phrases, and on generative (probabilistic) models. Given these different approaches, one way to improve segmentation might be to combine them. This is probably a direction for the further work in Text segmentation.

The other direction of research is the study of segmentation in spoken language (segmenting lectures, meetings or other speech events, etc.), particularly the study of monologue versus dialogue (Purver 2011).

Recently, work has started up again in story schemata, using narrative event chains (Chambers and Jurafsky 2008). The steps of structuring stories are: learning narrative relations between events sharing coreferring arguments, a temporal classification to partially order the connected events, and finally, pruning and clustering chains from the space of events. The narrative event chains could play in the further the same extended role as the lexical chains in earlier research.

Other further direction is the extension of early word-based algorithms with their topic-based variants. In (Riedl and Biemann 2012) the authors utilize semantic information from some topic models to inform text segmentation algorithms, and show that this approach substantially improves earlier algorithms TextTiling and C99. Similarly, some event-based variants of early word-based algorithms could be imagined for template filling in Information Extraction.

Starting from the evidence that texts within a given genre (news reports, scientific papers,etc) generally share a similar structure, that is independent of topic, the function played by their parts (functional segments) is studied: functional segments usually appear in a specific order, and they are not homogeneous. The problem of Functional segmentation versus Topic segmentation is a challenge for further research in Text segmentation.

CHAPTER 12

Text Summarization

12.1 Introduction

Text summarization (TS) has become the subject of intensive research in the last few years due to the explosion of the amount of textual information available and it is still an emerging field (Orasan 2006, Radev et al. 2002). A volume of papers edited by I. Mani and M. Maybury in 1999, a book of I. Mani in 2001, a Special Issue on Summarization of Computational Linguistics Journal (Radev et al. 2002), some special chapters in more recent handbooks, and a large number of tutorials and PhD theses provide good introductions and support to this fast evolving field.

A summary is "a shorter text, usually no longer than a half of the source text (mostly significantly less than that), which is produced from one or more original texts keeping the main ideas" (Hovy 2003).

Most of the work done so far produces two categories of summaries: *extracts* and *abstracts*. The extracts are summaries created by reusing important portions of the input verbatim, while abstracts are created by regenerating the extracted content (Hovy 2003). However, research in the field has shown that most of the sentences (80%) used in an abstract are those which have been extracted from the text or which contain only minor modifications (Orasan 2006, Hovy 2003). This, and the more difficult approach of abstracting, are the reasons why the majority of research is still focused on text extraction.

The idea of the first direction in TS, proposed by Luhn (Luhn 1958), was that, up to a point, the more frequently a word occurs the more likely it is to indicate salient material in the text. Luhn calculated for each sentence a score based on the number of frequent words (stems of words) contained, where the frequency is judged regarding a threshold. The summary consists of all sentences with scores above a cutoff point. Thus, the principle applicable for almost all extractive methods, for scoring the sentences and selecting the most scored sentences, is firstly given by Luhn.

In (Sparck Jones 2007, Tucker 1999) the stages of summarization are enumerated as *analysis (source representation or interpretation), condensation and synthesis:*

- The *analysis* stage establishes the linguistic level of representation, going from identifying sentence boundary to the identification of compound phrases, or even a semantic representation for each sentence using predicate logic. Granularity of the source representation is given by the size of its basic units, which could vary from a clause or a sentence until a segment of discourse.
- The second stage of summarization, *condensation* (*transformation*), has as result a summary representation. *Condensation* has two main processes (Tucker 1999): *selection* is choosing parts of the source representation judged as appropriate for inclusion in the summary, and *generalization* is the process of combining information from more than one part of the source representation to form the summary. *Generalization* is computationally hard to be realized, requiring some external knowledge and inferences, and thus *condensation* consists usually of *selection* alone.
- The *synthesis* (or *generation*) of the summary could be realized by a rigid or more flexible kind of output, depending on the addressability and the goal of the summary. Sentence extraction is considered as a moderately flexible kind of synthesis, even if the resulting summary is itself not very coherent; it is also the most easily to be obtained.

In terms of three stage model, the source representation in Luhn's model was simply the original text plus the frequencies of significant words (scoring the sentences), the condensation stage was the selection of the most scored sentences, and the synthesis stage was the output of the selected sentences (Tucker 1999).

At the most basic level, summaries could be classified into *extracts* and *abstracts* (as above) and within these two categories, a summary can be (Sparck Jones 1999):

- *Indicative*, which notes *that* the source text is about some topic(s) but not gives specific content;
- *Informative*, which conveys *what* the source text says about some topic(s) and covers this topic(s).
- *Critical* (vs.*neutral*), which is biased towards a particular view of the source text content.

From another point of view, summaries can be classified into *basic* (or *generic*) class and *user-focused*. A *basic* summary is likely to be useful for a variety of purposes, versus a special view (*user-focused*) summary. In this order, a *basic* summary must be *reflective* (Sparck Jones 2007) (representative of the source text as a whole), and *neutral*. A *basic* summary is text-driven and follows a bottom-up approach using IR techniques, while *user-focused* summary relies on user information need, and follows a top-down approach, using IE techniques (Lloret 2011, Radev et al. 2002). *Critical* summaries are difficult to be produced automatically, because they must include opinions, recommendations, etc. and therefore most systems generate either indicative or informative systems (Lloret 2011).

The stages of summarization according to (Hovy 2003) are somehow different from those introduced by Sparck Jones (as above). This view permits one to make a stronger distinction between *extracts* and *abstracts*, since the *extracts* are accomplished in only one (the first) stage of Hovy's description (Lloret 2011). The stages are topic identification, topic fusion and summary generation (Hovy 2003):

- *Topic identification*. It consists in assigning a score to each unit of input (word, sentence, phrases, etc). The criteria to assign this score include word and phrase frequency criteria (such as in Luhn's earliest work, who used only word frequency), positional criteria (first positions usually, scoring around 33 per cent from the total score (Hovy 2003)), cue phrase indicator criteria, title overlap criteria (the more words overlapping the title a sentence contains, the higher is the score of the sentence), cohesive or lexical connectedness criteria (the more connected sentences are assumed to be more important), discourse structure criteria (scoring the sentences by their centrality in a discourse). Combining scores improves the score significantly, even if there is not a best strategy. However in (Orasan 2006) it is shown that word frequency criteria, despite its simpleness, results in high quality summaries. (One prominent metric is *tf.idf* measure, which balances a terms frequency in the document against its frequency in a collection.)

This stage could be expressed by the following formula, where U is a unit of text, usually a sentence (Hahn and Mani 2000)

$$Score(U) = Freq(U) + Location(U) + CuePhrase(U)+$$

$$Head(U) + Connect(U) + Dicourse(U) + AddCriteria(U) \qquad (1)$$

Extractive summaries require only this stage (Hovy 2003), followed by the selection of sentences. There are three different ways of selecting summary sentences (Nenkova and McKeown 2012): a) the top n most scored sentences, where n is the length of the summary, b) using *maximal marginal relevance* principle in an iterative greedy procedure that disprefers sentences similar to already chosen sentences and c) using a global selection procedure that selects the optimal set of sentences so that the overall score maximized and the redundancy minimized.

- *Topic fusion (topic interpretation)*. In the case of Topic fusion, the topics identified as important are fused, and represented in new terms, concepts and words not found in the original text. This stage is presented only in abstractive summarization and remains difficult due to the problem of extraneous knowledge acquisition (Hovy 2003). However, extractive summaries (beside abstractive summaries) can take some benefits from a stage of pre-processing as segmentation, lexical chains or clustering, etc. These benefits will be highlighted in the next section.
- *Summary generation*. This stage requires techniques from natural language generation, namely text-planning, sentence-planning and sentence realization. The text-planner's task is to ensure that the information selected by topic identification and fused by the topic interpretation module, is phrased compactly and as briefly as possible. Text-planning is an area largely unexplored by computational linguists (Hovy and Lin 1999) and it is still true for the present. Another promising approach is *Text compression*, for example by compressing

syntactic parse tree. Usually, in this stage the systems must have access to external information such as an ontology or a knowledge base and even be able to perform inferences (Radev et al. 2002).

These stages are exemplified for the famous system SUMMARIST in the well known paper (Hovy and Lin 1999).

In (Nenkova and McKeown 2012) a distinction is made between Topic representation approaches and Indicator representation approaches of TS. While the first approach fits the first two stages in (Hovy 2003) (as above), in Indicator representation approaches the text is represented by a set of indicators which do not aim at discovering topicality. In sentence scoring by Indicator representation approach, the weight of a sentence is determined using machine learning techniques to discover different indicators and their weights.

The design and evaluation of a summarization system has to be related to three classes of context factors (Sparck Jones 1999, Sparck Jones 2007):

- *input* factors, that characterize properties of the source form (e.g., structure, scale, medium or genre), subject type (ordinary, specialized, restricted) and unit (single or multiple documents, etc.);
- *purpose* factors, which represent what summarizing is for, its intended use and audience;
- *output* factors, like degree of reduction and format, abstract or extract type, indicative or informative style, etc.

All these factors are interdependent, and the comparison of different systems can be made only if the context factors (input, purpose, output) are similar. The construction of a summarizer requires an explicit and detailed analysis of context factors. The taxonomy proposed in (Hovy and Lin 1999) comprises in the *input* characteristics of the source text(s) as for example: source size (single-document vs. multi-document), specificity (domain-specific vs. general), genre, scale (from book-length to paragraph length). The *output* factors are characteristics of the summary as a text: the way of derivation (extract vs. abstract), coherence (fluent vs. disfluent, a disfluent summary consists of individual words or text portions not structured), partiality (neutral vs. evaluative), conventionality (fixed vs. floating). The *purposes* factors, as characteristics of the summary usage are: audience (generic vs. query-oriented), usage (indicative vs. informative), and expansiveness (background vs. just-the-news). Different summarization techniques may apply to some combinations of factors and not to others (Hovy and Lin 1999).

All this discussion emphasizes the idea that summarizing is in general a hard task, "because we have to characterize a source text as a whole, we have to capture its important content, where (...) importance *is a matter of what is essential as well as what is salient*" (Sparck Jones 1999), and all these in a coherent manner. Moreover, due to the Web 2.0 evolving, the research in TS attempts to be extended to new types of websites, as blogs, forums, social networks. These conduct a new type of *sentiment* based summaries, which fall in the category of *critical* summaries (Lloret 2011). Also, to the previously mentioned issues, one can add those possed by different media summarization (images, video, audio) or by multi-lingual and cross-lingual summarization.

12.2 Methods and Algorithms

Some approaches of TS, falling into following categories, will be enumerated: summarization starting from linear segmentation, summarization by Lexical Chains or by Coreference Chains, summarization based on Discourse structure, Formal Concept Analysis (FCA) approach of summarization, summarization by sentences clustering, and other methods.

12.2.1 Summarization starting from linear segmentation

Extracting sentences (or paragraphs) without considering the relations among them, can result in incoherent summaries (for example there may be dangling anaphors, words or phrases that take their reference from another word or phrase), or in topical under-representation. Also, the problem of redundancy is not always solved optimally: the high scored sentences (with potential to be selected in the summary) could be very similar. Moreover, these sentences could cover only a part of source text, and not the text as a whole. Segmentation tends to diminish these drawbacks: the selected sentences are from *different* segments, and they are from *all* segments (uniformly representing all main topics in the source text). In (Boguraev and Neff 2000) it is mentioned that almost half of the summarization errors could have been avoided by using a segmentation component. In (Tatar et al. 2008a) it is proved that the precision of summaries in the presence of a previous segmentation achieves a 5.4% increase.

Let us remark that, since segmentation relies on the discourse properties, the method of summarization starting from segmentation falls in the discourse-based category. We will discuss in this section only linear segmentation, but the discussion could be extended to the hierarchical segmentation (see the Chapter Text Segmentation) as well.

In segmentation based summarization the positional criteria are widely used, and the simplest and most widely used method in summarization, namely to take the first sentence in a segment, often outperforms other methods. The positional score could be an explicit component of the score of a sentence (as in (Boguraev and Neff 2000)) or could be applied as a set of heuristics. In (Tatar et al. 2008a) we applied and compared three different criteria of selecting the most salient sentence of a segment (the first sentence, the best scored sentence, the least similar sentence) when the segmentation is accomplished by LTT method (see Chapter Text Segmentation). In logical segmentation (LTT) the score of a sentence is the number of sentences of the text which are entailed by it. The conclusion was that the strategy of extracting first sentence(s) from the segments is the best for the quality of the summary. The corresponding strategies applied to a text of length n, $T = \{S_1, \ldots, S_n\}$ are:

Definition 1

Given a segmentation of initial text, $T = \{Seg_1, \ldots, Seg_N\}$, a summary of the length X is calculated as:

$$Sum_1 = \{S'_1, \ldots, S'_X\} \tag{2}$$

where, for each segment Seg_i, $i = 1, \ldots, N$, first $NSenSeg_i$ sentences are selected. The computation of $NSenSeg_i$s is presented below.

The second strategy is that for each segment the sentence(s) which entail(s) a maximal number of other sentences are considered the most important for this segment, and hence they are included in the summary. The corresponding method is:

Definition 2

Given a segmentation of initial text, $T = \{Seg_1, \ldots, Seg_N\}$, a summary of the length X is calculated as:

$$Sum_2 = \{S'_1, \ldots, S'_X\} \tag{3}$$

where S'_k is a sentence S_{j*} such that $j* = argmax_j\{score(S_j) \mid S_j \in Seg_i\}$. For each segment Seg_i a number of $NSenSeg_i$ of different most scored sentences are selected. After selection of each sentence, Seg_i is modified in above formula.

The third way of reasoning is that from each segment the most informative sentence(s) (the least similar) relative to the previously selected sentences are picked up. The corresponding method is:

Definition 3

Given a segmentation of initial text, $T = \{Seg_1, \ldots, Seg_N\}$, a summary of the length X is calculated as:

$$Sum_3 = \{S'_1, \ldots, S'_X\}$$

where S'_k is a sentence S_{j*} such that $j* = argmin_j\{sim(S_j, Seg_{i-1}) \mid S_j \in Seg_i\}$. Again, for each segment Seg_i a number of $NSenSeg_i$ of different sentences are selected. The value $sim(S_j, Seg_{i-1})$ represents the similarity between S_j and the last sentence selected in Seg_{i-1} or between S_j and all sentences selected in $Seg_1, \ldots Seg_{i-1}$. The similarity between two sentences S, S' is calculated in (Tatar et al. 2008a) using $cos(S, S')$.

(Tatar et al. 2008a) shows that the strategy Sum_1 is the best by comparing the informativeness of the summaries obtained with the above methods with the original texts. The same measure of informativeness proves that the segmentation preceding the summarization could improve the quality of summaries obtained. In (Tatar et al. 2008a) the summarization without a previous segmentation is obtained by a Dynamic Programming method (see the description below).

Let it be remarked that the strategies Sum_1, Sum_2, Sum_3 could be applied after the segmentation realized by any method. In (Tatar et al. 2008b) a comparison of application of these strategies of summarization, when the segmentation is obtained by CTT and LTT (see Chapter Text Segmentation) is made, favoring the case of LTT.

Scoring the segments

The drawback of a segmentation-based summarization method (see Chapter on Text Segmentation) is that the number of obtained segments is fixed for a given text. For this reason, in this section several methods are presented to correlate dynamically the number of the obtained segments with the required length of the summary. In this way it is proved that summarization starting with a segmented text provides a better handling of the variation of length of a summary on one hand, and a dynamic correlation between the length of the source text and the length of the summary, on the other. The preprocessing step of segmentation gives the possibility to score each segment and thus to display the segments that contribute more to the output summary. The score of a sentence could be recalculated as a result of the score of the segment it belongs to.

(Kan et al. 1998) observed that treating the segments separately is an approach which loses information about the cohesiveness of the text as a whole unit. They created a framework for processing segments both as sub-documents of a whole and as independent entities. This enabled one to ask a set of general questions concerning how segments differ from each other and how a segment contributes to the document as a whole. The scoring of segments is made after the distribution of (proper and common) nounphrases and pronouns between segments and results in the classification of segments into three types: summary segments (the most important), anecdotal segments and support segments. A summarization system utilizes summary segments to extract key sentences.

In (Tatar et al. 2008a) an algorithm of scoring segments based on scores of sentences is elaborated. As a result, the numbers $NSenSeg_i$ involved in the formula Sum_1, Sum_2, Sum_3 are provided. After the score of a sentence is calculated, the score of a segment is the sum of the scores of contained sentences. The final score, $Score_{final}$, of a sentence is weighed by the score of the segment which contains it. The summary is generated from each segment by selecting a number of sentences proportional to the score of the segment.

The notations for this calculus are:

- $Score(Seg_j) = \dfrac{\Sigma_{S_i \in Seg_j} Score(S_i)}{|Seg_j|}$

- $Score_{final}(S_i) = Score(S_i) \times Score(Seg_j)$ where $S_i \in Seg_j$

- $Score(Text) = \Sigma_{j=1}^{N} Score(Seg_j)$

- *Weight of a segment* : $C_j = \dfrac{Score(Seg_j)}{Score(Text)}, 0 < C_j < 1$

The summarization algorithm with an arbitrary length of the summary is as follows:

Input: The segments $Seg_1, ...Seg_N$, the length of summary X (as parameter), $Score_{final}(S_i)$ for each sentence S_i ;

Output: A summary *SUM* of length X, where $NSenSeg_j$ sentences are selected from each segment Seg_j. The method of selecting the sentences is given by definitions Sum_1, Sum_2, Sum_3.

- Calculate the weights of segments (C_j), rank them in an decreased order, rank the segments Seg_j accordingly;
- Calculate $NSenSeg_i$: $NSenSeg_i = Integer(X \times C_j)$, if $Integer(X \times C_j) \geq 1$ or $NSenSeg_i = 1$ if $Integer(X \times C_j) < 1$;
- Select the best scored $NSenSeg_i$ sentences from each segment Seg_j ;
- Reord the selected sentences as in initial text.

12.2.2 Summarization by Lexical Chains (LCs)

To ensure the coherence of a summary, Lexical Chains are largely used (for LCs see Chapter Text Segmentation). According to (Sparck Jones 1999), the source text representation is a set of LCs, and the summary representation is the best scored LCs. In (Barzilay and Elhadad 1997), one of the early works in LCs summarization, the sentences are scored based on the degree of the connectedness (via LCs) of their words (nouns and noun compounds). An interpretation is a set of LCs in which each noun occurrence is mapped to exactly one chain. All possible interpretations are generated and then the best one is found. A chain score is determined by the number and weight of the relations between chain members. The weight of a relation correlates to their type: extra-strong (between a word and its repetition), strong (between two words connected by a WordNet relation) and medium strong (between the words from two WordNet synsets connected by paths longer than one). The LCs in (Barzilay and Elhadad 1997) are built in every segment and then the chains from different segments are merged if they contain a common word with the same meaning. The chains are scored according to the formula (Barzilay and Elhadad 1999):

$$Score(Chain) = Length(Chain) \times HomogenityIndex(Chain) \qquad (4)$$

where *Length(Chain)* is the number of occurrences of members of the *Chain*, and *HomogenityIndex(Chain)* is

$$1 - \frac{the\ number\ of\ distinct\ occurrences\ of\ members\ of\ the\ Chain}{Length(Chain)} \qquad (5)$$

A chain *Chain* is a *strong chain* if it satisfies the criterion:

$$Score(Chain) > Average(Scores) + 2 \times StandardDeviation(Scores) \qquad (6)$$

The strong chains are ranked and the highest scores are considered. For each chain one chooses the sentence that contains the first appearance of a chain member in it. A second (found better in (Barzilay and Elhadad 1997)) heuristic to extract sentences is: for each chain one chooses the sentence that contains the first appearance of a representative chain member (a representative word has a frequency in the chain no less than the average word frequency in this chain). The limitations of their method emphasized by the authors are: a) long sentences have significant likelihood to be selected, b) extracted sentences contain anaphora links to the rest of the text. However, the second problem occurs for every extracting method in TS.

Let it be remarked that in this work (Tatar et al. 2012) LCs are used as one of the possible segmentation method (see Chapter Text segmentation) and then summaries are obtained with strategies Sum_1, Sum_2, Sum_3, while in (Barzilay and Elhadad 1997) the source text is first segmented and then the summaries are obtained.

The complexity of Barzilay and Elhadad's algorithm (the calculus of all possible interpretations is exponential) appears to be a drawback when it is to be used effectively for Information Retrieval. In (Silber and McCoy 2002) Barzilay and Elhadad's research is continued with the major focus on efficiency. The base of their linear time algorithm

($0(n)$ in the number n of nouns) is the construction of an array of "meta-chains" (a meta-chain represents all possible chains in which the first word has a given sense). These meta-chains represent every possible interpretation of the text. Finally, for each word in the source text, the best chain to which the word belongs, is selected. An interesting point of view of the authors was the evaluating of the summaries by the examination of 1) how many of the concepts from strong lexical chains occur in the summary and 2) how many of the concepts appearing in the summary are represented in strong lexical chains. It is expected that the concepts identified as important by the LCs, are concepts found in the summary (Silber and McCcy 2002). The supposition which relies on this expectation is that the words are grouped into a single chain when they are "about" the same concept. This suggests a method of evaluation of FCA summarization (see the next Section) comparing the FCA-concepts of the source text with the FCA-concepts appearing in the summary.

Issues regarding generation of a summary based on LCs are the subject of many current works.

Summarization by Coreference chains

Identifying coreference chains (CRCs) can be an useful tool for generating summaries (Azzam et al. 1999), even if the coreference relations tend to require more complex techniques to compute. From the point of view of (Sparck Jones 1999), the source text representation is a set of CRCs, and the summary representation is by the best scored CRCs. The scoring criteria of CRCs are:

- Length of chain (the number of entries of a chain);
- Spread of chain (the distance between the first and the last entry of a chain);
- Start of a chain (the distance from the title or the first sentence of the text).

The best scored CRC will determine the summary as the set of sentences where an element of this chain occurs.

Another use of coreference in TS is studied by (Orasan 2007), who states that the integration of pronominal anaphora resolution into the term-based summarization process can substantially improve the informativeness of the produced summaries. Evaluation on the CAST corpus (Orasan 2006) shows that anaphora resolution improves the results of the summariser significantly at both 15%

and 30% compression rates. The reason of this improvement is that term-based summarization relies on word frequencies to calculate the score of a word. Because some of these words are referred to by pronouns, the frequencies of the concepts they represent are not correctly calculated. The proposed term-based summarization method takes the output of the anaphora resolver and increases the frequencies of words referred to by pronouns, thereby producing more accurate frequency counts. The measure used for computing the informativeness of a summary is the cosine similarity between it and the author produced summary as proposed by (Donaway et al. 2000). However, the conclusion in (Orasan 2006) is that for a term-based automatic summariser it is necessary to have an anaphora resolver which achieves a high success rate to have a noticeable improvement.

12.2.3 Methods based on discourse structure

(Sparck Jones 1999) emphasizes that discovering how to analyze the source text discourse is a major, long-term, research enterprise. An approach of TS in this direction is (Tucker 1999) where a sentence is parsed to quasi-logical form which captures discourse entities, their relations and their relative status. The sentence representations are then decomposed into their simple, atomic predications, and these are linked to build up the overall source representation as a cohesion graph. The transformation step, deriving the summary representation identifies the node set forming the summary content. It exploits weights for the graph edge types, applying a scoring function seeking centrality, representativeness, and coherence in the node set. The generation step synthesizes the output summary text from the selected predications. The output may consist of sentences or, for fragmentary predications, mere phrases (Tucker 1999).

Another direction following discourse structure is that from (Marcu 1997b) where it is shown that the concepts of rhetorical analysis and nuclearity can be used effectively for determining the most important units in a text. The approach relies on a rhetorical parser that builds RS-trees. When given a text, the rhetorical parser determines first the discourse markers and the elementary units. The parser then uses the information derived from a corpus in order to hypothesize rhetorical relations among the elementary units. In the end the parser applies a constraint-satisfaction procedure to determine the best text structure. The summarizer takes the output of rhetorical parser and selects the textual units that are most salient. An improvement of RST summarization by machine learning approach is (Chuang and Yang 2000) where each segment is scored by an average term frequency, the number of title words, the number of times it acts as a nucleus and the number of times it acts as a satellite. The score gives all segments in a text a total ordering and the most scored sentences of top segments are selected for a summary.

Another discourse structure theory, the Centering Theory, is also studied and used for automatic summarization. The transitions computed for utterances as in Centering Theory (Grosz et al. 1995) are examined in (Hasler 2004) for patterns which relate to the coherence of summaries. These transitions are also employed as an evaluation method of summarization (see Section Evaluation).

12.2.4 Summarization by FCA

In this section will be presented a method for selection of the sentences which are least similar to the rest of sentences and which are, at the same time, the most important from a certain point of view. The method belongs to the class of methods in which the summary is built up incrementally, sentence by sentence (Carbonell and Goldstein 1998). The frame which provides the instruments to realize these desiderata is the Formal Concept Analysis (FCA).

The basic idea of summarization by FCA

In the paper (Nomoto and Matsumoto 2001) the authors show that the exploiting of the diversity of topics in the text has not received much attention in the summarization literature. After them, an ideal summary would be one whose rank order in retrieved documents in IR is the same as that of its source document. They propose the different clusters as *different topics* (exactly as in older clustering methods) and for a reduced *redundancy*, a weighting scheme (for each sentence) which finds out the best scored sentences of each cluster. In this way, (Nomoto and Matsumoto 2001) pays much more attention to the principle of Maximal Marginal Redundancy (MMR) introduced in (Carbonell and Goldstein 1998). MMR selects a sentence in such a way that it is both relevant and has the least similarity to sentences selected previously. Our FCA approach (Tatar et al. 2011) satisfies both of these requirements of MMR, modifying all the time the characteristics of the rest of the selectable sentences.

In the paper (Filatova and Hatzivassiloglou 2004) the authors assert that the main step in text summarization is the identification of the most important "concepts" which should be described in the summary. By "concepts", they mean the named entities and the relationships between these named entities (a different vision from our FCA concepts (Tatar et al. 2011)).

In (Ye et al. 2007) the authors work with "concepts" which refer to an abstract or concrete entity expressed by diverse terms in a text. Summary generation is considered as an optimization problem of selecting a set of sentences with a minimal answer loss. Again, the notion of "concept" is different to that of the FCA concept.

In the method (Tatar et al. 2011) the FCA concepts are used and the idea applied is that the quality of a summary is determined by how many FCA concepts in the original text can be preserved with a minimal redundancy. The process of summarization is defined as extracting the minimal amount of text which covers a maximal number of "important" FCA concepts. The "importance" of a FCA concept is given by its generality in the Concept Lattice and by the number of the concepts "covered" by it (as below). The most important sentences are selected to be introduced in the summary, keeping a trace of the concepts already "covered".

The basic idea of the method is to associate with a text T formed by the sentences $\{S_1, \ldots, S_n\}$ a formal context (G, M, I) where:

- the objects are the sentences of the text: $G = \{S_1, \ldots, S_n\}$;
- the attributes are represented by the set M of the *most frequent terms* (nouns and verbs) in T;
- the incidence relation I is given by the rule: $(S_i, t) \in I$ if the term t occurs in the sentence S_i.

Let CL be the associated concept lattice of the formal context, where the set of all concepts (the nodes of CL) is denoted by $Conc$.

The "importance" (the weight) of a concept $c = (A, B)$ is defined as:

Definition 1 *The weight of a concept $c \in Conc$ is $w(c) = |\{m \mid c \leq \mu m\}|$.*

Definition 2 *An object concept S_i covers the concept c if $\gamma S_i \leq c$.*

The object concepts cover a bigger number of concepts if they are located in the lower part of the concept lattice *CL*. In other words, one is firstly interested in the sentences S_i so that γS_i are direct superconcepts of the bottom of the concept lattice *CL*. This kind of sentences are denoted by *Sentence*$_{bottom}$. The algorithm introduces sequentially in the summary *Sum* the sentences from *Sentence*$_{bottom}$ which cover a maximal number of attribute concepts at the introduction time.

Let some general desirable properties of the summary generated from a text be enumerated and the ways in which FCA method accomplishes them:

- *Text coverage*. Ideally, all the sentences of a text should be potentially represented, in one way or another, in a summary. Since all the sentences are descendents of the top of the concept lattice *CL*, this requirement is fulfilled.
- *Sibling Concepts distinctiveness*. At any level of the lattice *CL,* the sibling concepts are as different as possible from each other, the separation between the contribution of each sentence in summary being displayed.
- *General to specific*. In the lattice *CL* the most general concepts are situated in the lower part and the most specific ones in the upper part. Beginning with the sentences from *Sentence*$_{bottom}$ this desiderate is fulfilled.

Summarization by FCA - SFCA algorithm:

Input: A text $T = \{S_1, \ldots, S_n\}$, the concept lattice *CL* (where the set of objects are sentences and attributes are terms), the set of concepts *Conc*, the set of concepts *Sentence*$_{bottom}$, the length *L* of the summary.

Output: A summary *Sum* of the text *T* with the length *L*.

Step 1. The set of covered concepts is empty, $CC = \varnothing$; $Sum = \varnothing$.

Step 2. $\forall S_i \in Sentence_{bottom}$, $S_i \notin Sum$, calculate the weight $w(S_i)$:

$$w(S_i) = \sum_{\gamma S_i \leq c, c \in Conc \backslash CC} w(c)$$

Step 3. Choose the sentence with the maximum weight in the summary:

$$i^* = argmax_i \{w(S_i)\}, Sum = Sum \cup \{S_{i*}\}$$

Step 4. Modify the set of covered concepts: $CC = CC \cup \{c \mid \gamma S_{i*} \le c\}$.

Step 5. Repeat from **Step 2**, until the length of *Sum* becomes *L*.

In this paper the weight of a sentence has been defined as the number of covered concepts. In this way the condition of *coverage* is attempted first. It could be introduced in the definition of the weight, as well, a part of reflecting *distinctiveness* of the topics. Moreover, some weights associated with the factor of *coverage* and the factor of *distinctiveness* could be introduced.

Due to the definition of weight for a sentence S_i, **Step 2** in the SFCA algorithm is applied as:

Step 2. For each sentence $S_i \in Sentence_{bottom}$, $S_i \notin Sum$, calculate the weight:

$$w(S_i) = \mid \{m \mid \gamma S_i \le \mu m, \mu m \in Conc \setminus CC\} \mid$$

Experimental results

The method was tested on ten texts from DUC2002 documents. For the first text (Text1 or AP880314) (Figure 1) the concept lattice is given in Figure 2. One chose as attributes the verbs and nouns with a frequency ≥ 3.

S1. Hurricane Gilbert swept toward the Dominican Republic Sunday, and the Civil Defense alerted its heavily populated south coast to prepare for high winds, heavy rains and high seas.

S2. The storm was approaching from the southeast with sustained winds of 75 mph gusting to 92 mph.

S3. "There is no need for alarm," Civil Defense Director Eugenio Cabral said in a television alert shortly before midnight Saturday.

S4. Cabral said residents of the province of Barahona should closely follow Gilbert's movement.

S5. An estimated 100,000 people live in the province, including 70,000 in the city of Barahona, about 125 miles west of Santo Domingo.

S6. Tropical Storm Gilbert formed in the eastern Caribbean and strengthened into a hurricane Saturday night.

Figure1: contd....

S7. The National Hurricane Center in Miami reported its position at 2 am Sunday at latitude 16.1 north, longitude 67.5 west, about 140 miles south of Ponce, Puerto Rico, and 200 miles southeast of Santo Domingo.

S8. The National Weather Service in San Juan, Puerto Rico, said Gilbert was moving westward at 15 mph with a "broad area of cloudiness and heavy weather" rotating around the center of the storm.

S9. The weather service issued a flash flood watch for Puerto Rico and the Virgin Islands until at least 6 p.m. Sunday.

S10. Strong winds associated with the Gilbert brought coastal flooding, strong southeast winds and up to 12 feet feet to Puerto Rico's south coast. There were no reports of casualties.

S11. San Juan, on the north coast, had heavy rains and gusts on Saturday, but they subsided during the night.

S12. On Saturday, Hurricane Florence was downgraded to a tropical storm and its remnants pushed inland from the U.S. Gulf Coast.

S13. Residents returned home, happy to find little damage from 80 mph winds and sheets of rain.

S14. Florence, the sixth named storm of the 1988 Atlantic storm season, was the second hurricane.

S15. The first, Debby, reached minimal hurricane strength briefly before hitting the Mexican coast last month.

Figure 1: Text 1 (DUC2002 -AP880314)

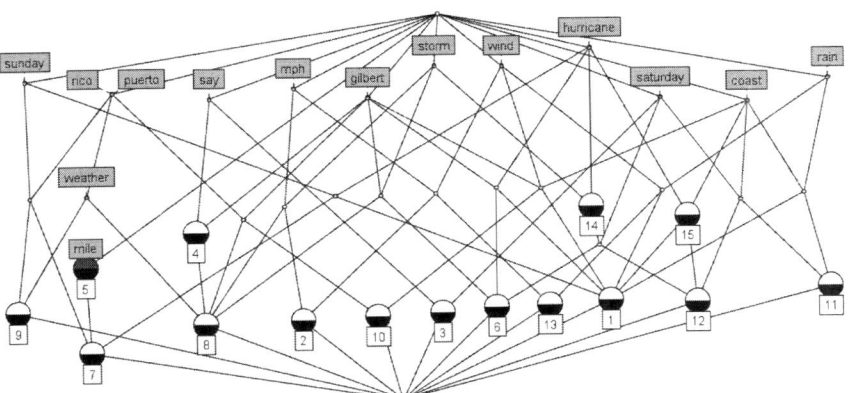

Figure 2: Concept lattice for Text1.

In the concept lattice for Text 1 all the concepts from $Sentence_{bottom}$ are γS_1, γS_2, γS_3, γS_6, γS_7, γS_8, γS_9, γS_{10}, γS_{11}, γS_{12}, γS_{13}, simply denoted in the concept lattice by 1, 2, 3, 6, 7, 8, 9, 10, 11, 12, 13. As the Text1 has the length 15, the length of the 30% summary is $L = 5 \leq |Sentence_{bottom}| = 11$. Thus the algorithm SFCA is applied only to the 11 sentences from $Sentence_{bottom}$. S_8 is the first sentence introduced in the summary, since γS_8 covers the concepts: $\mu puerto$, $\mu rico$, $\mu weather$, $\mu gilbert$, $\mu strom$, μmph, μsay and thus the maximal number of covered concepts is $w(S_8) = 7$.

Finally, the algorithm provides the summary $Sum = \{S_1, S_3, S_7, S_8, S_{10}\}$ with a precision of 60% (see Table below).

12.2.5 Summarization by sentence clustering

The idea of summarization by clustering has been applied for the first time in multi-document summarization. In (Hatzivassiloglou et al. 2001) the authors show that the method of multi-document summarization by clustering documents can also be applied for summarization of a single document.

The basic idea in summarization by clustering of sentences is that a cluster (a set of similar sentences) produces a reduced number of sentences in the final summary, since each cluster captures information directed to a particular aspect (topic) of input text. At the same time, the clustering offers an alternative to avoiding the redundancy problem, since only one (or a few sentences) from a cluster are selected in the final summary. The way the sentences can be chosen varies from using simple positional features (for example the sentences located earliest in the text are minimizing dangling references) to using the (near) centroids sentences. Some experiments about the best criteria of determining what sentences must be selected from clusters are given in (Radev et al. 2001) in connection with the summarizer MEAD, a centroid-based single-document and multi-document summarizer.

One of the first attempts to cluster sentences of a text was the paper (Yaari 1997). It applies lexical similarity measure (*cosine*) to identify the hierarchical discourse structure of a text and partition the text into segments. In (Tatar et al. 2011) are presented the results obtained by clustering the sentences of a text to obtain further the summary and the result is compared with that obtained by the SFCA algorithm. The algorithm is as follows:

Input: A text $T = \{S_1, \ldots, S_n\}$, the set of verbs and nouns from T, the length L of the summary.

Output: A summary *Sum* of the text T with the length L.

1. Calculate the frequency of the terms t (verbs and nouns) in each sentence S_i, denoted by $f(i, t)$;
2. Calculate the total frequency of the term t for all sentences as:

$$Total(t) = \Sigma^n_{i=1} f(i, t).$$

3. Choose the first m most frequent terms (nouns and verbs) t_1, $\ldots, t_m$ (which maximize $Total(t)$).
4. Represent each sentence S_i by an in vector:

$$V(i) = (f(i, t_1), \ldots, f(i, t_m)).$$

5. Apply the hierarchical agglomerative clustering algorithm for the set $\{S_1, \ldots, S_n\}$ using the similarity $sim(S_i, S_j) = cosine(V(i), V(j))$.
6. Build the summary: select from each cluster the sentence with the minimal index, which does not belong to the summary and re-traverse the clusters applying the same selection rule until L is reached.

In the experiment in (Tatar et. al 2011) 30% summaries were produced by different clustering methods, where n is the length of the text and L is the length of the summary. In the agglomerative hierarchical clustering algorithm a separate cluster is formed for each sentence and one continues by grouping the most similar clusters until a specific number of clusters are obtained. One has used as similarity measure between two sentences S_i and S_j:

$$cosine(V(i), V(j)) = \frac{\Sigma^m_{k=1} f(i, t_k) * f(j, t_k)}{\sqrt{\Sigma^m_{k=1} f^2(i, t_k) * \Sigma^m_{k=1} f^2(j, t_k)}}$$

and a new measure denoted as *com* (from *common* terms) and defined as follows:

$$sim(S_i, S_j) = com(V(i), V(j)) = \frac{\Sigma^m_{k=1} min(f(i, t_k), f(j, t_k))}{\Sigma^m_{k=1} max(f(i, t_k), f(j, t_k))}$$

In the language of FCA, the new proposed similarity measure $sim(S_i, S_j) = com(V(i), V(j))$ represents the ratio of the number of the common attributes of objects S_i and S_j and the total number of attributes of these.

As similarity between two clusters C1 and C2 one used in *single-link clustering* the similarity of two most similar members,

$$sim(C1, C2) = max\{sim(S_i, S_j) \mid S_i \in C1 \text{ and } S_j \in C2\}$$

and in *complete-link clustering* the similarity of two least similar members,

$$sim(C1, C2) = min\{sim(S_i, S_j) \mid S_i \in C1 \text{ and } S_j \in C2\}.$$

The algorithm works with all combinations for similarity of two sentences (*cosine* and *com*) and similarity between two clusters (*min* and *max*): *cosine + min, cosine + max, com + min* and *com + max*.

According to the results obtained on the same ten texts as input the new measure *com* behaves like *cosine* measure with a precision greater than 80%.

The results for 10 texts from DUC 2002 ($T1, \ldots T10$) obtained with SFCA algorithm and sentence clustering (SC) algorithm for *com+min*, when compared with manual summaries, are given in the Table below. Comparison of the results shows that both algorithms work with a good precision, but the precision is better for SFCA algorithm.

As examples of summaries for a text (T9) formed by 44 sentences (thus $L=15$) we obtained in (Tatar et. al 2011) by clustering and by FCA algorithm the following summaries:

- manual: {$S1, S2, S3, S5, S6, S8, S9, S15, S17, S18, S20, S24, S26, S27, S36$}
- FCA: {$S1, S3, S5, S8, S9, S15, S17, S18, S20, S21, S24, S27, S31, S32, S40$}
- Clustering (cosine+min):
  {$S1, S2, S4, S5, S8, S9, S10, S11, S12, S16, S18, S23, S35, S38, S41$}
- Clustering (com+min):
  {$S1, S2, S4, S5, S8, S9, S10, S11, S16, S17, S18, S23, S35, S38, S41$}

- Clustering (cosine+max):
  {$S1$, $S2$, $S4$, $S6$, $S9$, $S11$, $S12$, $S13$, $S15$, $S17$, $S21$, $S28$, $S31$, $S35$, $S38$}
- Clustering (com+max):
  {$S1$, $S2$, $S4$, $S6$, $S9$, $S11$, $S12$, $S13$, $S15$, $S17$, $S21$, $S28$, $S31$, $S35$, $S38$}

Let this section be finished with a comparison between sentences clustering method and FCA approach:

- The number of sentences per cluster could vary widely and it is difficult to estimate the number of clusters necessary for a summary of a given length. Instead, the FCA algorithm needs as an input the length of the summary.

- The clustering based methods select from each cluster more than one sentence, if the length of the summary is larger than the number of clusters. The only positional rule of the least index does not assure the need of non-redundancy. FCA approach, instead, will select at each step the sentences which cover not already covered concepts.

Table 1: SFCA *versus* SC *com + min* precisions with respect to manual summaries.

	T1	T2	T3	T4	T5	T6	T7	T8	T9	T10
n=	15	27	21	9	28	35	26	11	44	13
FCA	60%	44%	43%	66%	55%	45%	66%	75%	73%	75%
SC	40%	44%	43%	66%	22%	54%	33%	50%	46%	50%

12.2.6 Other approaches

Graph based approaches

The idea of graph-based summarization is only recently introduced in TS (Mihalcea 2005, Mihalcea and Radev 2011). In (Nenkova and McKeon 2012) the graph methods are classified in the class of indicator representation approaches, which do not attempt to interpret or represent the topics of the input text. Graph methods base summarization on the centrality of sentences in a graph representation of the input text. In the graph language, the nodes are text elements (mostly sentences, but they can be words or phrases) and the edges are links between these (representing the similarity, the overlap, the syntactic dependency, etc). The algorithm finds the most lexical central sentences (as the nodes visited most frequently), by a random walk on graph. To avoid duplicates, the final decision about a node can rely on maximal marginal relevance of node (Mihalcea and Radev 2011). In (Mihalcea 2005) a method of extractive summarization based on *PageRank* algorithm is introduced. The nodes of the graph are the sentences, the edges are the similarity relations between sentences (as a function of content overlap) with a weight associated with each edge. Finally, each node (sentence) has a score which integrates the weights of both incoming (*In*) and outgoing (*Out*) edges:

$$score(S_i) = (1 - d) + d \times \sum_{S_j \in I n(S_i)} \frac{score(S_j)}{\lceil Out(S_j) \rceil} \qquad (7)$$

where d is a parameter set between 0 and 1. The graph can be represented as an undirected graph or as a directed graph, with the orientation of edges from a sentence to sentences that follow (or are previous) in the text. The top scored sentences are selected for the summary. The method is evaluated on DUC 2002 documents using ROUGE evaluation toolkit and the best precision found is of 0.5121 (Mihalcea 2005).

Machine learning (ML) approaches

One of the first machine learning approach of TS was (Kupiek et al. 1995). They approach sentence extraction as a statistical classification problem. Given a training set of documents with hand-selected document extracts, they developed a classification function of some features that estimates the probability a given sentence is included in an extract. The studied features are: sentence length feature, fixed-phrase (or key words) feature, paragraph (or positional) feature, thematic word feature (the most frequent content words are defined as thematic words), uppercase (proper names) word feature. The idea is to introduce a simple Bayesian classification function that assigns for each sentence a score for inclusion in a generated summary. For each sentence s the probability to be included in a summary S, given the k features $F_1, F_2, \ldots F_k$, can be expressed using Bayes rule :

$$P(s \in S \mid F_1, F_2, \ldots F_k) = P(F_1, F_2, \ldots F_k \mid s \in S) \times P(s \in S)$$
$$/P(F_1, F_2, \ldots F_k). \qquad (8)$$

Assuming statistical independence of the features, the probability becomes:

$$P(s \in S \mid F_1, F_2, \ldots F_k) = \prod_{i=1}^{k} P(F_i \mid s \in S) \times P(s \in S) / \prod_{i=1}^{k} P(F_i) \qquad (9)$$

$P(s \in S)$ is a constant and $P(F_i \mid s \in S)$ and $P(F_i)$ can be estimated directly from the training set by counting occurrences.

One important drawback of this machine learning method is that the decision about inclusion in the summary is independently done for each sentence. The methods that explicitly encode dependencies between sentences outperforms other ML methods (Nenkova and McKeown 2012).

In the paper (Steinberger et al. 2007) the authors present a LSA-based sentence extraction summarizer which uses both lexical and anaphoric information. The first step in LSA method is the creation of a term by sentences matrix $A = [A1, A2, ..., An]$, where each column Ai represents the weighted term-frequency vector of sentence S_i in the document under consideration.

If there are m terms and n sentences in the document, then an $m \times n$ matrix A for the document is created. The next step is to apply Singular Value Decomposition (SVD) to matrix A. Thus the matrix A is defined as: $A = U\Sigma V^T$ where U is an $m \times n$ column-orthonormal matrix whose columns are called left singular vectors, Σ is an $n \times n$ diagonal matrix, whose diagonal elements are non-negative singular values sorted in descending order and V is an $n \times n$ orthonormal matrix, whose columns are called right singular vectors. The dimensionality of the matrices is reduced to r most important dimensions and thus, U is $m \times r$, Σ is $r \times r$ and V^T is $r \times n$ matrix. The original document represented by matrix A, is divided into r linearly-independent base vectors which express the main topics of the document. The summarization algorithm simply chooses for each topic the most important sentence for that topic. To improve the way of selection of the most important sentence for a topic (Steinberger et al. 2007) use the anaphoric chains (CRCs) identified by the anaphoric resolvers as additional terms in the initial matrix A. This is performed by generalizing the notion of *term* and treating anaphoric chains as another type of term that may occur in a sentence. The experiments show that adding anaphoric information yields an improvement in performance even if the performance of anaphoric resolver is low.

As always in Machine Learning, the advantage of using it is that it allows to test the performance of a high number of features. The disadvantage is that it needs a big training annotated corpus. Besides the common features, in most recent classification approaches of summarization, some new features based on Wikipedia are also used (Lloret 2011). Machine Learning approaches have proved to be much more successful in single document or domain specific summarization (Nenkova and McKeown 2012).

Dynamic Programming approach

In this section we describe the summarization method based on Dynamic Programming introduced by (Tatar et al. 2009c).

In order to meet the coherence of the summary, the algorithm selects the chain of sentences with the property that two consecutive sentences have at least one common word. This corresponds to the continuity principle in the centering theory which requires that two consecutive units of discourse have at least one entity in common (Orasan 2003). The assumption is made that, with each sentence a *score* is associated that reflects how representative that sentence is. In (Tatar et al. 2009c) the associated scores are the logical ones as used in LTT method, but any scoring scheme could be used. The score of a summary is the sum of the individual scores of contained sentences. The summary will be selected such that its score is maximum.

The idea of the Dynamic Programming Algorithm is the following.

Let it be considered that δ^k_i is the maximum score of the summary with length k that begins with the sentence S_i. Then δ^X_i is the maximum score of the summary with length X that begins with the sentence S_i, and $max_i(\delta^X_i)$ is the maximum score of the summary with length X. For a given length X one is interested in obtaining δ^X_1.

We have the next relations:

$$\delta^1_i = score(S_i)$$

$$\delta^k_i = score(S_i) + max_{j\ (i<j)}\ (\delta^{k-1}_j),\ k > 1$$

If a penalty is introduced for the case in which two consecutive sentences have no words in common, the recursive formula becomes:

$$\delta^k_{\ i} = max_{j\ (i<j)} \left\{ \begin{array}{ll} score(S_j) + (\delta^{k-1}_j) & \text{if } S_i, S_j \text{ have common words} \\ penalty * (score(S_j) + (\delta^{k-1}_j)) & \text{otherwise} \end{array} \right.$$

Penalty value chosen in the paper in order to inhibit the use of successive sentences without words in common is 1/10.

Thus, Dynamic Programming algorithm outputs for a sequence of scored sentences $S_1, \ldots, S_n$ and a given length X the optimal summary with the score δ^X_1.

Pure Entailment

The last method presented here relies on textual entailment to reduce the redundancy of a summary. It is called Pure Entailment method and is introduced in (Tatar et al. 2008a). The method is more efficient if instead of the text $\{S_1, S_2, \ldots S_n\}$ are used the best scored (by one of the above methods) sentences of the text.

> **Input**: Text= $\{S_1, S_2, \ldots S_n\}$
> **Output**: Summary S.
> $S = \{S_1\}; i = 2$
>   **while** $i < n$ **do**
>   **if** not $(S \rightarrow S_i)$
>   **then**
> $S := S \cup \{S_i\}$
>   **endif**
> $i:=i+1$
>   **endwhile**

12.3 Multi-document Summarization

Multi-document summarization is an automatic procedure of summarizing information from multiple texts written about the same topic. It is much more complex than summarizing a single document, due to inevitable diversity within a large documents set. The problems of dangling references and semantic redundancy are even more serious than in single document summarization as several documents are being dealt with at once.

In (Radev et al. 2002) are identified three major problems which arise in multi-summarization:

- Identifying important difference among documents and treating these accordingly;
- Recognizing and treating the redundancy;
- Ensuring summary coherence.

A success example of multi-document summarizer is MEAD, University of Michigan's summarization system, based on sentence extraction (Radev et al. 2000, Radev et al. 2001). For this tool, the authors developed a technique called centroid-based summarization, which uses the centroids of the clusters produced by a Topic Detection and Tracking (Allan et al. 1998) tool as input to identify which *sentences* are central to the topic of the cluster, rather than the individual articles. MEAD decides which sentences to include in the extract by ranking them according to a set of parameters. For each sentence in a cluster of related documents, MEAD computes three features and uses a linear combination of these to determine which sentences are most salient. Thus, for each sentence S_i a *Score* function is calculated with the formula:

$$Score(S_i) = w_c \times C_i + w_p \times P_i + w_f \times F_i \tag{10}$$

Here C_i is the centroid score which is a measure of the centrality of a sentence to the overall topic of a cluster, P_i is the position score which decreases linearly as the sentence gets further from the beginning of a document, and F_i is the "overlap with first" score which is the inner product of the tf*idf weighted vector representations of a given sentence and the first sentence (or title, if there is one) of the document (Radev et al. 2000, Radev et al. 2001). The coefficients w_c, w_p, w_f are between 0 and 1. P_i is for example $1/n_i$, where n_i is the relative position from the beginning of a document.

Reducing redundancy in multi-summarization becomes an even more significant problem. In *Score* equation the redundancy is measured as follows (Radev et al. 2000, Radev et al. 2001):

$$Score(S_i) = (w_c \times C_i + w_p \times P_i + w_f \times F_i) - w_R \times R(S_i). \tag{11}$$

Here $R(S_i)$ is a normalized cross sentence word overlap of S_i with all the rest of sentences and w_R is $max_{S_i}(Score(S_i))$.

The redundancy is solved for MEAD by cross-sentence informational subsumption, as well. If the information content of sentence S_i is contained within sentence S_j, then S_i becomes informationally redundant and the content of S_j is said to subsume that of S_i. Let us remark that in terms of entailment relation, in this case $S_j \rightarrow S_i$ (S_j *entails* S_i).

Moreover, (Radev et al. 2000) define the sentences S_i and S_j equivalent if they are subsuming each other (in this case they are said to belong to the same equivalence class). The *subsume* relation could be used in multi-summarization by determining first the classes of equivalence of sentences, choosing only a sentence from each class and then applying a single-document summarization method. In this way, each method of single-document summarization could be extended to multi-summarization.

One of the early approaches addressing the problem of measuring the redundancy is (Carbonell and Goldstein 1998), by introducing the MMR coefficient:

$$MMR(SS, A) = argmax_{S_i \in SS \setminus A}(\lambda \times Imp(S_i) - (1 - \lambda)max_{S_j \in A} sim(S_i, S_j))\ (12)$$

Here SS is the set of sentences to be summarized, A is the set of sentences already selected in the summary, $Imp(S_i)$ is the importance of the sentence S_i, Sim is the similarity between two sentences and λ is a parameter for the degree of redundancy (Mori et al. 2004). Given the above formula, *MMR* computes the ranked list of *important* sentences in SS when $\lambda = 1$, and the maximal diversity (novelty) list when $\lambda = 0$.

In the approach of (Carbonell and Goldstein 1998) summaries are created using greedy, sentence-by-sentence selection. Finding a global optimization method is very difficult but approximate solutions can be found using Dynamic Programming or Integer Linear Programming (Nenkova and McKeown 2012).

Ordering sentences in multi-summarization

Sentence ordering is much harder for multi-document summarization than for single-document summarization (Barzilay et al. 2001). The main reason is that unlike single document, multi-documents do not provide a natural order of texts.

The approaches of solving this problem are very different. Probabilistic ordering method treats the ordering as a task of finding the sentence sequence with the biggest probability (Lapata 2003). For a sentence sequence (text) $T = \{S_1, S_2, ..., S_n\}$, it is supposed that the probability of any given sentence S_i is determined only by its previous sentence. In this way the probability of a sentence sequence can be generated based on the condition probabilities $P(S_i | S_{i-1})$ of all adjacent sentence pairs in the sequence, and the problem of ordering sentences is solved as o problem of maximization of such of probability. Thus, the order is obtained by a complete graph, where the set of vertices V is equal to the set of sentences and each edge $u \rightarrow v$ has a weight, the probability $P(u | v)$. In (Lapata 2003) it is proposed a greedy algorithm of finding an approximate optimal ordering as an optimal path through a directed weighted graph.

When the multi-summarization is solved by clustering, a problem arises as to how the sentences are clustered into topics, and the other as to how the sentences belonging to the same topic are ordered. (Barzilay et al. 2001, Barzilay et al. 2002) combined topic relatedness and chronological ordering together to order sentences in the news genre: Majority Ordering (MO) and Chronological Ordering (CO). They show that MO performs well only when all input texts follow similar organization of the information, which is too strong a constraint. The CO algorithm can provide an acceptable solution for many cases, but it is not sufficient when summaries contain information that is not event based. The Majority Ordering algorithm is introduced in (Barzilay et al. 2002) using the themes, where a theme is a set of sentences conveying similar information drawn from different input texts. Consider two themes, Th_1 and Th_2; if sentences from Th_1 precede sentences from Th_2 in all input texts, then presenting Th_1 before Th_2 is likely to be an acceptable order, which is not the case when the order between sentences from Th_1 and Th_2 varies from one text to another. One way to define the order between Th_1 and Th_2 is to adopt the order occurring in the majority of the texts where Th_1 and Th_2 occur. In other words, the solution is to find a linear ordering between themes which maximizes the agreement between the orderings provided by the input texts. For this, for each pair of themes, Th_i and Th_j, the *counts* $C_{i,j}$ are determined, where $C_{i,j}$ is the number of input texts in which sentences from Th_i occur before sentences from Th_j. The weight of a linear order $(Th_{i_j}, \ldots, Th_{i_k})$ is defined as the sum of the *counts* for every pair Th_i and Th_j. In terms of a directed graph where nodes are themes, and a vertex from Th_i to Th_j has the weight $C_{i,j}$, one is looking for a path with maximal weight which traverses each node exactly once.

An improvement of the above algorithm consists in segmenting each input text (Barzilay et al. 2002). For each pair of themes A and B the number of pairs of sentences which appear in the same text $| AB |$, and the number of sentence pairs which appear in the same text *and* are in the same segment $|AB+|$ is calculated. Also, for each pair of themes A and B, the ratio $|AB+|/|AB|$ to measure the relatedness of two themes is calculated. If most of the sentences in A and B that appear together in the same texts are also in the same segments, it means that these themes are highly topically related and in this case, the ratio is close to 1. A and B are considered related if this ratio is higher than a predetermined threshold. This strategy defines pairwise relations between themes. A transitive closure of this relation builds blocks of related themes. In a second stage of the algorithm, a time stamp is assigned to each block of related themes using the earliest time stamp of the themes it contains.

12.4 Evaluation

Evaluation of summarization could be performed against the summarized text or against other summaries, that are obtained automatically or human made. A problem with this last method is the considerable subjectivity of human experts, since different experts might not arrive at the same summary of a given source text: human summarizers tend to agree only approximately 60 % of the time (Radev et al. 2002). In (Barzilay and Elhadad 1997) it is emphasized that the human subjects are consistent with respect to what they perceive as being the most important but less consistent with respect to what they perceive as being less important. However, even if human-reference approach is treated as a rough evaluation method, many works apply it for evaluation.

In (Sparck Jones 2007) a distinction is made between *intrinsic* evaluation (where a system is assessed with respect to its own declared objectives and quality) and *extrinsic* evaluation (where a system is assessed by how the produced summary is good enough in user assistance in performing a task). For the first kind of evaluation, fluency is judged, how well the summary covers stipulated key ideas, or how it compares with an ideal summary written by the author of the source text or a human abstractor.

The preference has moved from intrinsic to extrinsic evaluation, i.e. how useful TS is in order to help other tasks. For example, the questionnaire method of evaluation consists in establishing to what extent some questions about content that can be answered from the source text, can also be answered from the summary; or, in more concrete cases, to establish if the summary can satisfy some ultimate multiple purposes.

Other evaluation measures are those similar to Retention Ratio (RR) (Hovy 2003) defined as:

$$RR = \frac{information\ in\ Summary}{information\ in\ Initial\ text} \qquad (13)$$

In (Orasan 2006, Tatar et al. 2008a) the evaluation has been made by calculating the similarity (cosine) between the obtained summaries and the initial (summarized text), which is called the informativeness of summaries.

Ideally, one wants to measure interesting information content only. In (Hovy and Lin 1999) three ways of approximating the interesting information content are described: by the Shannon Game, the Question Game and the Classification Game. All these ways inspired many researchers in the evaluation of summarization approaches.

The quality of a summary could be described as the relation between RR and Compression ratio (CR) as well:

$$CR = \frac{length\ of\ Summary}{length\ of\ Initial\ text}.$$

A good summary is one in which CR is small while RR is large and the corresponding graph presents some characteristics (Hovy 2003).

In (Radev et al. 2002) it is stated that two classes of metrics have been developed for evaluation: *form* metrics and *content* metrics. *Form* metrics focus on grammaticality, overall text coherence and *content* metrics compare one or more human-made ideal abstracts and establish the percentage of correct information present in the systems summary (precision) and the percentage of important information omitted from the summary (recall). For the summarizer MEAD (Radev et al. 2001) one solution of evaluation was to construct an ideal summary by a group of human subjects asked to extract sentences and the sentences chosen by a majority of humans have been included in the ideal summary.

When the method of TS is based on LCs, the criterion of evaluation could be the percentage of the concepts common in the summaries and LCs. In (Silber and McCoy 2002) the system is evaluated by the percentage of noun senses of the summary that are in the strong lexical chains of the document (about 80 % in the case of the paper's experiment) and the percentage of strong lexical chains of the document that have a noun sense in the summary (also about 80 %).

An interesting point of view for evaluation is found in (Hasler 2004). The author asserts that evaluation is an important topic in summarization as there is often a disagreement on what constitutes a "good" summary and uses for evaluation some idea of Centering Theory (Grosz et al. 1995) (see Chapter Text Segmentation). In Centering Theory each utterance U_n (sentence or clause) in a text introduces a number of forward-looking centers (*Cfs*), which are NPs, ranked partially according to grammatical function. The more highly ranked a *Cf* is, the more likely it is to be a backward-looking center (*Cb*) of the next utterance U_{n+1}, providing a link between the two utterances U_n and U_{n+1}. The relationships between *Cfs* and *Cbs* of utterances result in transitions between utterances, which have a definite order of preference. This ordering of transitions reflects the idea that it is preferable for consecutive utterances to have the same *Cb* and also, for the most salient entity in one utterance, to be the *Cb* of the next utterance. By examining transitions in different sets of summaries of the same text (human or automatically produced) and comparing both sets of transitions and summaries, it should be possible to identify transitions typical of good summaries, which can be used to produce more locally coherent and readable summaries. These transitions can also be used to evaluate different summaries of the same text and select the best one (Hasler 2004).

Similar to machine translation metrics, a set of evaluation metrics known as ROUGE (for Recall-Oriented Understudy for Gisting Evaluation) has been used since 2004 in DUC Conferences (Document Understanding Conferences). Formally, ROUGE (namely ROUGE-N variant) measures the n-gram recall between a candidate summary and a set of reference summaries (Harabagiu and Lacatusu 2005). Another variant of ROUGE metric uses the longest common subsequence (LCS), based on the assumption that the pairs of summaries with longer LCSs will be more similar than those summaries with shorter LCSs. For n-grams, the metric ROUGE-N is (Lin 2004, Harabagiu and Lacatusu 2005):

$$ROUGE - N = \frac{\sum_{S \in Ref.summ} \sum_{n-gram \in S} Count_{match}(n - gram)}{\sum_{S \in Ref.summ} \sum_{n-gram \in S} Count(n - gram)} \qquad (14)$$

To assess the effectiveness of ROUGE measures, in (Lin 2004) is computed the correlation between ROUGE assigned summary scores and human assigned summary scores for a set of documents of DUC 2001, 2002, and 2003. The conclusion was that ROUGE package could be used effectively in automatic evaluation of one single document and multi-document summarization. However, these days more effort is being made in TS for researching on new metrics that outperform the previous standard for automatic text summarization evaluation (Conroy et al. 2011).

12.4.1 Conferences and Corpora

After the Summarization Evaluation Conference (SUMMAC, 1999), the series of conferences DUC (since 2001) has been the first sustained evaluation programme focused only on automatic summarizing, proposing year after year new challenges and new requirements for participating systems. The tasks of DUC have changed from simple to more complex ones, from generation of simple extracts of one single document (usually in English) to a generation of abstracts from a number of documents in a variety of languages and even producing summaries from a specific question or user-need (Lloret 2007). Along the editions of DUC, different methods of evaluation have progressed from the complete manual evaluation to fully automated evaluation with ROUGE. Since 2008 DUC conferences have become part of TAC (Text Analysis Conference). During all these years, the nature of summaries have changed from the simplest to the summaries which contain information about different aspects of topics, while the number of the participants grew (almost linear) from 15 at DUC 2001 until 32 at DUC 2007 and 32 at TAC 2010, showing the increasing interest in TS (Lloret 2011). Beside these conferences, specific workshops have been organized within the framework of other important conferences.

As corpora directly and freely usable in TS we must mention all data released at the DUC or TAC conferences. Moreover, we can signal here the CAST Project Corpora (Hasler et al. 2003) consisting of 163 documents with information about which sentences or parts of sentences can be removed from the document without affecting the sense.

12.5 Conclusions and Recent Research

Despite its long development and difficulty, TS attracts more and more researchers. Moreover, there is a fair amount of research going on to achieve high quality summaries, either extracts or abstracts, despite the fact that at present there is no theory as to how to construct the best (or at least a good) summary.

Firstly, let it be observed that the interest for good techniques of constructing *abstracts* continues to attract the researchers due to the challenges existing in this field:

- Unlike the extraction methods, abstraction requires using heavy machinery from natural language processing, including grammar and lexicons for parsing and generation (Hahn and Mani 2000);
- Abstracting requires some commonsense and domain-specific ontologies for reasoning during analysis and salience computation;
- Abstraction can be developed in two directions. The first one comprises compaction procedures that operate directly on syntactical trees to eliminate and regroup parts of those. After compaction, the original parse tree is considerably simpler becoming in essence a structural condensate (Hahn and Mani 2000). The second direction has its roots in artificial intelligence and focuses on natural language understanding. The support of methods are conceptual representation structures of the entire source content, which are assembled in a text knowledge base (Hahn and Mani 2000);
- Abstraction approaches provide more sophisticated summaries, which often contain material that enriches the source content and add new perspectives on the source (Sparck Jones 1999);
- The growing interest in abstraction will develop the studies about sentence compression and fusion of sentence parts. Also, this interest relates to other abstract representations of texts, as for example Information Items, as the smallest elements of coherent information in a text (Genest and Lapalme 2011).

As new directions in TS emerge, one can mention here the needs: to produce the summary which contains relevant information for a given user profile (so called personalized summaries), to produce updated summaries (interested in the most recent events in a given topic), to produce impact summaries from scientific articles (Nenkova and McKeown 2012), to produce opinion-based summaries (which have to consider the subjectivity of a person towards a topic (Lloret 2007)).

There are some areas of summarization which are becoming increasingly relevant. In all these, summarizers must be able to deal with a variety of document formats such as HTML and XML and access information in the tags associated with these documents. Developments in TS involving multiple languages are only at the beginning and a possible method for this type of summarization is a filtering mechanism which users could apply to produce a monolingual summary that contains content from multilingual sources (Hahn and Mani 2000). In general, the focus of many TS researchers will be MMM-summarization (multi-document, multi-language, and multi-media summarization) and summarization of new textualWeb 2.0 genres. In this case, in ranking the importance of sentences, other factors come into play such as the context in which the summary is generated and the links which point to a web page that has to be summarized. One of the earliest research on Web summarization (Amitay and Paris 2000) works by giving a search for a page, selecting all sentences containing a link to that page, and identifying the best sentences using some heuristics. Also Email summarization knows a fast development and relies on nature of email conversations as mean of getting answers to one's questions (Nenkova and McKeown 2012). To summarize an email mailbox multi-document, summarization techniques are used.

If one adds to these growing directions in TS, its applicability to other fields, the picture of the importance of the domain is completed. It is enough to mention here IR, QA and Text classification (see (Lloret 2011) for a detailed discussion).

CHAPTER 13

Named Entity Recognition

13.1 Introduction

Imagine you are given the following text and you are interested in recognizing key entity types.

> "Based on Altivar 61 drives, this solution allows you to reduce energy consumption by 20 to 30% on pumps depending on their characteristics. Energy savings is easily estimated via ECO8 software."

In these sentences, Altivar 61 and ECO8 should be recognized as computer hardware component and piece of software, respectively. Any other recognition, for instance, as person or location will be erroneous and spurious. Failing to recognize any of these entities at all is also an error.

Named Entity Recognition (NER) has attracted a growing interest in this field of Natural Language Processing and Information Extraction since the early 1990s. NER aims mainly to extract and classify rigid designators in the text by seeking to locate and classify atomic elements in the text into predefined categories such as the names of persons or biological species, organizations, locations, expressions of times, quantities, monetary values, percentages, etc. NER is also known as entity identification and entity extraction, which has been a subtask of information extraction.

This is due to some historical context of the origin of the term "Named Entity" as this was coined for the Sixth Message Understanding Conference (MUC-6) (Grishman and Sundheim 1996). By that time, MUC was focusing on Information Extraction (IE) tasks where structured information of company activities and defence related activities are extracted from unstructured text, such as newspaper articles. People noticed, however, that it is essential to recognize information units such as names, including those of persons, organizations and locations. Also, numeric expressions including time, date, money and percentage have been added. Identifying references to these entities in the text was recognized as one of the important sub-tasks of IE, which has been called "Named Entity Recognition and Classification (NERC)" ever since.

In this chapter, a reference will be made to the main trends and methods in the field, i.e., hand-crafted rules (also known as rules based approaches) and machine learning approaches. Methods based on hand-crafted rules are usually linguistic, grammar-based techniques, whereas approaches based on machine learning are usually based on statistical models. Hand-crafted, grammar-based systems are more precise, but at the cost of lower recall and months of work by experienced computational linguists. Statistical NER systems, on the other hand, require a large amount of manually annotated training data. Nevertheless, both trends and methods are still suffering from serious shortcomings when it comes to building and maintaining large-scale NER systems.

13.2 Baseline Methods and Algorithms

13.2.1 Hand-crafted rules based techniques

Named Entity (NE) extraction is considered part of the bigger picture depicted by Information Extraction (IE) as a discipline concerned with extracting structured information from unstructured text. It has become increasingly important in recent years. In order to identify regions of text that are of interest, most NE and IE systems make use of *extraction rules.*

Some typical examples of rule-based systems are CIMPLE (Doan et al. 2008), GATE (Cunningham et al. 2002), System T (Chiticariu et al. 2010, Krishnamurthy et al. 2008). These systems make use of rules throughout the entire NE/IE extraction flow. An example rule might read, in plain English, "identify a match of a dictionary of salutations followed by a match of a dictionary of last names and mark the entire region as a candidate person". Such a rule would be expressed formally using the rule language of the system such as XLog in CIMPLE, JAPE in GATE, or AQL in System T.

The relationship with machine learning-based systems, to be discussed in the following section, is exemplified by the fact that machine learning-based techniques also make use of rules. However, this happens in the first stage of a NE extraction flow in order to identify basic features such as capitalized words or phone numbers. Subsequently, these basic features are fed as input to various machine learning algorithms that implement the rest of the extraction process.

The development of highly accurate information extraction rules (or extractors, in short) is a laborious process. It is extremely time consuming, and prone to error. The developer usually starts by building an initial set of rules and subsequently proceeds with an iterative process consisting of three main steps:

- execute the rules over a set of test documents and identify mistakes in the form of wrong results (false positives), as well as missing results (false negatives);
- analyze the rules to understand the causes of the mistakes;
- and finally determine refinements that can be made to the extractor to correct the mistakes.

This 3-step process is repeated until the developer is satisfied with the accuracy of the extractor. In practice, however, things become easily complex, since an extractor might contain hundreds of rules, which might pose complex interactions between these rules. Therefore, rule refinement is a "trial and error" process, in which the developer has to implement multiple candidate refinements, which are evaluated individually in order to understand their effects.

An "ideal" refinement would eliminate as many mistakes as possible, while minimizing effects on existing results which are correct.

The challenges faced by the developer when it comes to rule refinement is exemplified by the following paradigm. Let us assume that a developer is interested in eliminating false positives from the result in order to improve the precision of the extractor.

For instance, while it identifies all mentions of correct persons, or true positives (e.g., John Stanley, Mr. Stanley), it also identifies incorrect mentions, or false negatives (e.g., Smith Barney, Morgan Stanley), which are meant to be organizations or brand names. To this end, the goal of the developer will be to refine the rules in order to remove as many false positives from its output, while preserving true positives, as much as possible. In order to do so, there are several optional modifications, which have different effects and side effects.

For example, a modification that filters mentions containing the term Stanley will eliminate two of the three false positives, however, it will also remove two correct results (i.e., John Stanley, and Mr. Stanley). Another modification might have the sideeffect of removing the single correct result (John Stanley). Moreover, a rule modification where Stanley is no longer identified as a last name might fail to identify correct candidates John Stanley and Mr. Stanley, respectively as a side effect. A rule modification targeting to eliminate Morgan Stanley being recognized as a person could take the form that Morgan is no longer identified as first name. However, this modification will cause failure to mark an entity named Morgan Doe as a candidate person.

Possibly, another more viable solution is to assume that an extractor is available for identifying mentions of organizations as well, the developer might use it to add a subtraction on top of the applicable rule. This way, an entity recognized as a person, though the entity stands for an organization, e.g., a brand name, is subtracted from the final result. To this end, Smith Barney and Morgan Stanley will be identified as organizations to be removed from all recognized named entities as persons, while not affecting any correct ones.

In addition to these challenges, another major challenge is imposed by the sheer complexity and the steep learning curve of the languages used for the specification of extraction rules. On the example of JAPE (Java Annotation Pattern Language) as a constituent of GATE, a grammar is defined, which consists of a set of phases, each of which consists of a set of pattern/action rules. The phases run sequentially and constitute a cascade of finite state transducers over annotations.

The left-hand-side (LHS) of the rules consist of the description of an annotation pattern. The right-hand-side (RHS) consists of annotation manipulation statements. Annotations matched on the LHS of a rule may be referred to on the RHS by means of labels that are attached to pattern elements. Consider the following example:
Phase: Jobtitle

Input: Lookup
Options: control = appelt debug = true

Rule: Jobtitle1
(
{Lookup.majorType == jobtitle}
(
{Lookup.majorType == jobtitle}
)?
)
:jobtitle
-->
:jobtitle.JobTitle = {rule = "JobTitle1"}

In this example, the LHS is the part preceding the '-->' and the RHS is the part following it. How things can escalate and become more complex is exemplified by the following rule, which is meant to extract football players as named entities in GATE/JAPE.

```
/*
 * playercontext.jape
 *
 * Dhavalkumar Thakker, Nottingham Trent University/PA Photos
10 March 2009
 *
 * $Id: playercontext.jape 0001 2009-03-10 dhaval $
 *
 */
Phase: playercontext
Input:  Lookup Token Team
Options: control = brill debug = true

//rules for identifying name of the player based on Team names
Rule: playercontext1
Priority:50
(
            {Team}
 {Token.string=="players"}
)
(
{Token.kind==word, Token.category==NNP, Token.
```

```
orth==upperInitial}
    {Token.kind==word, Token.category==NNP, Token.
orth==upperInitial}
    ):player
    -->
    :player.Player = {rule= "playercontext-playercontext1" }

    Rule: playercontext2
    Priority:100
    (
    {Team}
     {Token.string=="players"}
    )
    (
    {Token.kind==word, Token.category==NNP, Token.
orth==upperInitial}
    {Token.kind==word, Token.category==NNP, Token.
orth==upperInitial}
    ):player1
    {Token.string=="and"}
    (
    {Token.kind==word, Token.category==NNP, Token.
orth==upperInitial}
    {Token.kind==word, Token.category==NNP, Token.
orth==upperInitial}
    ):player2
    -->
    :player1.Player = {rule= "playercontext-playercontext2" } ,
    :player2.Player = {rule= "playercontext-playercontext2" }

    Rule: playercontext3
    Priority:80
    (
    {Team}
      {Token.string=="players"}
    )
    (
```

```
{Token.kind==word, Token.category==NNP, Token.
orth==upperInitial}
    {Token.kind==word, Token.category==NNP, Token.
orth==upperInitial}
    ):player1
    {Token}
    (
      {Token.kind==word, Token.category==NNP, Token.
orth==upperInitial}
      {Token.kind==word, Token.category==NNP, Token.
orth==upperInitial}
    ):player2
    -->
    :player1.Player = {rule= "playercontext-playercontext3" } ,
    :player2.Player = {rule= "playercontext-playercontext3" }
```

13.2.2 Machine learning techniques

Supervised NE Learning (SL): Supervised NE learning is the current dominant technique for addressing the named entity recognition problem. These techniques comprise of Hidden Markov Models (HMM), Decision Trees, Maximum Entropy Models (ME), Support Vector Machines (SVM) and Conditional Random Fields (CRF). Despite all variants, the typical SL approach consists of a system that reads a large annotated corpus, memorizes lists of entities, and creates disambiguation rules based on discriminative features.

The main idea often proposed refers to tagging words within a test corpus, when they are annotated as entities in the training corpus. The performance of the system usually depends on the vocabulary transfer, which is the proportion of words, without repetitions, appearing in both training and testing corpus.

Generally speaking, the larger the training data in the supervised methods, the more accuracy can the system achieve. However, annotating a large corpus is not easy. An attractive alternative idea is to annotate only the data which can help to improve the overall accuracy. The favorite practice is to annotate the data which is tagged with uncertainty by the current system.

Given this shortcoming of supervised learning techniques for name entity recognition,one will embark, in the following, on semi-supervised and unsupervised learning techniques with exemplar algorithmic approaches.

Semi-supervised NE Learning: Given that, as reported recently, recent machine learning approaches have a problem with annotated data availability, which in turn is a serious shortcoming in building and maintaining large-scale NER systems, it has been attempted to build NER systems with very little supervision. This is also based on the presumption that human supervision is indeed limited to listing a few examples of each named entity (NE) type, usually up to four to ten NE types.

Semi-supervised learning techniques, however, promise to enhance recognition capacity up to 100 NE types. It is, therefore, believed that semi-supervised learning techniques are about to break new ground in the area of machine learning in that limited supervision can build complete NER systems. In addition, performances that compare to baseline supervised systems in the task of annotating NEs in texts have been reported. In the following sector, one will get to know these techniques a bit more closer in terms of the design of a baseline, semi-supervised NER system (called BaLIE) that performs at a level comparable to that of a simple supervised learning-based NER system. The architecture of this system was published at Canadian AI 2006 (Nadeau et al. 2006).

BaLIE is meant to solve two common limitations of *rule-based* and *supervised* NER systems. Firstly, it requires no human intervention such as manually labelling training data or creating gazetteers. Secondly, the system can handle more than the three classically named-entity types (person, location, and organization). In order to gain NER capabilities, besides a NER module, BaLIE also features a tokenizer, a sentence boundary detector, a language guesser, and a part-of-speech tagger.

Regardless of the very details of the architecture, BaLIE, as of most NER systems around, one needs to meet two main targets: create large lists of NEs and resolve ambiguity in a given document. The distinction between these tasks is important. It might seem that having a list of entities on hand makes NER trivial, however, one can extract city names from a given document merely by searching it for each city name in a city list. This strategy, however, often fails because of ambiguity. For example, consider the words "It" (a city in the state of Mississippi and a pronoun) and "Jobs" (a person's surname and a common noun). Without resolving such ambiguity, a system cannot perform robust, accurate NER. Moreover, the combination between NE extraction and a simple form of NE disambiguation is usually accompanied by highly effective heuristics. In the following, one will embark on exemplary descriptions how a gazetteer, a list of named entities, is generated, as well as how simple heuristics are being applied to identify, classify and record entities in the context of a given document.

Generating lists of entities: This is a task that has been investigated extensively in the past. One of the first ideas has been to study lexical patterns, which could be used to identify nouns from the same semantic class. For instance, a noun phrase that follows the pattern "the capital city of" is usually a country. Another idea has been to examine words with similar syntactic dependency relationships in order to create large clusters of semantically related words (Lin and Pantel 2001). However, this technique does not discover the labels of the semantic classes, which is a common limitation of clustering techniques.

An example of an algorithm: In the following, an algorithm is discussed (Etzioni et al. 2005), which has been claimed to outperform all previous methods for creating a large list of a given type of entity or semantic class: the task of automatic gazetteer generation. This exemplary algorithm is meant to explain how to generate a list of thousands of cities from only a few seed examples.

First step: A Web search engine such as Yahoo! can be selected by interfacing with the corresponding developer API. Then, a query can be composed of k manually chosen entities (e.g., "Montreal" AND "London" AND "Paris" AND "Athens"). The main idea here is that when k is set to 4 and the seed entities are common city names, the Web pages will be retrieved from which many names of cities can be harvested, in addition to the seed names. Subsequently, these additional names of cities from each retrieved Web page can be extracted for the list of cities from the gazetteer. It is also worth mentioning that less than four entities might result in lower precision, whereas more than four entities might result in lower recall.

It is also worth noting that the same strategy can be applied to name of persons, name of companies, car brands, and many other types of entities.

Second step: During this step, a Web page wrapper is being applied whose goal is to isolate desired information. It acts as an abstract layer over HTML and it is provided with the location of a subset of the desired information within a page. The wrapper isolates the entire set of desired information and hides the remainder of the page. The goal of the wrapper is, therefore, to hide everything in the page but the named entities that are likely to be in HTML structures similar to that of the seed names.

More details can be found in (Nadeau et al. 2006) where the following example is given, since learning to isolate desired information on a Web page starting with a few seed examples is considered as an instance of learning from positive and unlabelled data. For instance, in the following HTML code, the <a> node contains the city name "Ottawa", which is the desired information. This, in turn, is labelled "positive" for the purpose of training the wrapper.

```
    <tr>
        <td>Day5</td>
        <td><img src="bullet.gif">
        <a href="vacation.htm" label="positive">Ottawa</a> </td>
        <td>Ottawa, Museum of Civilization: Morning drive to Canada's
capital city,
            Ottawa. This afternoon visit the Canadian Museum of
Civilization...</td>
    </tr>
```

Web page wrapper attributes can also be asserted such as cell row in innermost table *(numeric)*, cell column in innermost table *(numeric)*, class: {Positive, Negative}. Subsequently, the description of a typical HTML node could take the form:

a,6,0,0,4002,26,0.684211,8,0.222222,1,0.027778,td,104,0.514851,0, 0,2,1, Positive

Third step: The two steps above are repeated with the aim to bring new entities that are added to the final gazetteer. During each iteration, k new randomly chosen entities are used to refresh the seed for the system from the gazetteer under construction. Preference is given to seed entities that are less likely to make noise, such as those appearing in multiple Web pages.

Resolving Ambiguity: This is a complementary task towards the creation of a list of NEs, not necessarily existing in many NER systems. In a nutshell, this is the task anchored in a list look-up strategy as a method of performing NER by scanning a given input document to look for terms that match a list entry. The list look-up strategy, however, has three main problems: entity-noun ambiguity errors, entity boundary detection errors and entity-entity ambiguity errors. This is the reason why the gazetteer-generating module, presented previously, is not in itself adequate for reliable NER. Usually, these problems are resolved with some sort of heuristics.

Entity-noun ambiguity is adhered to an entity as the homograph of a noun. For instance, the plural word "jobs" and the surname "Jobs" is an example of this occurrence. In order to avoid this problem, the following heuristic can be taken as an example. In a given document, we assume that a capitalized word or phrase (e.g., "Jobs") is a named-entity, unless the following condition applies:

- The word appears in the document without capitals (e.g., "jobs");
- The word appears at the beginning of a sentence or at the beginning of a quotation (e.g., "Jobs that pay well are often boring.");
- The word is in a sentence in which all words with more than three characters start with a capital letter (e.g., a title or section heading).

Entity-boundary detection is adhered to the problem of recognizing where a NE begins and ends in a document, for instance, finding only "Boston" in "Boston White Sox"). This usually happens under the following circumstances:

- A NE is composed of two or more words (e.g., "Jean Smith"), each listed separately (e.g., "Jean" as a first name and "Smith" as a last name).
- A NE is surrounded by unknown capitalized words (e.g., "New York Times" as an organization followed by "News Service" as an unlisted string).

An exemplar heuristic to tackle this problem has been the so-called longest match strategy. Accordingly, all consecutive entities of the same type are merged with every entity with any adjacent capitalized words. However, consecutive entities of different types are not being merged, since they would have resulted in unknown types.

The heuristic above is general enough to be applied independently of the entity type. Other merging rules could improve the precision of a system, such as the creation of a new 'organization' type entity by merging a location with a following organization. However, such rules must be avoided because this kind of manual rule engineering results in brittle, fragile systems that do not adapt well to new data.

Entity-entity ambiguity underpins the problem when the string standing for a NE belongs to more than one type. For instance, if a document contains the "France" NE, it could be either the name of a person or the name of a country. For this problem, it is proposed that at least one occurrence of the NE should appear in a context, where the correct type is clearly evident. For example, in the context "Dr. France," it is clear that "France" is the name of a person. Under such circumstances, one could use several cues, such as professional titles (e.g., farmer), organizational designators (e.g., Corp.), personal prefixes (e.g., Mr.) or personal suffixes (e.g., Jr.), but this should be avoided, since this kind of manual rule engineering is inflexible.

Instead, the following heuristic is being applied. When an ambiguous entity is found, its aliases are used in two ways. Firstly, if a member of an alias set is unambiguous, this can be used to resolve the whole set. For instance, "Pacific ocean" is clearly a location, but "Pacific" can be either a location or an organization. If both belong to the same alias set, it is assumed that the whole set is a "location" type.

Also, one could use the alias resolution to include unknown words in the model. For instance, if an entity (e.g., "Steve Hill") is formed from a known entity (e.g., "Steve") and an unknown word (e.g., "Hill"), occurrences of this unknown word shall be allowed to be added in the alias group.

Unsupervised NE Learning: *Clustering* is the typical approach in unsupervised learning. One can try, for instance, to recognize named entities from clustered groups based on the similarity of context. These techniques, however, rely on some sort of lexical resources such as WordNet, lexical patterns or statistics computed on a large corpus, which is not annotated. Some ideas and algorithmic approaches reported in the literature are introduced in the following.

Given an input word, it is attempted to find an appropriate NE type. NE types can be taken, for instance, from WordNet, e.g., location>country, animate>person, animate>animal. The approach is to assign a topic signature to each WordNet synset merely by listing words that co-occur frequently with it in a large corpus. Then, given an input word in a given document, the word context (words appearing in a fixed-size window around the input word) is compared to type signatures and classified under the most similar one (Alfonseca and Manandhar 2002).

Another idea has been the identification of hyponyms / hypernyms described in order to identify potential hypernyms of sequences of capitalized words appearing in a document. For instance, when X is a capitalized sequence, the lexical pattern "such as X" is searched on the Web and in the retrieved documents. The noun that immediately precedes the query can be chosen as the hypernym of X. For instance, the lexical patterns "city such as X", "organization such as X", could easily lead to the conclusion that X is either a city or an organization, respectively (Cimiano and Völker 2005, Evans 2003, Hearst 1992).

Another cue has been delivered by the observation that named entities often appear simultaneously in several news articles, whereas common nouns do not, since a strong correlation between being a named entity and appearing punctually, in time, and simultaneously in multiple news sources has been evidenced. This technique allows identifying rare named entities in an unsupervised manner and turns out to be useful in combination with other NE recognition methods (Shinyama and Sekine 2004).

13.3 Summary and Main Conclusions

Of the major trends and methods documented earlier in this chapter, a recent trend away from hand-crafted rules towards machine learning approaches has been identified in the NER field. Hand-crafted systems provide a good performance at a relatively high system engineering cost. As with supervised learning techniques for the extraction of knowledge, a general prerequisite within the NER domain is the availability of a large collection of annotated data. Despite the fact that such collections are available from the evaluation forums, they still remain rather rare and limited in domain and language coverage.

Therefore, researchers turned to exploring semi-supervised and unsupervised learning techniques that promise fast deployment for many entity types without the prerequisite of an annotated corpus. However, semi-supervised and unsupervised learning techniques make use of an expressive and varied set of features, which turn out to be just as important as the choice of machine learning algorithms.

Despite the controversial aspects of machine learning techniques, a general belief that semi-supervised learning techniques are superior in the machine learning community emerged. It has been shown that limited supervision can build complete NER systems. In addition, performances have been reported on standard evaluation corpora, which are similar or outperform baseline supervised systems in the task of annotating NEs in texts.

Another emerging feature is the brittleness of state-of-the-art NER systems, meaning that NER systems developed for one specific domain do not perform well on other domains. Therefore, considerable effort must be involved in tuning NER systems to perform well in a new domain. This turned out to be true for both rule-based and trainable statistical systems.

Bibliography

Agirre, E. and P. Edmonds eds. 2006. *WSD: Algorithms and Applications*, Springer, New York.

Alfonseca, E. and S. Manandhar. 2002. "An Unsupervised Method for General Named Entity Recognition and Automated Concept Discovery". *In* Proc. International Conference on General WordNet.

Allan, J. 2002. "Challenges in Information Retrieval and Language Modeling". Report on a Workshop for Intelligent Information Retrieval, University of Massachusetts, Amherst.

Allan, J and J. Carbonell, G. Doddington, J. Yamron and Y. Yang. 1998. "Topic Detection and Tracking Pilot Study". Final Report. Proceedings of the DARPA Broadcast News Transcription and Understanding Workshop, Lansdowne, VA.

Allen, J. 1995. *Natural language understanding*. Benjamin/Cummings Publ. 2nd ed. Wokingham, UK.

Alonso i Alemany, L. 2005. "Representing discourse for automatic text summarization via shallow NLP techniques". PhD Thesis. Universitat de Barcelona. Spain.

Amer-Yahia, S., P. Case, T. Rölleke, J. Shanmugasundaram and G. Weikum. 2005. Report on the DB/IR Panel at SIGMOD 2005, SIGMOD Record 34(4): 71–74.

Amitay, E. and C. Paris. 2000. "Automatically Summarising Web Sites—Is There A Way Around It?" Proceedings of the Ninth ACM International Conference on Information and Knowledge Management (CIKM 2000), Washington, DC.

Androutsopoulos, I and P. Malakasiotis. 2010. "A Survey of Paraphrasing and Textual Entailment Methods". Journal of Artificial Intelligence Research 38: 135–187, Moento Park, CA.

Apel, K.-O. In: Begründung H. Seifert and G. Radninzky eds.1989. Handlexikon der Wissenschaftstheorie, Ehrenwirth, M\"unchen 14–19.

Armstrong, M.A. 1988. *Groups and Symmetry, Undergraduate Texts in Mathematics*, Springer Verlag.

Artin M. 1993. *Algebra*, Birkhäuser Advanced Texts.

Azzam, S., K. Humphreys and R. Gaizauskas. 1999. "Using coreference chains for text summarization", 77–84. Proceedings of the ACL'99 workshop on coreference and its applications, College Park, MD.

Baker, M.C. 2001. *The Atoms of the Language*, Oxford Press, Oxford, UK.

Banarjee, S. and T. Pedersen. 2002. An adapted lesk algorithm for WSD using WordNet. Proceedings of the Third Int. Conf. in Intelligent Text Processing and Comp. Linguistics, Mexico City, pp. 136–145.

Banarjee, S. and T. Pedersen. 2003. "Extended Gloss Overlaps as a Measure of Semantic Relatedness". Proceedings of the Eighteenth International Joint Conference on Artificial Intelligence, Acapulco, Mexico.

Banea, C. and R. Mihalcea. 2011. "Word Sense Disambiguation with multilingual features", 25–34. Proceedings of ICSC, Oxford, UK.

Bar-Haim, R., I. Dagan, B. Dollan, L. Ferro, D. Giampiccolo, B. Magnini and I. Szpektor. 2006. "The Second PASCAL Recognizing Textual Entailment Challenge". Proceedings of the Second PASCAL RTE challenge, Venice, Italy.

Barwise, John and John Perry. 1983. *Situations and Attitudes*. A Bradford Book. MIT Press.

Barwise, J. and J. Seligman. 1993. "Imperfect information flow", *In* Proceedings of the Eighth IEEE Symposium on Logic in Computer Science, 252–261 Montreal.

Barzilay, R. and M. Elhadad. 1997. "Using Lexical Chains for Text Summarization", 10–17 Proceedings of the Association for Computational Linguistics and the European Chapter of the Association for Computational Linguistics (ACL-97/EACL-97), Workshop on Intelligent Scalable Text Summarization, Madrid, Spain.

Barzilay, R., N. Elhadad and K.McKeown. 2001. "Sentence ordering in multidocument summarization", 149–156. Proceedings of the First International Conference on Human Language Technology Research (HLT-01), San Diego, CA.

Barzilay, R., N. Elhadad and K. McKeown. 2002. "Inferring strategies for sentence ordering in multi document news summarization". In Journal of Artificial Intelligence Research, 17: 35–55.

Bateson, G. 2000. *Steps to an Ecology of Mind*, University of Chicago Press, Chicago.

Beeferman, D., A. Berger and J. Lafferty. 1997. "Text segmentation using exponential models", 35–46 Proceedings of the 2nd Conference on Empirical Methods in NLP, Providence, RI.

Belohlavek R., Introduction to Formal Concept Analysis, Dept. of Computer Science, Palacky University, Olomouc.

Beretti, S.,A. Del Bimbo and P. Pala. 2004. "A Graph Edit Distance Based on Node Merging", *In* Image and Video Retrieval: Third International Conference, CIVR 2004, 464–472, Dublin, Ireland

Berners-Lee, T., J. Hendler and O. Lassila. 2001. The Semantic Web. Scientific American, http: //www.sciam.com/2001/0501issue/0501berners-lee. html.

Birkhoff, G. 1967. *Lattice Theory*. American Mathematical Society, 3rd edition.

Birkhoff, G. and J. von Neumann. 1936. "The Logic of Quantum Mechanics". Annals of Mathematics 37: 823–843.

Boguraev, B. and M. Neff. 2000. "Lexical Cohesion, Discourse Segmentation and Document Summarization". Proceedings of the 33rd Hawaii International Conference on System Sciences, Mani, HI.

Bos, J. and K. Markert. 2006. "Recognising Textual Entailment with logical inference", 628–635. Proceedings of HLT/EMNLP, Vancouver, CA.

Boyer, C. B. and U. C. Merzbach. 1991. *A History of Mathematics.* Wiley.

Brachman, R.J. and J.G. Schmolze. 1985. "An overview of the KL–ONE knowledge representation system". Cognitive Science 9(2): 171–216

Brun, C. 2000. "A Client/Server Architecture for Word Sense Disambiguation", 132–138. Proceedings of the 17th conference on Computational Linguistics, Saarbrücken.

Burton-Jones, A., V. Storey, V. Sugumaran and S. Purao. 2003. "A Heuristic-based Methodology for Semantic Augmentation of User Queries on the Web", *In* 22nd International Conference on Conceptual Modeling, 476 – 489 Chicago, Illinois.

Budanitsky, A. and G. Hirst. 2006. "Evaluating WordNet-based Measures of Lexical Semantic Relatedness". Computational Linguistics 32: 14–47.

Callan, J. P., W. B. Croft and S. M. Harding. 1992. "The INQUERY retrieval System", 78–83. *In* Proceedings of DEXA-92, 3rd International Conference on Database and Expert Systems Applications.

Carbonell,J. and J. Goldstein. 1998. "The use of MMR, diversity based reranking for reordering documents and producing summaries", 335–336. Proceedings of 21th ACM/SIGIR conference, Melbourne.

Carpineto C. and G. Romano. 2004. *Concept Data Analysis, Theory and Applications*, Wiley. New Jersey.

Carpuat, M. and D. Wu. 2007. "Improving Statistical Machine Translation using WSD", 61–72. Proceedings of the Joint Conference on Empirical Methods in NLP and Computational Natural Language Learning, Prague, CZ.

Chambers, N. and D. Jurafsky. 2008. "Unsupervised learning of narrative event chains". Proceedings of ACL-08, Hawaii, USA.

Cheng, P.J., J.W. Teng, R.C. Chen, J.H. Wang, W.H. Lu and L.F. Chien. 2004. "Translating Unknown Queries with Web Corpora for Cross Language Information Retrieval", 146–153. *In* Proceedings SIGIR'04, Sheffield, UK.

Chiticariu, L., R. Krishnamurthy, Y. Li, S. Raghavan, F. Reiss and S. Vaithyanathan. 2010. "SystemT: An Algebraic Approach to Declarative Information Extraction". *In* Proceedings Association Computational Linguistics (ACL).

Choi, F. Y. 2000. "Advances in domain independent linear text segmentation", 26–33. Proceedings of 6th Applied Natural Language Processing Conference, NAACL, Seattle, Washington.

Chomsky, Noam. 1957. *Syntactic Structures*. Mouton, The Hague.

Chomsky, Noam. 1965. *Aspects of the theory of syntax*. MIT Press, Cambridge MA.

Chuang, W. and J. Yang. 2000. "Extracting Sentence Segments for Text Summarization: A Machine Learning Approach", 152–159. Proceedings of the 23rd International Conference on Research in Information Retrieval (SIGIR '00), Athens, Greece.

Cimiano, P., A. Hotho and S. Staab. 2004. "Clustering ontologies from text" 1721–1724. Proceeding of LREC, Lisbon, Portugal.

Cimiano, P.,A. Hotho and S. Staab. 2005. "Learning Concept hierarchies from text corpora using Formal Concept Analysis". Journal of AI Research, 24: 305–339.

Cimiano, P. and J. Völker. 2005. "Towards Large-Scale, Open-Domain and Ontology-Based Named Entity Classification". pp. 166–172. *In* Proceedings Conference on Recent Advances in Natural Language Processing, Borovets, BG

Cohn P.M. 2000. *Classic Algebra*. J. Wiley.

Conroy, J.M., J.D. Schlesinger and D.P. Leary. 2011. "Nouveau-ROUGE: A novelty metric for update summarization". Computational Linguistics. 37: 1–8.

Corley, C. and R. Mihalcea. 2005. "Measuring the semantic similarity of texts", 13–18. Proceedings of the ACL Workshop on Empirical Modeling of Semantic Equivalence and Entailment, Ann Arbor, MI.

Cunningham, H., D. Maynard, K. Bontcheva and V. Tablan. 2002 "GATE: A Framework and Graphical Development Environment for Robust NLP Tools and Applications". pp. 168–175. *In* Proceedings Association Computational Linguistics (ACL), Philadelphia, PA.

Curran, James Richard. 2003. From Distributional to Semantic Similarity. Ph.D. Thesis, School of Informatics, University of Edinburgh, Institute for Communicating and Collaborative Systems, Edinburgh.

Dagan, I., O. Glickman and B. Magnini. 2005. "The PASCAL Recognising Textual Entailment Challenge". Proceedings of the First PASCAL Challenges Workshop on Recognising Textual Entailment, Southampton, UK.

Dagan, I., B. Dolan, B. Magnini and D. Roth. 2009. "Recognizing textual entailment: Rational, evaluation and approaches". Journal of Natural Language Engineering 15(4): i–xvii.

Deitel, A., C. Faron and R. Dieng. 2001. "Learning Ontologies from RDF Annotations". *In* Proceedings International Joint Conf. on Artificial Intelligence, Workshop Ontology Learning.

Delmonte, R., S. Tonelli and R. Tripodi. 2009. "Semantic Processing for Text Entailment with VENSES". Proceedings of the TAC 2009 Workshop on Textual Entailment. Gaithersburg, Maryland.

Devlin, K. 1999. *Infosense - Turning Information into Knowledge*, Freeman, New York.

Doan, A., J. F. Naughton, R. Ramakrishnan, A. Baid, X. Chai, F. Chen, T. Chen, E. Chu, P. DeRose, B. Gao, C. Gokhale, J. Huang, W. Shen and B.-Q. Vuong. 2008. "Information Extraction Challenges in Managing Unstructured Data". *SIGMOD Record*, 37(4): 14–20.

Donaway, R., K. W. Drummey and L. A. Mather. 2000. "A comparison of rankings produced by summarization evaluation measures" 69–78. Proceedings of NAACL-ANLP 2000 Workshop on Text Summarisation, Seattle, Washington.

Emmeche, C. and J. Hoffmeyer. 1991. From language to nature, Semiotica 84: 1–42.

Etzioni, O., M. Cafarella, D. Downey, A.-M. Popescu, T. Shaked, S. Soderland, D.S. Weld and A. Yates. 2005. Unsupervised Named-Entity Extraction from the Web: An Experimental Study. Artificial Intelligence 165: 91–134.

Evans, R. 2003. "A Framework for Named Entity Recognition in the Open Domain". pp. 137–144. *In* Proc. Recent Advances in Natural Language Processing, Borovets, BG.

Fellbaum, C. [ed]. 1998. *"WordNet: an electronic lexical database"*. MIT Press, Cambridge, MA.

Ferret, O. 1998. "How to thematically segment texts by using lexical cohesion?" 1481–1483. Proceedings of ACL-COLING'98, Montreal, Quebec, CA.

Filatova, E. and V. Hatzivassiloglou. 2004. "Event-based extractive summarization". pp. 104–111. Proceedings of the ACL Text Summarization Workshop, Barcelona.

Fogarolli, A. 2011. Word Sense Disambiguation based on Wikipedia Link Structure. Studies in Computational Intelligence, 329: 1–26.

Gärdenfors, P. 2000. *Conceptual Spaces: The Geometry of Thought*. The MIT Press, Cambridge, Massachusetts.

Gale, A., K. Church and D. Yarowsky. 1992. "One Sense Per Discourse". Proceedings of the DARPA Speech and Natural Language Workshop, NewYork.

Ganguly, D.,J. Leveling and G. J. F. Jones. 2011. "Query expansion for language modeling using sentence similarities", 62–77. Proceedings of the IRFC.

Ganter, B. and R. Wille. 1999. *Formal Concept Analysis: Mathematical Foundations*. Springer.

Ganter, B., Stumme, G., Wille, R (Eds.) 2005: "Formal Concept Analysis, Foundations and Applications". Lecture Notes in Computer Science 3626, Springer, Heidelberg.

Genest, P.E and G. Lapalme. 2011. "Framework for Abstractive Summarization using Text-to-Text Generation", 64–73. Proceedings of the Workshop on Monolingual Text-To-Text Generation, Portland, USA.

Grassmann, H. 1862. "Extension Theory. In History of Mathematical Sources". American and London Mathematical Society.

Giampiccolo, D., B. Magnini, I. Dagan and B. Dolan. 2007. "The Third PASCAL Recognizing Textual Entailment Challenge", 1–9. Proceedings of the Workshop on Textual Entailment and Paraphrasing, Prague.

Glickman, O., I. Dagan and M. Koppel. 2005. "Web Based Probabilistic Textual Entailment". Proceedings of the First PASCAL Challenges Workshop on Recognising Textual Entailment, Southampton, UK.

Grishman, R. and B. Sundheim. 1996. "Message Understanding Conference - 6: A Brief History". *In* Proceedings International Conference on Computational Linguistics, Copenhagen, Denmark.

Grosz, B. and C. Sidner. 1986. "Attention, Intentions and the Structure of Discourse". Computational Linguistics 12: 175–204.

Grosz, B., B. Joshi and S. Weinstein. 1995. Centering: a framework for modelling the local coherence of discourse. Computational Linguistics 21: 203–225.

Hahn, U. and I. Mani. 2000. "The challenges of automatic summarization". IEEE-Computer 33(11): 29–36.

Harabagiu, S. 1999. "From lexical cohesion to textual coherence: a data driven perspective". Journal of Pattern Recognition and Artificial Intelligence, 13: 247–265.

Harabagiu, S. and D.Moldovan. 1999. "A parallel system for Textual Inference". IEEE Transactions parallel and distributed systems 10: 254–270.

Harabagiu, S. and F. Lacatusu. 2005. "Topic Themes for Multi-Document Summarization", 202–209. Proceedings of SIGIR'05, Salvador, Brazil.

Harnad, Stevan. 1990. "The symbol grounding problem". Physica D 42: 335–346

Hasler, L. 2004. "An Investigation into the Use of Centering Transitions for Summarisation", 100–107. Proceedings of CLUK'04, Birmingham, UK.

Hatzivassiloglou, V., J. Klavans, M. Holcombe, R. Barzilay, M. Kan and K. McKeown. 2001. "SimFinder: A flexible Clustering Tool for Summarization". Proceedings of the Workshop of Summarization, NAACL-01, Pittsburgh, PA.

Hearst, M. 1992. "Automatic Acquisition of Hyponyms from Large Text Corpora". *In* Proceedings International Conference on Computational Linguistics, Nantes, France.

Hearst, M. 1997. "TextTiling: Segmenting Text into Multi-paragraph Subtopic Passages". Computational Linguistics 23: 33–76.

Hobbs, J. 1985. "On the coherence and structure of discourse. Technical Reports". CSLI-85-37. Center for the study of Language and Information. Stanford University.

Hovy, E. and C. Lin. 1999. "Automated Text Summarization in SUMMARIST", pp 81–98. *In* I. Mani and M. Maybury. [eds.]. Advances in Automatic Text Summarization. MIT Press, MA, USA.

Hovy, E. 2003. "Text summarization", pp 583–598. *In* R. Mitkov ed]. *The Oxford Handbook of Computational Linguistics*. Oxford University Press, UK.

Ide, N. and J. Veronis. 1998. "Introduction to the special issue on WSD: the state of the art". Computational Linguistics 24: 1–40.

Ide, N. and Y. Wilks. 2006. "Making sense about sense". pp. 47–73. *In* E. Agirre and P. Edmonds (eds). *WSD: Algorithms and Applications*, Springer, New York, NY.

Inkpeen, D., D. Kipp and V. Nastase. 2006. "Machine Learning Experiments for textual entailment". Proc. of the Second PASCAL Challenges on Recognising Textual Entailment, Venice, Italy.

Ioannidis, I. and G. Koutrika. 2005. "Personalized Systems: From an IR & DB Perspective", Tutorial at VLDB Conference, Trondheim, Norway

Jackson, P. and I. Moulinier. 2007. *Natural Language Processing for Online Applications*. John Benjamins Publishing Company, Amsterdam.

Jaenich, K., *Lineare Algebra*, Springer Verlag, 2004.

Jiang, J.J. and D.W. Conrath. 1998. "Semantic Similarity Based on Corpus Statistics and Lexical Taxonomy". *In* Proceedings of the International Conference on Research in Computational Linguistic, Taiwan.

Jobbins, A. and L. Evett. 1998. "Text Segmentation Using Reiteration and Collocation", 614–618. Proceedings of ACL-COLING'98, Montreal, CA.

Johannesson, P. 1994. "A Method for Transforming Relational Schemas into Conceptual Schemas", 115–122. *In* Proceedings 10th Inter. Conf. on Data Engineering

Joshi, A.K., B. L. Webber and I. A. Sag (eds.). 1981. "Elements of Discourse Understanding, Chapter Procedural Semantics as a Theory of Meaning", 300–334. Cambridge University Press, Cambridge, UK.

Jurafsky, D. and J. Martin. 2000. "Speech and Language Processing". *An introduction to Natural Language Processing, Computational Linguistics, and Speech recognition*. Prentice Hall, NJ, USA.

Kapetanios, E., D. Baer and P. Groenewoud. 2005. "Simplifying syntactic and semantic parsing of NL-based queries in advanced application domains", Data and Knowledge Engineering, 55(1): 38–58.

Kapetanios, E., V. Sugumaran and D. Tanase. 2006. "Multi-Lingual Web Querying: A Parametric Linguistics Based Approach". *In* Proceedings International Workshop on Applications of Natural Language to Information Systems (NLDB2006), Klagenfurt, Austria

Kan, M-Y., J. Klavans and K. McKeown. 1998. "Linear segmentation and segment significance", 197–205. Proceedings of the 6th International Workshop of Very Large Corpora (WVLC-6), Montreal, CA.

Kaufmann, S. 1999. "Cohesion and collocation: Using context vectors in text segmentation", 591–595. *In* Proceedings of the 37th Annual Meeting of the Association of for Computational Linguistics (Student Session), College Park, USA.

Keimel K., Einführung in die Algebra, TU Darmstadt, 2003.

Kietz, Jörg-Uwe,Raphael Volz and Alexander Maedche. 2000. "Extracting a domain-specific ontology from a corporate intranet", 167–175. *In* Proc. of 4th Conf. on Computational Natural Language Learning and of the 2nd Learning Language in Logic Workshop, Claire Cardie, Walter Daelemans, Claire N´edellec, and Erik Tjong Kim Sarg, [eds.], Somerset, New Jersey. Association for Computational Linguistics.

Kobayashi, M. and K. Takeda. 2000. "Information Retrieval on the Web". ACM Computing Surveys, 32(2): 146–173

Kouylekov, M. and B. Magnini. 2005. "Tree Edit Distance for Recognizing Textual Entailment: Estimating the Cost of Insertion". Proceedings of the Second PASCAL Challenges Workshop on Recognising Textual Entailment, Venice, Italy.

Krishnamurthy, R., Y. Li, S. Raghavan, F. Reiss, S. Vaithyanatham and H. Zhu. 2008. "SystemT: A System for Declarative Information Extraction". *SIGMOD Record* 37(4): 7–13

Kuhns, Robert. 1996. "A Survey of Information Retrieval Vendors", Technical Report SMLI TR-96-56, Sun Microsystems Laboratories, Mountain View, CA.

Kupiec, J., J. Pedersen and F. Chen. 1995. A Trainable Document Summarizer, Proceedings of SIGIR', Seattle, WA.

Labadie, A. and V. Prince. 2008a. "Intended Boundaries Detection in Topic Change Tracking for Text Segmentation". Proceedings of 5th International Workshop on Natural Language Processing and Cognitive Science, Barcelona, Spain.

Labadie, A. and V. Prince. 2008b. "Finding text boundaries and finding topic boundaries: two different tasks?" Proceedings of GoTAL'08, Gothenburg, Sweden.

Lamprier, S.,T. Amghar, B. Levrat and F. Saubion. 2007. "ClassStruggle: a clustering based text segmentation", 600–604. Proceedings of SAC'07. ACM Press, Seoul, Korea.

Lamprier, S., T. Amghar, B. Levrat and F. Saubion. 2007. "On evaluation methodologies for text segmentation algorithms", 19–26. Proceedings of ICTAI 2007, Patras, Greece.

Lang S. 2004. *Linear Algebra*, Springer Verlag.

Lapata, M. 2003. "Probabilistic text structuring: Experiments with sentence ordering", 545–552. Proceeding of the ACL, Sapporo, Japan.

Leacock, C. and M. Chodorow. 1998. "Combining local context and WordNet similarity for word sense identification". pp. 265–283. *In* Christianne Fellbaum, editor, *WordNet*: An Electronic Lexical Database. MIT Press, MA.

Lesk, M. 1986. "Automatic Sense Disambiguation using MRD: How to tell a Pine Cone from an Ice Cream Cone?" Proceedings of the 5th annual international conference on Systems documentation, NY, USA.

Li, Yuhua, Zuhair A. Bandar and David McLean. 2003. "An Approach for Measuring Semantic Similarity between Words Using Multiple Information Sources". IEEE Transactions on Knowledge and Data Engineering, 15(4): 871–882.

Lin, Dekang and P. Pantel. 2001. "Induction of Semantic Classes from Natural Language Text". *In* Proceedings of ACM SIGKDD Conference on Knowledge Discovery and Data Mining.

Lin, C-Y. 2004. "ROUGE: A Package for automatic evaluation of summaries", 74–81. Proceedings of ACL, Workshop of Text Summarization, Barcelona, Spain.

Lloret, E. 2007. "Text Summarization: An Overview", Available: http: //www. dlsi.ua.es/~elloret/publications/TextSummarization.pdf.

Lloret, E. 2011. "Text Summarization based on HLT and its applications". PhD Thesis, Universidad de Alicante, Spain.

Lord, P.W., R.D. Stevens, A. Brass and C.A. Goble. 2003. "Investigating Semantic Similarity Measures across the Gene Ontology: the Relationship between Sequence and Annotation". Bioinformatics, 19(10): 1275–83.

Lu, W.H.,L.F. Chien and H.J. Lee. 2004. "Anchor Text Mining for Translation of Web Queries: A Transitive Translation Approach". ACM Transactions on Information Systems, 22(2): 242–269.

Luhn, H.P. 1958. "The automatic creation of literature abstracts". IBM Journal, 2: 159–165.

Maedche, A. and S. Staab. 2000a. "Discovering conceptual relations from text". pp. 321–325. *In* Proceedings of ECAI'2000, Berlin, Germany.

Maedche, A. and S. Staab. 2000b. "Mining ontologies from text". *In* Proceedings of Knowledge Engineering and Knowledge Management (EKAW 2000), LNAI 1937. Springer.

Maedche, A. and S. Staab. 2000c. "Semi-automatic engineering of ontologies from text". *In* Proceedings of 12th Int. Conf. on Software and Knowledge Engineering, Chicago.

Maedche, A. and S. Staab. 2001. "Ontology Learning for the Semantic Web". IEEE Inteligent Systems, 16(2).

Mann, W.C. and S.A. Thompson. 1988. "Rhetorical Structure Theory: Toward a functional theory of text organization". TEXT 8: 243–281.

Manning, C. and H.Schutze. 1999. *Foundation of statistical natural language processing*. MIT Press, Cambridge, MA.

Marathe, M. and G. Hirst. 2010. "Lexical Chains Using Distributional Measures of Concept Distance". pp. 291–302. Proceedings 11th International Conference CICLing 2010, Iasi, Romania.

Marcu, D. 1997a. "The rhetorical parsing of natural language texts", 96–103. Proceedings of 35th Annual Meeting of the Association for Computational Linguistics, (ACL'97/EACL'97), Madrid, Spain.

Marcu, D. 1997b. "From discourse structure to text summaries", 82–88. Proceedings of the ACL/EACL '97 Workshop on Intelligent Scalable Text Summarization, Madrid,Spain.

Marcus, M.P. 1984. "Some Inadequate Theories of Human Language". *In* T.G. Bever, J.M. Carroll, and L.A. Miller, [eds.]. Talking Minds: The Study of Language in Cognitive Science, 253–278. MIT Press, Cambridge, MA.

Melton, J. and A. Eisenberg. 2001. SQL "Multimedia and Application Packages (SQL/MM)". SIGMOD Record 30(4).

Meyberg K. 1979. *Algebra*. Carl Hanser Verlag.

Mihalcea, R. and D. Moldovan. 2000. "An iterative approach of WSD", 219–223. Proceedings of FLAIRS, Orlando, Florida.

Mihalcea, R. 2005. "Language Independent Extractive Summarization", 49–52. Proceedings of the ACL Interactive Poster and Demonstration Sessions, Ann Arbor, MI.

Mihalcea, R. and D. Radev. 2011. "Graph-based natural Language Processing and Information retrieval". Cambridge University Press, New York, NI.

Mihalcea, R. 2007. "Using Wikipedia for automatic Word Sense Disambiguation", 196–203. Proceedings of NAACL HLT, Rochester, NY.

Miller, G. 1995. "WordNet: a lexical database for English". Communications of the ACM. 38: 39–41

Mirkin, S., J. Berant, I. Dagan, and E. Shnarch. 2010. "Recognising Entailment within Discourse", 770–778. Proceedings of the 23rd International Conference on Computational Linguistics (Coling), Beijing, China.

Mitkov, R. 2002. *Anaphora Resolution*. Pearson Education, Longman, London, UK.

Miyamoto, S. 1990. *Fuzzy Sets in Information Retrieval and Cluster Analysis*, Kluwer Pubishers.

Monz, C. and M. de Rijke. 2001. "Light-Weight Entailment Checking for Computational Semantics", pp. 59–72. *In* Blackburn, P. and Kohlhase, M. [eds]. Proc. of the third workshop on inference in computational semantics. (ICoS-3), Siena, Italy.

Morris, J. and G. Hirst. 1991. "Lexical Cohesion Computed by Thesaural Relations as an Indicator of the Structure of Text". Computational Linguistics 17: 21–48.

Morris, J. 2006. "Readers' subjective perceptions of lexical cohesion and implications for computers' interpretations of text meaning". Proceedings of CaSTA Conference on Breadth of Text, University of New Brunswick, CA.

Nadeau, D., P. Turney and S. Matwin. 2006. "Unsupervised Named Entity Recognition: Generating Gazetteers and Resolving Ambiguity". Proceedings of Canadian Conference on Artificial Intelligence, Quelec, CA.

Navigli, R. 2009. "Word Sense Disambiguation: A survey". ACM Computing Surveys. 41(2): 1–69

Navigli, R and S.P. Ponzetto. 2012. "Multilingual WSD with just a few lines of code: the BabelNet API", 67–72. Proceedings of the 50th Anual Metting of the ACL, Jeju, Republic of Korea.

Negri, M., A. Marchetti, Y. Mehdad, L. Bentivogli and D. Giampiccolo. 2012. "Semeval-2012 Task 8: Cross-lingual Textual Entailment for Content Synchronization". Proceedings of the 6th International Workshop on Semantic Evaluation (SemEval 2012), Montreal, CA.

Nenkova, A. and K. McKeown. 2012. "A survey of text summarization techniques", 43–76. In C.C.Aggarwal and C.X. Zhai [eds.]. *Mining Text Data*. Springer Science, New York, NY.

Ng, T. H., B. Wang and Y. S. Chan. 2003. "Exploiting parallel texts for Word sense Disambiguation: An empirical study", 455–462. Proceedings of the 41st Annual Meeting of ACL, Sapporo, Japan.

Nomoto, T. and Y. Matsumoto. 2001. "A new approach to unsupervised Text summarization", 26–34. Proceedings of SIGIR, New Orleans, LA.

Okumura, M. and T. Honda. 1994. "WSD and text segmentation based on lexical cohesion", 755–761. Proceedings of COLING-94, Kyoto, Japan.

Orasan, C. 2003. "An evolutionary approach for improving the quality of automatic summaries", 37–45. Proceedings of the Multilingual Summarization and Question Answering - Machine Learning and Beyond Workshop. Sapporo, Japan.

Orasan, C. 2006. "Comparative evaluation of modular automatic summarization systems using CAST". PhD Thesis. University of Wolverhampton, UK.

Orasan, C. 2007. "Pronominal anaphora resolution for text summarization", 430–436. Proceedings of Recent Advances in NLP (RANLP), Borovets, Bulgaria.

Owei, V. 2002. "An Intelligent Approach to Handling Imperfect Information in Concept Based Natural Language Queries". ACM Transactions on Information Systems, 20(3): 291–328.

Passonneau, R and D. Litman. 1997. "Discourse Segmentation by Human and Automated Means". Computational Linguistics, 23: 103–139.

Pedersen, T., S. Patwardhan and J. Micheelizzi. 2004. "Wordnet:: similarity-measuring the relatedness of concepts". pp. 267–270. Proceedings of 5th NAACL, Boston, MA.

Perini, A and D. Tatar. 2009. "Textual entailment as a directional relation revisited". pp. 69–72. Proceedings of KEPT2009, Knowledge Engineering Principles and Techniques, Cluj-Napoca, Romania.

Perini, A. 2012. "DirRelCond3: Detecting Textual Entailment Across Language with Conditions on Directional Text Relatedness Scores". Proceedings of the 6th International Workshop on Semantic Evaluation, Montreal, CA.

Pevzner, L. and M. Hearst. 2002. "A Critique and Improvement cf an Evaluation Metric for Text Segmentation". Computational Linguistics 28: 19–36.

Power, R.,D. Scott and N. Bouayad-Agha. 2003. "Document Structure". Computational Linguistics 29: 211–260.

Prince, V. and A. Labadie. 2007. "Text segmentation based on document understanding for information retrieval", 295–304. Proceedings of NLDB'07, Paris, France.

Priss, U. 2005. "Linguistic application of Formal Concept Analysis", 149–160. *In* Ganter, Stumme, Wille. [eds.]. Formal Concept Analysis, Foundations and Applications, LNCS 3626, Springer, Heidelberg.

Priss, U. 2009. "Linguistic Data Exploration", 177–198. *In* P. Hitzler, H. Scharfe. [eds.]. *Conceptual Structures in Practice,* CRC Press.

Purdea, I. 1977. *Pic, Gh., Tratat de algebră modernă,* vol. 1, ed. Academiei RSR.

Purver, M. 2011. "Topic segmentation", 291–317. *In* Tur, G. and de Mori, R., [eds.]. *Spoken Language Understanding: Systems for Extracting Semantic Information from Speech.* Wiley, New Jersey.

Rada, R., H. Mili, E. Bicknell and M. Blettner. 1989. "Development and Application of a Metric on Semantic Nets". IEEE Transactions on Systems, Man, and Cybernetics, 19(1): 17–20.

Radev, D., E. Hovy and K. McKeown. 2002. Introduction to the Special Issues on Summarization. Computational Linguistics. 28: 399–408.

Radev, D., H. Jing and M. Budzikowska. 2000. "Centroid-based summarization of multiple documents: sentence extraction, utility-based evaluation, and user studies". Proceedings ANLP-NAACL Workshop on summarization, Seattle, USA.

Radev, D., S. Blair-Goldensohn and Z. Zhang. 2001. "Experiments in Single and Multi-Document Summarization using Mead". First Document Understanding Conference, New Orleans LA.

Raina, R.,A. Ng and C. Manning. 2005. "Robust textual inference via learning and abductive reasoning". Proceedings of the Twentieth National Conference on AI. AAAI Press, Pittsburgh, PA.

Resnik, Philip. 1999. "Semantic Similarity in a Taxonomy: An Information-Based Measure and its Application to Problems of Ambiguity in Natural Language". Journal of Artificial Intelligence Research, 11: 95–130.

Reynar, J. 1998. "Topic Segmentation: algorithms and applications". PhD Thesis. Univ. of Pennsylvania, PA.

Riedl, M. and C. Biemann. 2012. "How Text Segmentation Algorithms Gain from Topic Models". 553–557. Proceedings of Conference of the North American Chapter of the ACL: Human Language Technologies, Montreal, Canada.

Rijsbergen, Keith van. 2004. *The Geometry of Information Retrieval.* Cambridge University Press, Cambridge, United Kingdom.

Rodriguez, M.A. and M.J. Egenhofer. 2003. "Determining Semantic Similarity Among Entity Classes from Different Ontologies". IEEE Transactions on Knowledge and Data Engineering, 15(2): 442–456.

Rus, V. 2001. Logic form transformation for WordNet glosses and its applications. PhD Thesis. Southern Methodist University, CS and Engineering Department, Dallas, TX.

Sahlgren, M. 2001. "Vector-based semantic analysis: representing word meanings based on random labels". Proceedings of ESSLI Workshop on SKAC, Helsinki, Finland.

Salvo Braz, R., R. Girju, V. Punyakanok, D. Roth and M. Sammons. 2005. "An Inference Model for Semantic Entailment in Natural Language". pp. 29–32. Proceedings of the First PASCAL Challenges Workshop on Recognising Textual Entailment, Southampton, UK.

Sanderson, M. 1994. "Word sense disambiguation and information retrieval", 142–150. Proceedings of the 17th ACM/SIGIR Conference, Dublin, Ireland.

Schaal, M., R. Mller, M. Brunzel and M. Spiliopoulou. 2005. "RELFIN - Topic Discovery for Ontology Enhancement and Annotation". *In* Proceedings of European Conference on the Semantic Web (ECSW 2005), Heraklion, Crete, Greece.

Schank, R. 1973. "Identification of conceptualizations underlying natural language". *In* Computer Models of Thought and Language, 187–247. Freeman, San Francisco, CA

Schütze, H. 1998. "Automatic word sense discrimination". Computational Linguistics, 24(1): 97–124

Sebeok, T. and M. Danesi. 2000. *Forms of Meaning*. DeGuyter, Berlin

Seligman, J. and L.S. Moss. 1997. "Situation theory", 239–309. *In*: Logic and Language, J. van Bentham, A. Ter Meulen [eds.]. The MIT Press, Cambridge, MA

Serban, G. and D. Tatar. 2004. "UBB system at Senseval3". 226–229. Proceedings of Workshop in Word Disambiguation, ACL, Barcelona, Spain.

Shinyama, Y. and S. Sekine. 2004. "Named Entity Discovery Using Comparable News Articles". *In* Proceedings International Conference on Computational Linguistics.

Sidorov, G. and A. Ghelbukh. 2001. "Word Sense Disambiguation in a Spanish explanatory dictionary". 398–402. Proceedings of TALN, Tours, France.

Silber, G. and K. McCoy. 2002. "Efficiently computed lexical chains as an intermediate representation for automatic text summarization". Computational Linguistics. 28: 487–496.

Sowa, J.F. 1984. "*Conceptual Structures: Information Processing in Mind and Machine*". Addison–Wesley, The system programming series.

Sparck Jones, K. 1999. "Automatic summarising: factors and directions", 1–14. *In* I. Mani and M. Maybury [eds]. *Advances in automatic text summarization*. MIT Press, Cambridge, MA.

Sparck Jones, K. 2007. "Automatic summarising: a review and discussion of the state of the art". Technical Report Nr. 679, University of Cambridge, UK.

Steinberger, J.,M. Poesio, M. Kabadjov and K. Jezek. 2007. "Two uses of anaphora resolution in summarization". Information Processing and Management 43: 1663–1680.

Steinberger, J. and M. Kristan. 2007. "LSA-Based Multi-Document Summarization", 87–91. The 8th International PhD Workshop on Systems and Control, a Young Generation Viewpoint, Balatonfured, Hungary.

Stern, A. and I. Dagan. 2012. "BIUTEE: A Modular Open-Source System for recognizing Textual Entailment", 73–78. Proceedings of the 50th Annual Meeting of ACL, Jeju, Republic of Korea.

Stokes, N., J. Carthy and A.F. Smeaton. 2004. "Select: a lexical cohesion based news story segmentation system". AI Communications, 17(1): 3–12.

Stokes, N. 2004. "Applications of Lexical Cohesion Analysis in the Topic Detection and Tracking Domain". PhD. Thesis, National University of Ireland, Dublin.

Stokoe, C., M. J. Oakes and J. I. Tait. 2003. "Word sense disambiguation in information retrieval revisited", 159–166. *In* Proceedings of the 26th Annual International ACM SIGIR Conference on Research and Development in Information Retrieval, Toronto, Canada.

Streicher Th. 2001. Allgemeine Algebra für Informatiker, TU Darmstadt.

Tatar, D. and G. Serban. 2001. "A new algorithm for WSD". Studia Universitatis "Babes-Bolyai", Informatica 2: 99–108.

Tatar, D. and G. Serban. 2003. "Word clustering in QA systems". Studia Universitatis "Babes-Bolyai", Informatica 1: 23–33.

Tatar, D. and M. Frentiu. 2006. "Textual inference by theorem proving and linguistic approach". Studia Universitatis "Babes- Bolyai", Seria Informatics 2: 31–41.

Tatar, D., G. Serban, A. Mihis, M. Lupea and M. Frentiu. 2007a. "A chain dictionary method for Word Sense Disambiguation and applications", 41–49. Proceedings of Knowledge Engineering Principles and Techniques (KEPT), University Press, Cluj-Napoca, Romania.

Tatar, D., G. Serban and M. Lupea. 2007b. "Text entailment verification with text similarities", 33–40. Proceedings of Knowledge Engineering Principles and Techniques (KEPT), University Press, Cluj-Napoca, Romania.

Tatar, D., A. Mihis and D. Lupsa. 2008a. "Text Entailment for Logical Segmentation and Summarization", 233–244. *In* Kapetanios, E., Sugumaran, V., Spiliopoulou, M. [eds.] Proceedings of 13th International Conference on Applications of Natural Language to Information Systems, Springer, London, UK. (LNCS 5039).

Tatar, D., A. Mihis and G. Serban. 2008b. "Lexical Chains Segmentation in Summarization", 95–101. Proceedings of SYNASC 2008, IEEE Computer Society, Timisoara, Romania.

Tatar, D., E. Tamaianu-Morita and G. Serban-Czibula. 2009a. "Segmenting text by lexical chains distribution", 41–44. Proceedings of Knowledge Engineering Principles and Techniques (KEPT), University Press, Cluj-Napoca, Romania

Tatar, D., G. Serban, A. Mihis and R. Mihalcea. 2009b. "Textual Entailment as a Directional Relation". Journal of Research and Practice in Information Technology 41: 53–64.

Tatar, D., E. Tamaianu-Morita, A. Mihis and D. Lupsa. 2009c. "Entailment-based linear segmentation in summarization". IJSEKE, International Journal of Software Engineering and Knowledge Engineering. 19(9): 1023–1038

Tatar, D., M. Lupea and Z. Marian. 2010a. "Learning Taxonomy for Text Segmentation by Formal Concept Analysis", 223–228. Proceedings of SYNASC 2010, IEEE Computer Society, Timisoara, Romania.

Tatar, D., E. Kapetanios, C. Sacarea and D. Tanase. 2010b. "Text Segments as Constrained Formal Concepts". Proceedings of SYNASC 2010, IEEE Computer Society, Timisoara, Romania.

Tatar, D.,M. Lupea and Z. Marian. 2011. "Text summarization by Formal Concept Analysis approach". Proceedings of KEPT 2011, University Press, Cluj-Napoca, Romania.

Tatar, D., D. Inkpen and G. Czibula. 2013. "Text Segmentation Using Roget-based weighted Lexical Chains". Computing and Informatics 32(2): 393–410.

Thanh Tho, Quan, Siu Cheung Hui, A.C.M. Fong and Tru Hoang Cao. 2006. "Automatic Fuzzy Ontology Generation for Semantic Web". IEEE Transactions on Knowledge and Data Engineering, 18(6): 842–856.

Thomason, R.H. 1974. "Formal Philosophy: Selected Papers of Richard Montague", 247–270. Yale University Press.

Tucker, R. 1999. "Automatic summarizing and the CLASP system". PhD thesis, University of Cambridge, Cambridge, U.K.

Turtle, H. and W. B. Croft. 1989. "Inference networks for document retrieval". *In* Proceedings of the 13th Annual Conference on Research and Development in Information Retrieval (SIGIR-89), 1–24.

Tversky, A. 1977. "Features of Similarity". Psychological Review 84: 327–352.

Walker, D. "Knowledge resource tools for accesing large text files". In S. Nirenberg [ed.]. 1987. *Machine Translation: Theoretical and Methodological Issues*. Cambridge University Press, Cambridge, England.

Wang, M. and C.D. Manning. 2010. "Probabilistic tree-edit models with structured latent variables for textual entailment and question answering". pp. 1164–1172. Proceedings of COLING, Beijing, China.

Wang, J.H.,J.W. Teng, P.J. Cheng, W.H. Lu and L.F. Chien. 2004. "Translating Unknown Cross-Lingual Queries in Digital Libraries Using a Web-based Approach", 108–116. *In* Proceedings JCDL'04, Tucson, Arizona, USA

Widdows, D. 2004. *Geometry and Meaning*. CSLI publications, Stanford, California.

Wille, R. 2008. "Communicative Rationality, Logic, and Mathematics". *In* LNAI 4933, ICFCA.

Wille, R. 1995. "Begriffsdenken: Von der griechischen Philosophie bis zur künstlichen Intelligenz heute". Dilthey-Kastanie, Ludwig-Georgs-Gymnasium, Darmstadt, 77–109.

Wille, R. 1994. "Plädoyer für eine philosophische Grundlegung der Begrifflichen Wissensverarbeitung". *In*: R. Wille, M. Zickwolff (eds.) Begriffliche Wissensverarbeitung—Grundfragen und Aufgaben, B.I.-Wissenschaftsverlag, Mannheim, 11–25.

Wille, R. 1997. "Conceptual landscapes of knowledge: a pragmatic paradigm for knowledge processing". *In*: G. Mineau, A. Fall (eds.). Proceedings of the International Symposium on Knowledge Representation, Use, and Storage Efficiency. Simon Fraser University, Vancouver, 2–13.

Wille, R. 1999. "Conceptual landscapes of knowledge: A pragmatic paradigm for knowledge processing". *In* W. Gaul and H. Locarek-Junge (eds.), Classification in the Information Age, 344–356, Springer.

Wille, R. 2000. "Begriffliche Wissensverarbeitung: Theorie und Praxis. Informatik Spektrum 23",357–369.

Wille, R. 2006. "Methods of Conceptual Knowledge Processing, Formal Concept Analysis", 4th International Conference ICFCA 2006, Dresden, Germany, LNAI 3874, Springer, 1–29.

Wittgenstein, L. 1953. *Philosophical Investigations*. Blackwell, Oxford.

Woods, W.A. 1973. "Progress in Natural Language Understanding: An Application to Lunar Geology". *In* AFIPS Conference Proceedings, National Computer Conference, 42: 441–450.

Yaari, Y. 1997. "Segmenting of expository text by hierarchical agglomerative clustering". Proceedings of RANLP'97. Borovets BG.

Yaari, Y. 1998. "Texplore—exploring expository texts via hierarchical representation", 25–31. Proceedings of COLING-ACL Workshop on CVIR, Montreal, CA.

Yarowsky, D. 1992. "Word sense disambiguation using statistical models of Roget's categories trained on large corpora". pp. 454–460. Proceedings of COLING'92, Nantes, France.

Yarowsky, D. 1995. "Unsupervised Word Sense Disambiguation rivaling supervised methods", 189–196. Proceedings of ACL, Cambridge, MA.

Yarowsky, D. 1999. *Hierarchical Decision Lists for WSD*. Kluwer Acadmic Publishers.

Ye, S., T. Chua, M. Kan and L. Qiu. 2007. "Document concept lattice for text understanding and summarization". Information Processing and Management, 43: 1643–1662.

Zhong, Z. and H.T. Ng. 2012. "Word Sense Disambiguation improves Information Retrieval", 273–282. Proceedings of the 50th Annual Meeting of the ACL, Jeju, Republic of Korea.

Zhou, Y., J. Qin, H. Chen and J.F. Nunamaker. 2005. "Multilingual Web Retrieval: An Experiment on a Multilingual Business Intelligence Portal", 43a, *In* HICSS'05, Track 1.

Index